Advanced Processing and Manufacturing Technologies for Structural and Multifunctional Materials IV

Advanced Processing and Manufacturing Technologies for Structural and Multifunctional Materials IV

A Collection of Papers Presented at the 34th International Conference on Advanced Ceramics and Composites
January 24–29, 2010
Daytona Beach, Florida

Edited by
Tatsuki Ohji
Mrityunjay Singh

Volume Editors
Sanjay Mathur
Tatsuki Ohji

The American Ceramic Society

WILEY

A John Wiley & Sons, Inc., Publication

Published by John Wiley & Sons, Inc., Hoboken, New Jersey.
Published simultaneously in Canada.

Library of Congress Cataloging-in-Publication Data is available.

ISBN 978-0-470-59473-5

Printed in the United States of America.

10 9 8 7 6 5 4 3 2 1

Contents

RAPID PROCESSING

JOINING AND MACHINING

Preface

The Fourth International Symposium on Advanced Processing and Manufacturing Technologies for Structural and Multifunctional Materials and Systems (APMT) was held during the 34th International Conference on Advanced Ceramics and Composites, in Daytona Beach, FL, January 24–29, 2010. The aim of this international symposium was to discuss global advances in the research and development of advanced processing and manufacturing technologies for a wide variety of non-oxide and oxide based structural ceramics, particulate and fiber reinforced composites, and multifunctional materials. A total of 96 papers, including invited talks, oral presentations, and posters, were presented from more than 10 countries (USA, Japan, Germany, France, Italy, Slovenia, Serbia, Belgium, Turkey, Sweden, Canada, China, Korea, India and Israel). The speakers represented universities, industry, and research laboratories.

This issue contains 25 invited and contributed papers, all peer reviewed according to the American Ceramic Society Review Process. The latest developments in processing and manufacturing technologies are covered, including green manufacturing, smart processing, advanced composite manufacturing, rapid processing, joining, machining, and net shape forming technologies. These papers discuss the most important aspects necessary for understanding and further development of processing and manufacturing of ceramic materials and systems.

The editors wish to extend their gratitude and appreciation to all the authors for their cooperation and contributions, to all the participants and session chairs for their time and efforts, and to all the reviewers for their valuable comments and suggestions. Financial support from the Engineering Ceramic Division and the American Ceramic Society is gratefully acknowledged. Thanks are due to the staff of the meetings and publication departments of the American Ceramic Society for their invaluable assistance.

We hope that this issue will serve as a useful reference for the researchers and technologists working in the field of interested in processing and manufacturing of ceramic materials and systems.

Tatsuki Ohji, Nagoya, Japan
Mrityunjay Singh, Cleveland, USA

Introduction

This CESP issue represents papers that were submitted and approved for the proceedings of the 34th International Conference on Advanced Ceramics and Composites (ICACC), held January 24–29, 2010 in Daytona Beach, Florida. ICACC is the most prominent international meeting in the area of advanced structural, functional, and nanoscopic ceramics, composites, and other emerging ceramic materials and technologies. This prestigious conference has been organized by The American Ceramic Society's (ACerS) Engineering Ceramics Division (ECD) since 1977.

The conference was organized into the following symposia and focused sessions:

Symposium 1	Mechanical Behavior and Performance of Ceramics and Composites
Symposium 2	Advanced Ceramic Coatings for Structural, Environmental, and Functional Applications
Symposium 3	7th International Symposium on Solid Oxide Fuel Cells (SOFC): Materials, Science, and Technology
Symposium 4	Armor Ceramics
Symposium 5	Next Generation Bioceramics
Symposium 6	International Symposium on Ceramics for Electric Energy Generation, Storage, and Distribution
Symposium 7	4th International Symposium on Nanostructured Materials and Nanocomposites: Development and Applications
Symposium 8	4th International Symposium on Advanced Processing and Manufacturing Technologies (APMT) for Structural and Multifunctional Materials and Systems
Symposium 9	Porous Ceramics: Novel Developments and Applications
Symposium 10	Thermal Management Materials and Technologies
Symposium 11	Advanced Sensor Technology, Developments and Applications

Focused Session 1	Geopolymers and other Inorganic Polymers
Focused Session 2	Global Mineral Resources for Strategic and Emerging Technologies
Focused Session 3	Computational Design, Modeling, Simulation and Characterization of Ceramics and Composites
Focused Session 4	Nanolaminated Ternary Carbides and Nitrides (MAX Phases)

The conference proceedings are published into 9 issues of the 2010 Ceramic Engineering and Science Proceedings (CESP); Volume 31, Issues 2–10, 2010 as outlined below:

- Mechanical Properties and Performance of Engineering Ceramics and Composites V, CESP Volume 31, Issue 2 (includes papers from Symposium 1)
- Advanced Ceramic Coatings and Interfaces V, Volume 31, Issue 3 (includes papers from Symposium 2)
- Advances in Solid Oxide Fuel Cells VI, CESP Volume 31, Issue 4 (includes papers from Symposium 3)
- Advances in Ceramic Armor VI, CESP Volume 31, Issue 5 (includes papers from Symposium 4)
- Advances in Bioceramics and Porous Ceramics III, CESP Volume 31, Issue 6 (includes papers from Symposia 5 and 9)
- Nanostructured Materials and Nanotechnology IV, CESP Volume 31, Issue 7 (includes papers from Symposium 7)
- Advanced Processing and Manufacturing Technologies for Structural and Multifunctional Materials IV, CESP Volume 31, Issue 8 (includes papers from Symposium 8)
- Advanced Materials for Sustainable Developments, CESP Volume 31, Issue 9 (includes papers from Symposia 6, 10, and 11)
- Strategic Materials and Computational Design, CESP Volume 31, Issue 10 (includes papers from Focused Sessions 1, 3 and 4)

The organization of the Daytona Beach meeting and the publication of these proceedings were possible thanks to the professional staff of ACerS and the tireless dedication of many ECD members. We would especially like to express our sincere thanks to the symposia organizers, session chairs, presenters and conference attendees, for their efforts and enthusiastic participation in the vibrant and cutting-edge conference.

ACerS and the ECD invite you to attend the 35th International Conference on Advanced Ceramics and Composites (http://www.ceramics.org/icacc-11) January 23–28, 2011 in Daytona Beach, Florida.

Sanjay Mathur and Tatsuki Ohji, Volume Editors
July 2010

Green Manufacturing and Smart Processing

SECURING THE SUPPLY OF PRECIOUS AND SPECIAL METALS - THE NEED OF CLOSING THE LOOP

Christian Hagelüken
Umicore Precious Metals Refining
D 63457 Hanau, Germany

ABSTRACT

Special and precious metals are key ingredients for high tech applications like information technology, electronics, or car-catalysts, and emerging clean technologies such as photovoltaics, fuel cells or electric car batteries. With a relatively recent use in mass applications, 80% or more of the cumulative mine production for e.g. the Platinum Group Metals (PGM), gallium, or indium took place just in the last 30 years.

A future sufficient access to these 'technology metals' is essential. Their primary production is often coupled to base metals and concentrated in few mining regions of the world, causing a complex demand-supply balance and high price volatility. Hence it is necessary to secure their resource efficient use along the lifecycle and to establish effective recycling systems to "close the loop", preserving limited resources. Appropriate management at the product's end-of-life however is challenging. Most consumer products are characterized by an "open cycle" with non transparent global flows and multiple owners along the lifecycle, a complex material composition, and demanding collection and recycling chains.

The contribution addresses these issues from a holistic perspective and elaborates the system interdependencies and potential ways of improvement. On the example of precious metals it will be shown that in many cases effective recycling technologies exist but that the majority of consumer products do not enter in such recycling chains so far.*

INTRODUCTION

Metals are classical examples of non-renewable resources, and their extraction from Earth by mining of ores cannot be seen as sustainable in the strict sense of the word. Mining, by definition, depletes the ore reserves. Through mineral processing and subsequent smelting and refining, ores are disintegrated, and the desired metals are isolated for use in the technosphere. Special and precious metals play a key role in modern societies as they are of specific importance for clean technologies and other high tech equipment. Important applications are information technology (IT), consumer electronics, as well as sustainable energy production such as photovoltaic (PV), wind turbines, fuel cells and batteries for hybrid or electric cars. They are crucial for more efficient energy production (in steam turbines), for lower environmental impact of transport (jet engines, car catalysts, particulate filters, sensors, control electronics), for improved process efficiency (catalysts, heat exchangers), and in medical and pharmaceutical applications. Figure 1 provides an overview of these main applications areas for selected metals and illustrates their significance for modern life. For example, electronic products can contain up to 60 different elements and in their entity are major demand drivers for precious and special metals: Just the annual sales of mobile phones and computers account e.g. for about 3% of the world mine production of gold and silver, 15% of palladium and over 20% of cobalt.[1]

Driving forces for the booming use of these "technology metals" (used here as a synonym for precious and special metals) are their extraordinary and sometimes exclusive properties, which make many of these metals essential components in a broad range of applications. Building a more sustainable society with the help of technology hence depends to a large extent on sufficient access to technology metals.

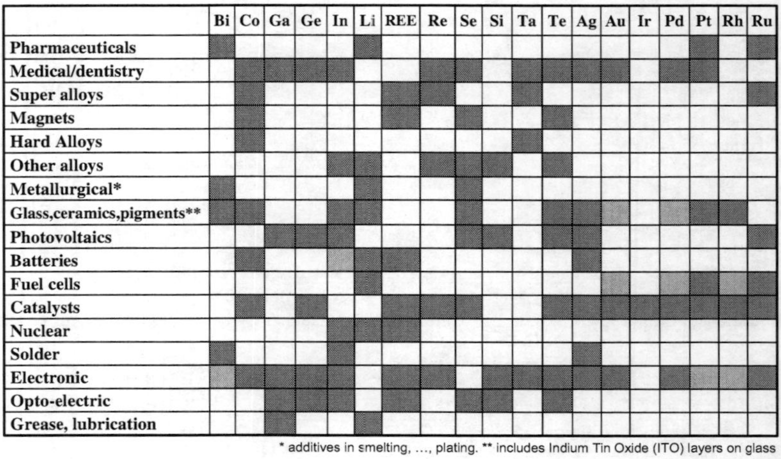

Figure 1: Important applications for technology metals[2]

POTENTIAL METAL SCARCITIES AND SECURITY OF SUPPLY

In the context of raising metal prices and the boom in demand for many technology metals, a discussion on potential metal scarcities restarted about 4 years ago.[3,4,5] More than 30 years after the Club of Rome's "The Limits to Growth" publication from 1972, it put again more emphasis on the finite character of our natural resources, a debate which had calmed down for almost two decades in between. Since the 1970s, a lot has happened specifically with respect to the use of the "technology metals". 80% or more of the cumulative mine production of platinum group metals (PGM), gallium, indium, rare earth elements, and silicon, for example, has occurred over the last 30 years. For most other special metals, more than 50% of their use took place in this period, and even for the "ancient metals" gold and silver use from 1978 onward accounts for over 30% (Fig. 2). In many cases the booming demand especially from consumer mass applications drove up metal prices significantly. For example, the significant increase in demand of platinum and palladium was mainly caused by automotive catalysts (50% of today's platinum/palladium demand) and electronics (Fig. 3). So more often the question is raised: "How soon will we run out of key element resources?" and occasionally: "Are severe shortages of certain critical metals within the next decade threatening?" Governments in the US, Japan and since recently also in Europe undertake efforts to define which metallic resources are specifically critical for their economies and which measures should be taken to improve their long-term supply security.[6,7]

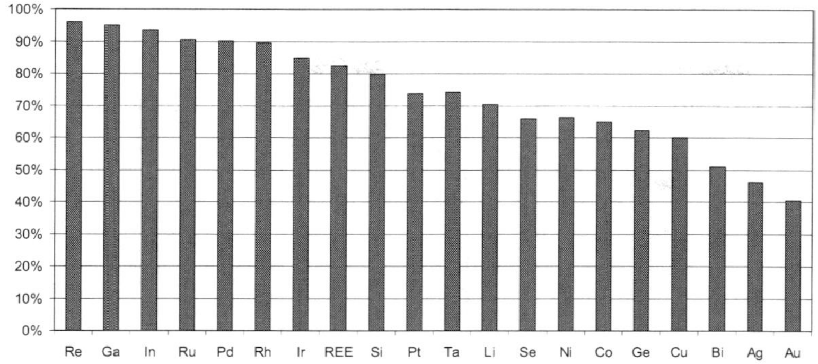

Figure 2: Share of technology metals mined since 1978 compared to cumulated production between 1900–2007; copper and nickel included for comparison (modified after [2]).

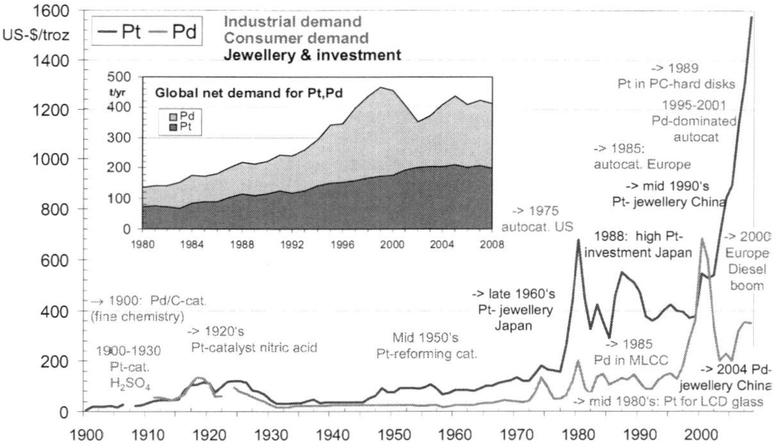

Figure 3: Long term development of demand and nominal prices for platinum (Pt) and palladium (Pd)

The current debate takes place between two extremes - resource optimists versus resource pessimists. Optimists argue that in principal market mechanisms will help to overcome supply shortages. Increased metal prices will lead to new exploration and mining (of so far uneconomic deposits) and technical substitution will be able to replace scarce metals by others with similar properties, or by thrifting and innovative technologies. Pessimists start with information about ore resources, compiled by the US Geological Services (USGS)[8] among others, and then divide these numbers by the current and projected annual demand. For some metals such as indium this leads indeed to rather short "static lifetimes". While the scientific debate is open to the many facets of the matter, media sometimes tend to bring this in rather black and white statements. This contribution follows a pragmatic "resource realist" approach, without diving into detailed discussions on statistics and single metals. The aim is to discuss the main parameters and mechanisms that impact metal scarcities and what can be done to prevent them.

DIMENSIONS OF RESOURCE SCARCITY

Three types need to be distinguished, namely absolute, temporary and structural resource scarcity, and in this context the understanding of the primary supply chain is crucial.[2]

Absolute scarcity would mean the depletion of economically mineable ore resources. In this case all ore deposits of a certain metal – including the ones which have not yet been discovered by exploration – would have been widely mined out, and the total market demand for a metal would exceed the remaining mine production. This would first lead to extreme price increases and finally force substitution of that metal (or technology) in certain applications, or would put severe limits to the further technology distribution (as worst case a good technology, e.g. for energy generation, is endangered because a key metal is not available). However, within the foreseeable future such an absolute scarcity is rather unlikely, and here the arguments of the resource optimists count. Extremely high prices would make deep level mining and mining of low grade deposits, which are currently left aside, economically feasible. Also it would trigger more exploration, leading to the discovery of new ore bodies. Exploration is very costly and time consuming, so as long as mining companies have enough accessible deposits for the next two decades there is not much incentive for them to conduct additional exploration. Accordingly, the data reported by USGS[8] and other geological services do not report the absolute availability of metals on the planet but compile the known deposits that can be extracted economically already today (reserves), or where it is expected to be potentially feasible (resources). If exploration and mining efforts extent deeper into the earth's crust or oceans and cover a wider geographical area, maybe even into arctic regions, substantial new metal resources are very likely to be accessible, however this will not come without trade offs as we will see below[9].

In contrast, *temporary or relative scarcity* is a phenomenon which has been already experienced. In this case, metal supply is for a certain period in time not able to meet the demand. Reasons can be manifold. New technological developments, strong market growth in existing applications, or speculative buying of investors can drive up the demand significantly within a short time so that mine supply lags behind. Also the supply can be disrupted by political developments, armed conflicts, natural disasters or other constraints in the mining countries itself, within the transport of ore concentrates, or also at major smelters/refineries. Temporary scarcities are a main reason for the sometimes extreme price volatility in metal markets. The risk on temporary scarcities increases with increasing concentration of the major mines or smelters in few and/or unstable regions, or in few companies. Also a low number of applications in which the metal is used increases the risk. Often, different factors come together and then accelerate the development. For instance, in the first quarter of 2008 a soaring demand for PGMs from automotive catalysts and (speculative) investment coincided with a reduced supply from South African mines due to shortages in electric power. The prices of platinum and rhodium went to record heights within a short time as South Africa produces over 75% of platinum and rhodium supply.

Speculation about potential depletion of indium resources started when from 2003 onwards the sales boom of LCD devices (monitors, TVs, mobile phones etc), which use indium-tin oxide (ITO) as transparent conductive layer, drove up indium prices significantly. The development of indium prices (Fig. 4) is a good example to illustrate the effect of temporary scarcities. The supply could not follow the sudden jump in demand and indium prices went up by factor 10 between 2003 and 2006. An important impact on this development has the manufacturing technology used for LCD applications, which is a sputter process. From the indium contained in the ITO targets, less than 20% end up as conductive layer on LCD screens, while the rest goes into production scrap at various stages of the process. Hence the gross indium demand is significantly higher than what is finally used for the product. The production scrap, however, is not lost, the biggest part of it are spent targets which can be recycled very efficiently. With the limited recycling capacities available before the boom, the huge new scrap arising could not be handled. But due to the increasing indium prices, recycling became attractive, spent targets were stockpiled, and new recycling capacities were build up. After 2006 an increased primary supply (also triggered by higher prices) and a significant secondary supply (working down the target stockpiles) drove down the indium prices again, which was further amplified end of 2008 by the economic crisis. It is important to understand that the high recycling rates (> 50%) reported for indium only refer to this recycling of high grade production scrap. So far, hardly any indium recycling takes place for end-of-life products, which will be much more challenging from a technical and economic point of view (the In concentration in the final product is very low). It can be assumed that in the meantime sufficient recycling capacity for production scrap is available and that the stocks have been largely eaten up. Hence, a further increase in indium demand, supported as well by thin film solar cells, is likely to drive up the prices again. On the long run, for the indium supply the structural scarcity as described below will become an important factor (In is a by-product from zinc mining).

In future, a take off in thin film photovoltaics would boost the demand for tellurium, indium, selenium, and gallium, mass applications of electric vehicles will require large amounts of lithium, cobalt and some rare earth elements, and fuel cell cars would need significantly more platinum than is used today in a catalytic converter. Developing and expanding mining and smelting capacities is highly capital intensive, risky, and it takes many years to materialize. Hence, temporary scarcities are likely to happen more often in future.

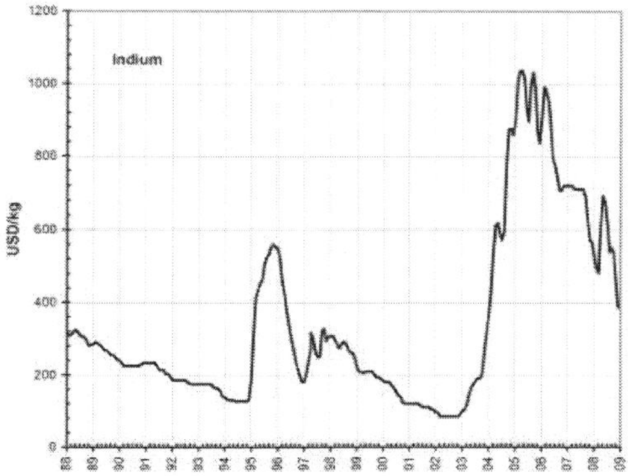

Figure 4: Development of indium prices (monthly averages) since 1988

The *structural scarcity* is most severe for many technology metals, which are often not mined on their own but occur only as by-products from so-called major or carrier metals[9]. Indium and germanium, for example, are mainly by-products from zinc mining, gallium from aluminum, and selenium, tellurium from copper (and lead). The PGMs occur as by products from nickel- and copper mines, and as coupled products in own mines. Within the PGMs ruthenium and iridium are by-products from platinum and palladium (Fig. 5). Since the by-product ("minor metal") is only a very small fraction of the carrier metal, here the usual market mechanisms do not work. An increasing demand will certainly lead to an increasing price of the by-product metal, but as long as the demand of the major metal does not rise correspondingly, mining companies will not produce more, because this would erode the major metal's price. In this respect, the supply of by-product metals is price-inelastic, even a "ten-fold increase" in its price could usually not compensate the negative impact on total revenues when there is oversupply of the major metal. Moreover, many technology metals are important ingredients for several emerging technologies simultaneously (Fig. 1), so a competition between applications becomes likely and increasing demand from various segments will intensify the pressure on supply.

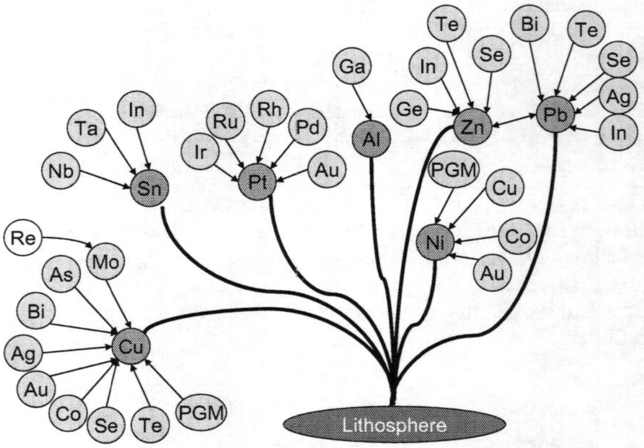

Figure 5: Coupling of minor and major metals production. The figure indicates, which minor metals are produced as by-products of major metals[2]

Substitution is not likely to become the solution for many of these metals either since the required functional properties can often be met only by metals from the same metal family. For example, substituting platinum by palladium in catalytic applications will just shift the problem from one temporary/structural scarce metal to the other, which was experienced in the second half of the 1990s, pushing the before cheaper palladium to record heights in 2000/2001 (Fig. 3). In emerging opto-electronics the crucial metals are silicon, tellurium, gallium, selenium, germanium, and indium. They can partially substitute each other, though this will not really mitigate the problem (Fig. 6). It can only be overcome by increasing the efficiencies in the primary supply chain (possibly leading to considerable gains) and, above all, by comprehensive recycling efforts as pointed out hereafter. Omitting the fact that many technology metals are by-products and that structural scarcity is possible is thus the weak point in the resource optimists' argumentation.

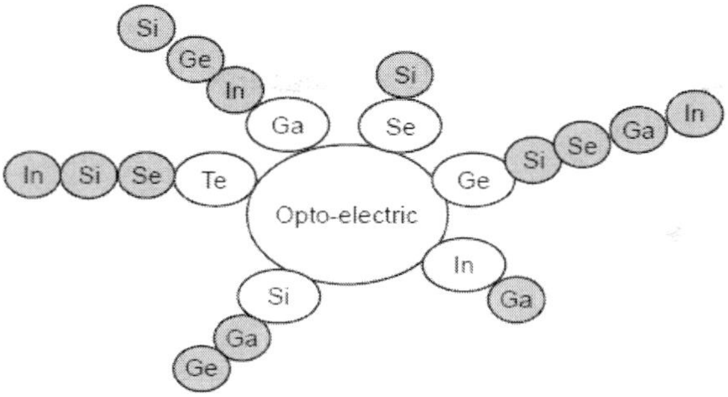

Figure 6: Potential substitution of metals in opto-electrics. Inner spheres show the elements used in the application, outer spheres depict possible replacement elements.[2]

Independent of whether or not supply constraints are likely, the impact of mining of lower grade ores and from more challenging locations must not be overlooked. It will inevitably lead to increasing costs, energy demand, and raising emissions, it will impact the biosphere (rain forest, arctic regions, oceans), and it can increase the dependence on certain regions ("battle for resources"). This can imply significant constraints on emerging technologies, unless effective life cycle management enables the use of recycled (secondary) metals in the forthcoming years.

THE NEED OF CLOSING THE LOOP

Metals are not consumed, they are only transferred from one manifestation into another, moving in and between the lithosphere and the technosphere. Thus the latter becomes our future 'renewable' resource in society. Thoroughly extracting "urban mines" is the only sustainable solution to overcome supply disruptions. Metal combinations in products often differ from those in primary deposits, which results in new technological challenges for their efficient recovery. In products such as electronics or catalysts, the precious metals (Au, Pt, Pd, ...) have become the economic drivers for recycling ("paying metals"), while many special metals (Se, Te, In, ...) can be recovered as by-products when state-of-the art treatment and refining operations are used. Very low concentration of technology metals in certain products or dissipation during product use sets economic and technical limits in many cases, and technical challenges exist especially for complex products like vehicles, computers, etc. Effective recycling requires a well tuned recycling chain, consisting of different specialized stakeholders: Starting with collection of old products, followed by sorting/dismantling and preprocessing of relevant fractions, and finally recovery of technology metals. The latter requires sophisticated, large scale metallurgical operations like the Umicore integrated smelter-refinery in Antwerp, Belgium where currently seven precious metals (Ag, Au, Pt, Pd, Rh, Ru, Ir), as well as eleven base and special metals (Cu, Pb, Ni, Sn, Bi, As, Sb, Se, Te, In, Ga) are recovered and supplied back to the market. Most of these metals are recovered with high yields, in the case of the precious metals yields are well over 95% of what was contained in the feed material to the plant. The plant input of approximately 1000 metric tons per day comprises over 200 different categories, the majority of which consists of recyclables (car catalysts, various process catalysts, cell phones, circuit boards, photographic residues, fuel cells, etc.) and smelter by-products (slags, flue dust, anode slimes, effluent treatment sludges, etc.).[10]

Recycling technology has made significant progress and further improvements extending the range and yield of metals are underway. When required, recycling technologies are adapted to new products, as has been successfully the case e.g. for certain petrochemical process catalysts, fuel cells, diesel particulate filters, or high grade residues from thin film solar cell manufacturing. Design for sustainability based on a close dialogue between manufacturers and recyclers can further support effective recycling as it starts already in the design and manufacturing phase and proceeds all along its lifecycle.

However, the biggest challenge to overcome is the insufficient collection of consumer goods, and inefficient handling within the recycling chain. As long as goods are discarded with household waste, stored in basements or ending up in environmentally unsound recycling operations, the total recovery rates will remain disappointingly low, as it is the case today for most consumer goods. Legislation can be supportive but monitoring of the recycling chain as well as tight enforcement of the regulations are crucial for success. For example, in spite of a comprehensive European legislative framework ("Directive on waste electrical and electronic equipment/WEEE-Directive"; "Directive on end-of-life vehicles/ELV Directive"), a significant share of end-of-life computer, cell phones, cars, etc. are currently not recycled properly. Instead they are discarded or (illegally) exported to Asia or Africa under the pretext of "reuse" to circumvent the Basel Convention regulations on transboundary shipments of waste. The same happens in North America and Japan.[11] This leads to a situation where state-of-the-art, high financial investment recycling facilities in industrialized countries are underutilized because 'recycling' and the associated environmental burden of environmentally unsound treatment is 'outsourced' to the developing world. Except some inefficient gold and copper recovery, technology metals are lost in such primitive "backyard recycling processes", the "urban mine is wasted irreversibly.[12]

A striking example is the automotive catalyst. Due to the high prices of the contained PGMs platinum, palladium and rhodium its recycling is economically highly attractive (several tens of US-dollars per piece paid to the scrap yard). An autocatalyst is easy to identify and remove from an end-of-life car, a comprehensive collection infrastructure exists, and state-of-the-art metallurgical treatment operations achieve PGM recovery yields of 98%. Nevertheless, on a global scale only about 50% of the PGMs originally used for automotive catalysts are finally recovered, the rest is lost inevitably. The main reasons are global flows of end-of-life cars (e.g. in Germany from a little over 3 million annual car deregistrations only about 0.5 million cars are recycled within Germany, the remainder is exported largely out of Europe) and a high degree of intransparency and "informal" business practices in the early parts of the recycling chain (even in industrialized countries).[13] Figure 7b shows the typical structure of so called "open cycles" for consumer goods. The insufficient cooperation along the life cycle and recycling chain (although "extended producer responsibility" has been implemented), combined with insufficient tracking of product and material streams along the entire chain explain why inefficient open cycles continue to exist.

To effectively close the loop for consumer products, new business models need to be introduced that provide strong incentives to hand in products at their end-of-life into professional recycling systems. This can include deposit fees on new products, or product service systems (PSS) like leasing or other approaches. Especially for emerging technologies (electric vehicles, fuel cells, photovoltaics etc.) setting up "closed loop structures" will be essential and manufacturers who put successful models in place can thus secure their own supply of technology metals in the future.

Such closed loop structures exist successfully already in most industrial applications of precious metals. For example, PGM-catalysts used in fine chemistry or oil refining are turned around very efficiently at their end of life. Usually well over 90% of the PGMs used in the fresh catalysts are finally recovered, even at long catalyst use times (up to 10 years in some applications), several regeneration cycles, and difficult operating conditions in the chemical reactor or oil refinery. The metallurgical steps to recover the PGMs from the spent catalysts are similar to the ones used for automotive catalysts. The decisive differences lie in the lifecycle structure and the steps prior to metallurgical recovery. Here, for industrial process catalysts the complete lifecycle in handled very transparently in a highly professional way between the industrial actors involved (Fig. 7a). Catalyst manufacturers, users, and recyclers work closely together, the location of the catalyst is always well defined, and a profound knowledge exists about the properties and use history of a specific catalyst. Usually catalyst users (e.g. chemical plants) maintain the property of the PGMs throughout the entire lifecycle. Recycling is contracted with a precious metals refinery as a "toll refining operation" with a physical credit of the PGMs back to the user who provides them directly to a catalyst manufacturer for the production of a fresh catalyst. From there, a new lifecycle starts. As a consequence, the net demand for PGMs from the (petro)chemical industry as a whole is just below 10% of the total global PGM net demand. It is used to cover market growth, new application and the small losses that occurred during the catalyst lifecycle. The gross demand, however, i.e. the annual new use of PGMs for process catalysts, is as high as for automotive catalysts, but the latter makes up for about 50% of the global PGM net demand. A transformation of "open cycles" in consumer applications into "closed cycles" as prevail in many industrial applications would be a big step towards a secured supply of technology metals.[14]

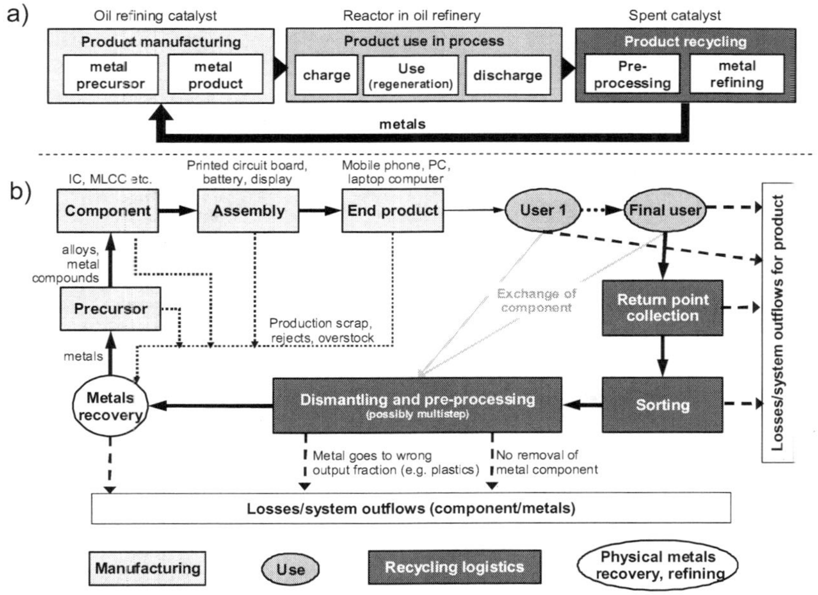

Figure 7: (a) Closed loop systems for industrial applications (example process catalyst) versus (b) Open loop systems for consumer goods (example consumer electronics)[14]

CONCLUSION

In an ideal system, the sustainable use of metals could indeed be achieved by avoiding spillage during each phase of the product life cycle. As illustrated in figure 8, such losses occur at various stages and it needs to analyze the specific impact factors to identify the most appropriate means for each stage. It is important to understand that universal means to improve recycling do not exist, if material properties or technology constraints have the main impact, then completely different measures will be required than if societal or life cycle issues are the main loss driver. Mining and recycling thus need to evolve as a complimentary system, where the primary metals supply is widely used to cover inevitable life cycle losses and market growth, and secondary metals from end-of-life products contribute increasingly to the basic supply. Effective recycling systems would thus make a significant contribution to conserve natural resources of scarce metals and secure sufficient supply of technology metals for future generations.

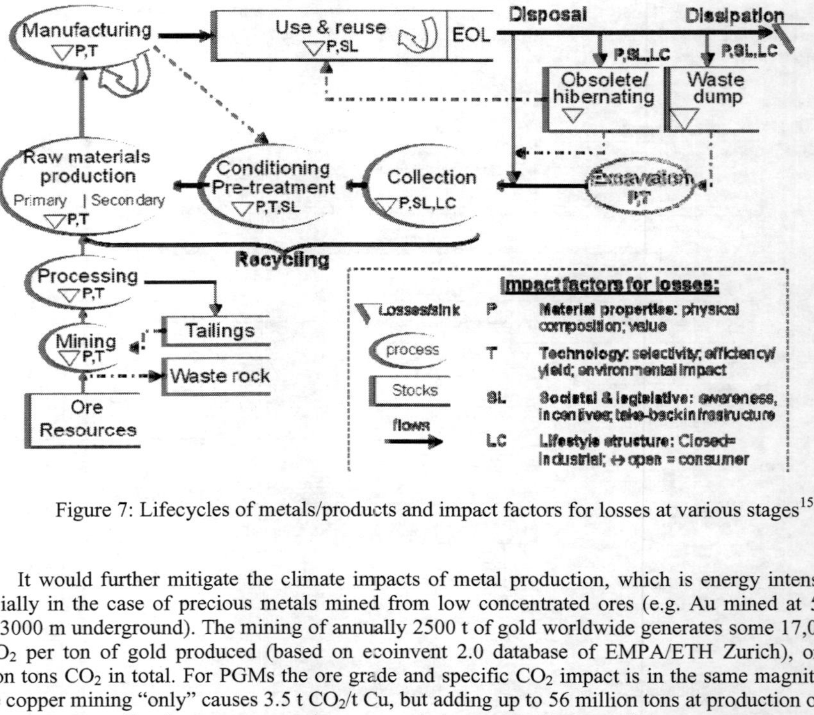

Figure 7: Lifecycles of metals/products and impact factors for losses at various stages[15]

It would further mitigate the climate impacts of metal production, which is energy intensive, especially in the case of precious metals mined from low concentrated ores (e.g. Au mined at 5 g/t from 3000 m underground). The mining of annually 2500 t of gold worldwide generates some 17,000 t of CO_2 per ton of gold produced (based on ecoinvent 2.0 database of EMPA/ETH Zurich), or 42 million tons CO_2 in total. For PGMs the ore grade and specific CO_2 impact is in the same magnitude, while copper mining "only" causes 3.5 t CO_2/t Cu, but adding up to 56 million tons at production of 16 million tons annually. Some mass products are relatively rich "bonanzas" in comparison, e.g., a computer motherboard with ca. 250 g/t of gold, a mobile phone handset with 350 g Au/t, or an automotive catalytic converter with some 700 g/t of PGM. If effective collection systems and state-of-the-art recovery processes are used, the secondary metal production from such products requires only a small fraction of energy/CO_2 compared to mining.[1,2]

Such products carry a high intrinsic metal value which makes recycling attractive under an economical point of view as well. Recovering pure metals from a PC circuit board costs only about 20% of its intrinsic metal value, leaving sufficient margins to pay for logistics and dismantling, in case of car catalysts the cost share is even less. This similarly applies for large multi-metal products such as cars. But for other technology metals containing products like TVs, audio equipment and household appliances, the intrinsic metal value is usually not sufficient to pay the total costs of the recycling chain, and incentives by legislation, manufacturers, or distributors are needed for stimulation. However, if the true costs of landfill and environmental damage caused by non-recycling would be accounted for, then on a macroeconomic level proper recycling most probably is viable for such products as well. In this sense efficiently recycling our end-of-life products today is insurance for the future: It will prevent/smoothen metal price surges and secure a sustainable and affordable supply of metals needed for our products of tomorrow.

FOOTNOTE

The article is an updated and extended version of a contribution by the author entitled "Precious Element Resources" to the McGraw-Hill 2010 Yearbook of Science & Technology.

REFERENCES

[1]C. Hagelüken, C.E.M. Meskers, Mining our computers – Opportunities and challenges to recover scarce and valuable metals, in: Proceedings of Electronics Goes Green Conference 2008, H. Reichl, N. Nissen, J, Müller, O. Deubzer (eds.), Fraunhhofer IRB, Stuttgart, 585-590 (2008).

[2]C. Hagelüken, C.E.M. Meskers, Complex lifecycles of precious and special metals, in: Graedel, T., E. van der Voet (eds): Linkages of Sustainability, Strüngmann Forum Report, vol. 4. Cambridge, MA: MIT Press, (2010).

[3]R.Gordon, M. Bertram, T. Graedel, Metal stocks and sustainability, PNAS 103 (5): 1209-1214 (2006).

[4]J. Tilton, G. Lagos, Assessing the long-run availability of copper, Resources Policy 32, 19–23 (2007).

[5]M. Wolfensberger, D. Lang, R. Scholz, (Re)structuring the field of non-energy mineral resource scarcity. ETH working paper 43. ETH Zürich: R. Scholz, Natural and Social Science Interface (NSSI) (2008).

[6]NRC 2008, Minerals, Critical Minerals, and the U.S. Economy, Committee on Critical Mineral Impacts of the U.S. Economy, Committee on Earth Resources, National Research Council. The National Academies Press, Washington DC (2008).

[7]EU COM 699, The raw materials initiative – meeting our critical needs for growths and jobs in Europe, Brussels, SEC (2008) 2741 (2008).

[8]U.S. Geological Survey (USGS), Mineral commodity summaries 2009, Washington DC (2009).

[9]F.W. Wellmer, Reserves and resources of the geosphere, terms so often misunderstood. Is the life index of reserves of natural resources a guide to the future? Z. dt. Ges. Geowiss. 159/4, 575-590 (2008).

[10]C.E.M. Meskers, C. Hagelüken, G, Van Damme, Green recycling of EEE, EPD Congress 2009 at the TMS Annual Meeting. S.M. Howard (ed.) (2009).

[11]J. Puckett, S. Westervelt, R. Gutierrez, Y. Takamiya, The digital dump – exporting re-use and abuse to Africa. Seattle: Basel Action Network (2005).

[12]D. Rochat, C. Hagelüken, M. Keller, R. Widmer, Optimal recycling for printed wiring boards in India. R'07, World Congress - Recovery of materials and energy for resource efficiency. L.M. Hilty, X. Edelmann, A. Ruf (eds.) (2007).

[13]C. Hagelüken, Closing the loop – recycling of automotive catalysts, Metall 61 (1-2), 24-39 (2007).

[14]C. Hagelüken, M. Buchert, P. Ryan, Materials flow of Platinum Group Metals, GFMS, London (2005).

[15]McLean, H., F. Duchin, C. Hagelüken, K. Halada, S.E. Kesler,Y. Moriguchi, D- Mueller, T.E. Norgate, M.A. Reuter, E. van der Voet (2009), Mineral Resources: Stocks, flows, and prospects, in: Graedel, T., E. van der Voet (eds): Linkages of Sustainability. Strüngmann Forum Report, vol. 4. Cambridge, MA: MIT Press (2010).

MECHANICAL PROPERTIES OF Cr-Si-N-O THIN FILMS DEPOSITED BY RF REACTIVE UNBALANCED MAGNETRON SPUTTERING

Jun Shirahata[1], Tetsutaro Ohori[1], Hiroki Asami[2], Tsuneo Suzuki[1], Tadachika Nakayama[1], Hisayuki Suematsu[1], and Koichi Niihara[1]

[1]Extreme Energy-Density Research Institute, Nagaoka University of Technology, 1603-1 Kamitomioka-cho, Nagaoka 940-2188, Japan

[2]Department of Mechanical Engineering, Tomakomai national college of Technology, 433 Nishikioka, Tomakomai 059-1275, Japan

ABSTRACT

Cr-Si-N-O thin films were prepared by radio frequency (RF) reactive unbalanced magnetron sputtering. Composition of the thin films was analyzed by utilizing Rutherford backscattering spectroscopy. It was found that these thin films contained approximately up to 10at.% silicon. Film thicknesses were measured by using a scanning electron microscope. It was found that specimens have various thicknesses due to each sample position and sputtering rates of targets. Phases in the samples were identified by X-ray diffraction. Diffraction peaks due to Cr_2N, chromium oxides, silicon nitride, and silicon oxide were not observed. All samples showed only broad peaks based on CrN. From the results of Fourier transform infrared spectroscopy, peaks attributed to the Cr-N bond of CrN, peaks attributed to the Si-O bond of SiO_2 and peaks attributed to Silicon Nitride were observed. No peaks attributed to chromium oxide were found. A nano-indentation method was used to measure indentation hardness and elastic modulus of thin films. In accordance with a change of Si content, the hardness increases up to a maximum value of 40GPa.

INTRODUCTION

In recent years, machining industries prefer dry and high speed processes rather lubricated processes to reduce amount of substances of concern (SOC). Therefore, cutting tools are used at severe conditions and hard coatings which have superior characteristics than the conventional materials are required. Chromium nitride (CrN) thin film has many superior characteristics including low friction coefficient and high thermal stability comparing to coatings based on titanium nitride. In previous studies, it was found that Cr(N,O) thin films prepared by pulsed laser deposition (PLD)[1,2,3] have high hardness and good oxidation resistance than those of common CrN. Furthermore, to improve these properties, Asami *et al.* have doped various metal elements into Cr(N,O) thin film[4]. One of them, (Cr,Mg)(N,O) [4] thin films showed excellent hardness of 4300HV. Recently, we have been tried to prepare Cr(N,O) thin films by using RF reactive unbalanced magnetron sputtering[5], and succeed in the preparation of Cr(N,O) thin film.

Since Cr-Si-N[6,7] thin films are well known for high hardness. One hypothesis of the hardening was for amorphous Si_3N_4 layer surrounding CrN fine grain. By replacing CrN by Cr(N,O), we expect that hardness of Cr-Si-N-O thin films can be further increased. In this work, we report mechanical properties of the Cr-Si-N-O thin films.

EXPERIMENTAL PROCEDURE

Cr-Si-N-O thin films were deposited on crystalline silicon (100) wafer of 20 × 20mm^2. Substrates were set liner in front of targets with a working distance of 80mm. Figure 1 shows a schematic drawing of the experimental setup used in this study, and the target configuration was shown in Figure 2. A combinatorial method has been used in this process. Square plates of chromium with 88mm in wide and 150mm in length (purity 99.9%) and silicon with 88mm in wide and 50mm in length (purity 99.999%) were used as targets and set in order as shown in Figure 2. For this reason, specimens have graded compositions of Si.

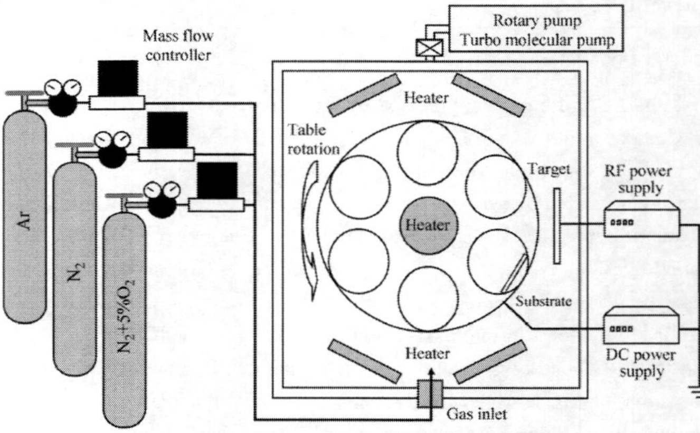

Figure 1. Schematic drawing of the experimental setup.

In this experiment, Ar gas as sputtering gas, and N_2 and N_2+5%O_2 gas as reactive gases were used. Gas flow rates were controlled carefully by using mass flow controllers. O_2 gas flow rate was varied by using N_2+5%O_2 gas flow rate from 0 to 7sccm since we need limited quantity of oxygen. The working pressure of 6.0×10^{-1} Pa was kept by controlling N_2 gas flow rate to minimize change in condition of thin film growth. At that time, Ar and N_2 gas flow rates were 200sccm and approximately 400sccm, respectively. Before the deposition, the chamber was pumped down to less than 1.0×10^{-3} Pa using a rotary pump and a turbo molecular pump. After this evacuating, substrates were heated to approximately 450°C for 1 hour, the evacuation was also kept during the heating process to reduce an influence of residual oxygen. Si substrates were cleaned by Ar ion bombardment with a substrate bias of -650V for 20 minutes. During the deposition, Ar, N_2, and O_2 gases were mixed, and they passed an ion source for ionization. Power of 1000W by an RF source was applied trough the targets and substrates. Substrate bias of -100V was applied and table was rotated at 1rpm. The deposition time was changed to prepare the specimens which have a suitable film thickness for each measurement. The typical experimental conditions are shown in Table I.

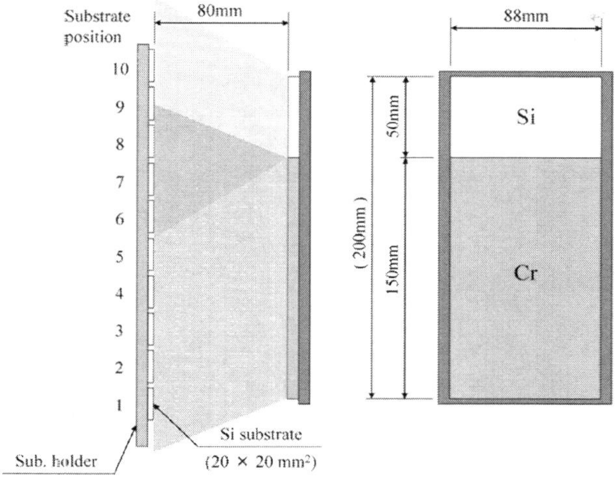

Figure 2. Schematic drawing of target configuration.

Table I Typical experimental conditions for preparation of the Cr-Si-N-O thin films.

Substrate	Si(100), Glassy carbon
Base pressure [Pa]	$<1\times10^{-3}$
Substrate temperature [℃]	450
Target material	Cr(99.9%), Si(99.999%)
Cathode power [W]	1000(RF type, 13.56MHz)
Sub. bias voltage [V]	-100
Ar gas flow rate [sccm]	200
N_2 gas flow rate [sccm]	approximately 400
O_2 gas flow rate [sccm]	0, 7
Pressure in process [Pa]	6×10^{-1}
Film thickness [μm]	0.2, 2

Compositions of the thin films were analyzed by Rutherford backscattering spectroscopy (RBS; NHV Corporation, NT-1700HS) using He^+ ions of 2MeV. By fitting the spectrum, compositions of Cr, Si, N, and O (C_{Cr}, C_{Si}, C_N, and C_O, respectively) were determined. The film thickness was measured by using scanning electron microscope (SEM; JEOL, JSM6700F). Phases in the samples were identified by X-ray diffraction (XRD; Rigaku, RINT2000) using a Cu Kα radiation of 0.15418nm under the operation conditions of 50kV and 300mA. The chemical bonding state was examined by Fourier transform infrared spectroscopy (FT-IR, JASCO, FT/IR-4200). The data were recorded from 400 to 4000cm^{-1} with a resolution of 4cm^{-1}. The FT-IR spectrum was corrected by subtracting the absorbance of the Si substrate. The indentation hardness H_{IT} and elastic modulus E_{IT} of the Cr-Si-N-O thin films were calculated by utilizing a nanoindenter (FISCHER instruments, HM2000). In this measurement, Vickers indenter was used as a load of 5mN.

RESULTS AND DISCUSSION

Figure 3 shows a result of compositional analysis by RBS. C_{Si} was changed <1at.% to 10at.%. C_{Cr} of the thin films was decreased with increasing C_{Si}. Since no differences in samples at substrate positions of #1-#5, only thin films on substrate positions of #5-#10 are used for measuring.

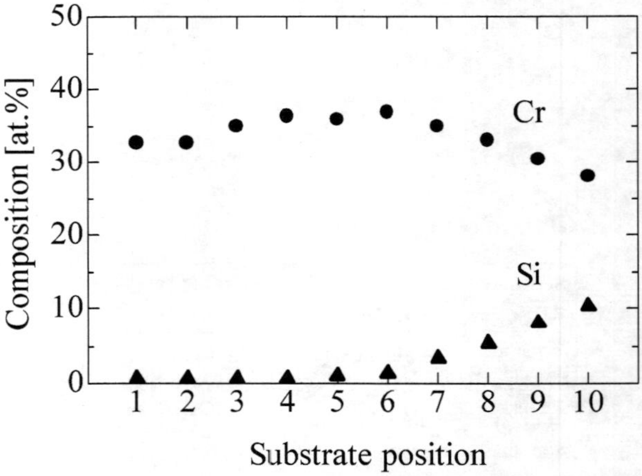

Figure 3. Cr content C_{Cr} and Si content C_{Si} of Cr-Si-N-O thin films.

Cross-section images of the thin films which were set at substrate positions of #5, #7, and #10 were displayed in Figure 4. Change of the film thicknesses is shown in Figure 5. Film thicknesses were influenced strongly by the substrate positions and sputtering rates of targets since the sputtering rate of Si is lower than that of Cr.

XRD patterns of the thin films which have various C_{Si} are shown in Figure 6. For discussion propose, CrN data according to the International Center for Diffraction Data (ICDD)[8] are referenced at bottom of Figure 6. The XRD patterns of Cr-Si-N-O thin films matched with the referenced data of CrN. The diffraction peaks due to Cr_2N, chromium oxide, silicon nitride, and silicon oxide were not observed. It suggests that the Cr-Si-N-O thin films are mainly of NaCl structure based on CrN. Si exists in either an amorphous phase of silicon nitride and/or silicon oxide or the crystalline phase to form (Cr,Si)(N,O). Until 3at.% of C_{Si}, Cr-Si-N-O thin films orientated CrN-111. Above 5at.% of C_{Si}, the orientations were changed to CrN-200.

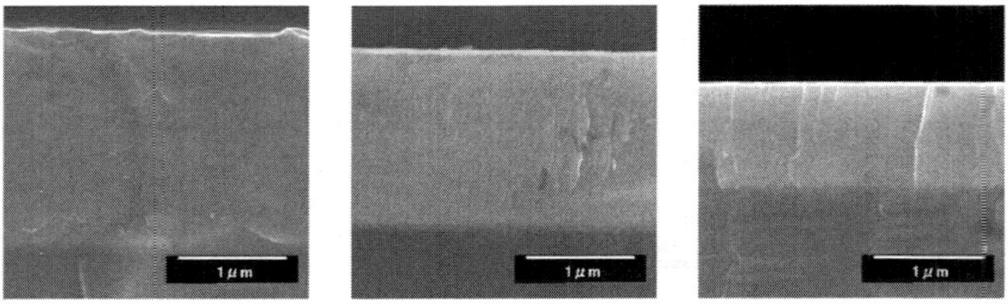

(1) #5 (2) #7 (3) #10

Figure 4. Cross-section images of the Cr-Si-N-O thin films which set at different substrate position.

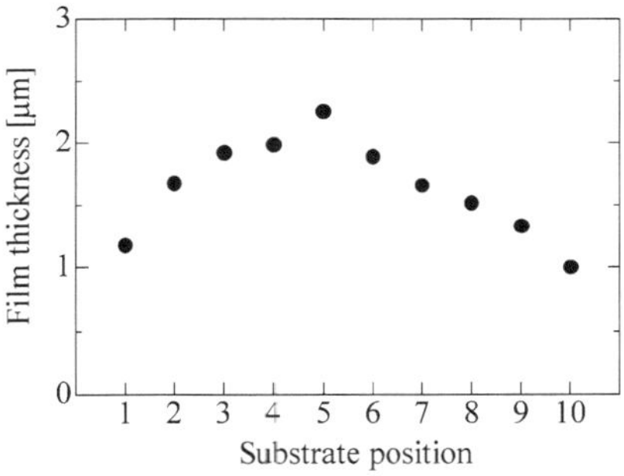

Figure 5. Change of film thickness.

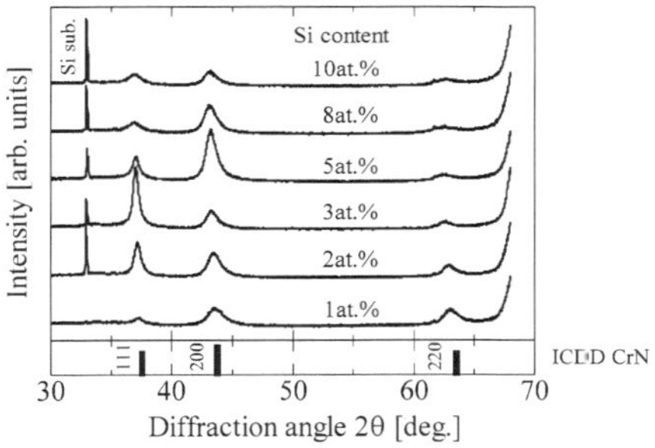

Figure 6. XRD patterns of Cr-Si-N-O thin films.

FT-IR spectra of the Cr-Si-N-O thin films are shown in Figure 7, and the spectra of some related compounds[9-12] are displayed at the bottom of Figure 7. From the results, peaks around $500cm^{-1}$ and the broad peaks around $1100cm^{-1}$ were observed. In the case of Cr-Si-N, it is well known that Si exists in amorphous Si_3N_4 and/or Si dissolved into CrN[6] at low Si content. Therefore, we suppose that an amorphous Si_3N_4 is formed until 3at.% of C_{Si}, and an amorphous SiO_2 is made above 5at.% of C_{Si}.

Indentation hardness and elastic modulus of the thin films are shown in Figure 8. The hardness was increased to maximum value of 40GPa with increasing C_{Si} up to 3.5at.%, subsequently it was decreased. This result is similar to change of the orientation. It is well known that mechanical properties are affected by orientation of the thin films. In the Cr-Si-N-O thin films, the thin films being orientated to CrN-111 is harder than the films being orientated to CrN-200. On the other hands, the elastic modulus was changed similar with the change of hardness. This implies that hardening is caused by the elastic modulus increase due to the compositional change and the effect of the orientation to the hardness change in minor.

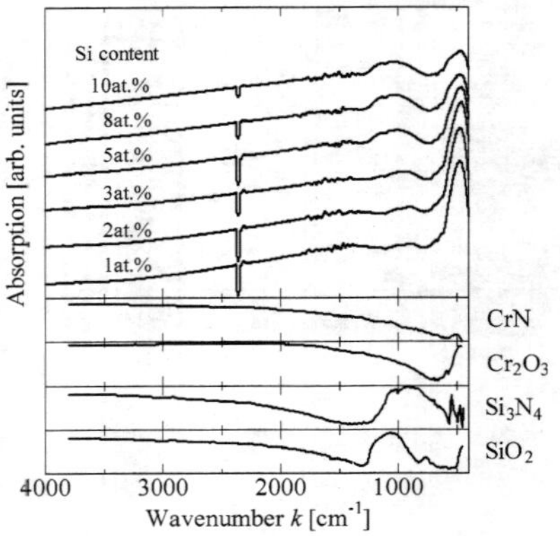

Figure 7. Absorption spectra of samples.

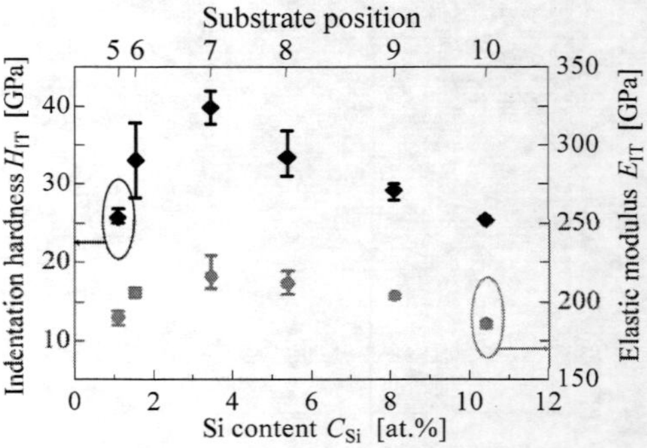

Figure 8. Indentation hardness of the thin films.

CONCLUSION

We prepared the Cr-Si-N-O thin films, and measured phases, microstructure and mechanical properties. It was found that those thin films has up to 10at.% of C_{Si}, crystal structure of all samples are NaCl structure based on CrN. From the results of XRD and FT-IR, it was suggested that Si was existing in an amorphous phase or (Cr,Si)(N,O) phases. The indentation hardness and elastic modulus were increased with increasing C_{Si} up to 3.5at.% because of elastic modulus increase, subsequently it was decreased above 5at.% of C_{Si}.

ACKNOWLEDGMENT

The authors want to thank Mr. Obata, Mr. Katagiri, and Mr. Sekimoto of Nagaoka University of Technology for RBS analysis. We thank all students in Extreme Energy–Density Research Institute for discussing about this study.

REFERENCES

[1]T. Suzuki, H. Saito, M. Hirai, H. Suematsu, W. Jiang and K. Yatsui, Preparation of $Cr(N_x,O_y)$ thin films by pulsed laser deposition, *Thin Solid Films*, **407**, 118-121 (2002).

[2]J. Inoue, H. Saito, M. Hirai, T. Suzuki, H. Suematsu, W. Jiang and K. Yatsui, Mechanical Properties and Oxidation Behavior of Cr-N-O Thin Films Prepared by Pulsed Laser Deposition, *Trans. Mat. Res. Soc. Jpn.*, **28**, 421-424 (2003).

[3]T. Suzuki, J. Inoue, H. Saito, M. Hirai, H. Suematsu, W. Jiang and K. Yatsui, Influence of oxygen content on structure and hardness of Cr–N–O thin films prepared by pulsed laser deposition, *Thin Solid Films*, **515**, 2161-2166 (2006).

[4]H. Asami, J. Inoue, M. Hirai, T. Suzuki, T. Nakayama, H. Suematsu and K. Niihara, Hardness optimization of (Cr,Mg)(N,O) Thin Films Prepared by Pulsed Laser Deposition, *Advanced Materials Research*, **11**, 311-314 (2006).

[5]J. Shirahata, T. Ohori, H. Asami, T. Suzuki, T. Nakayama, H. Suematsu, Y. Nakajima and K. Niihara, Preparation of Cr(N,O) thin films by RF reactive unbalanced magnetron sputtering, in submission.

[6]H. Y. Lee, W.S. Jung, J. G. Han, S. M. Seo, J. H. Kim and Y. H. Bae, The synthesis of CrSiN film deposited using magnetron sputtering system, *Surf. Coat. Technol.*, **200**, 1026-1030 (2005).

[7]A. Thobor-Keck, F. Lapostolle, A. S. Dehlinger, D. Pilloud, J. F. Pierson and C. Coddet, Influence of silicon addition on the oxidation resistance of CrN coatings, *Surf. Coat. Technol.*, **200**, 264-268 (2005).

[8]Powder Diffraction File, ICDD International Center for Diffraction Data : CrN (11-0065).

[9]R. A. Nyquist and R. O. Kagel, Infrared Spectra of Inorganic Compounds, Academic Press, 116-117 (1971). : Chromium(III) nitride CrN

[10]R. A. Nyquist and R. O. Kagel, Infrared Spectra of Inorganic Compounds, Academic Press, 216-217 (1971). : Chromium(III) sesquioxide Cr_2O_3

[11]R. A. Nyquist and R. O. Kagel, Infrared Spectra of Inorganic Compounds, Academic Press, 114-115 (1971). : Silicon nitride Si_3N_4

[12]R. A. Nyquist and R. O. Kagel, Infrared Spectra of Inorganic Compounds, Academic Press, 210-211 (1971). : Silicon dioxide SiO_2

ROOM-TEMPERATURE DEPOSITION AND MAGNETO-OPTICAL PROPERTIES OF TRANSPARENT COBALT/LEAD ZIRCONATE TITANATE (PZT) NANOCOMPOSITE FILMS BY AEROSOL DEPOSITION

Jae-Hyuk Park and Jun Akedo
National Institute of Advanced Industrial Science and Technology (AIST),
1-2-1 Namiki, Tsukuba, Ibaraki 305-8564, JAPAN

ABSTRACT

Highly transparent magnetic nanocomposite thick films consisting of cobalt nanoparticles embedded in a host matrix of lead zirconate titanate (PZT) were prepared by aerosol deposition (AD) at room temperature. Transparent thick nanocomposite cobalt/PZT films display very dense without any pores as well as any cracks. As the wt% of nanocobalt increased, magneto-optic effect measured in the transmission mode, especially Faraday rotation effect of nanocomposite layers gradually increases with good linearity. The 0.1 wt% nanocobalt -containing nanocomposite film acquired the Faraday rotation angle of approximately 0.14 degree and the transmittance of more than 30%.

INTRODUCTION

Nanocomposite systems, materials containing particles of nanometer dimensions have shown interesting properties related to its extremely small size. Some of their optical, magnetic, electronic, mechanical, and chemical properties are different from those exhibited by the same composition in bulk material [1, 2]. Of these magnetic nanocomposite systems have become a subject of growing interest. In particular, transparent magnetic materials have been the subject of many investigations because of their novel properties and potential applications. The attractive properties of the transparent magnetic compounds are related to the magneto-optical effects. The scientific as well as industrial applications for this technology include optical fiber sensors, optical isolators, information storage, magnetooptical switches, modulators, magnetic field and electric field sensors etc [3-7].

Nanocomposite systems have been prepared using various conventional method including sputtering, sol–gel, colloidal solutions, ion implantation, Chemical Vapor Deposition (CVD) and other hybrid methods [1-7]. Since these conventional methods are based on deposition of atom and cluster size with two or several kinds of element, it is very difficult that the methods may control the size and the shape of the nanomagnetic particles in the host matrix. For example, it may be difficult to acquire a good spatial distribution and control a desirable concentration of the nanomagnetic particles in the host matrix. Further, these methods require a high temperature annealing processes of more than 600 ℃ in order to make grain growth of several tens nanometer from nanoparticles with several nanometers. Also, it may be difficult to prepare highly transparent nanocomposite films because of high concentration of nanoparticles with several nanometers in host matrix.

Our group have proposed novel concept for materializing room-temperature deposited functional ceramic films using aearosol deposition (AD) [8-12] and has studied transparent optical nanoocmposite systems, including Au / lead zirconate titanate (PZT) [13]. In this study, we report a room-temperature deposition of highly transparent cobalt/lead zirconate titanate (PZT) nanocomposite thick films by the aerosol deposition (AD) and demonstrate its optical, magnetooptical, and structural properties.

EXPERIMENTAL PROCEDURES

Recently, as a novel ceramic film-forming method, an aerosol deposition (AD) has been developed by Akedo. et al [8-12]. Figure 1 shows an illustration of an AD system for preparing magnetic naoocomposite layer. AD has been attracting much attention for the fabrication of thin and thick complex oxide films with a high deposition rate and a low process temperature on different kinds of substrates. This novel method makes use of crystalline fine particles with multi-composition as raw materials and has no change of original composition from starting materials before and after coating, compared to conventional coating method including sputter, sol-gel and CVD etc. AD is good candidate to prepare nanocomposite films with various kinds of host dielectric matrix from metal oxide to complex oxide.

Figure 2 shows a schematic of novel concept for the preparation of magnetic nanocomposite systems with AD. Since the AD can make a itself structure of composite powder as a dense thick layer as described above, our new concept for the preparation of nanocomposite systems with AD has many merits. For example but not by way of limitation, this concept can further apply various sizes of nanomagnetic particles, and achieves desirable distribution of nanomagnetic particles in dielectric matrix because the structure of deposited layer is similar to that of composite powder. The AD can tailor desired concentration of nanomagnetic particles in host matrix by just controlling an amount of the given nanomagnetic particles mixing with host matrix particles. These properties strongly indicate that AD has enough possibilities of preparing new nanocomposite systems with wide range of practical applications.

For preparing nanocomposite films, a composite metal-dielectric powder is prepared from PZT (Zr/Ti=52/48) powder and nano-particles of cobalt (20- 50 nm), as shown in Fig.3. The size of PZT is approximately 200-500 nm. The concentrations of nanocobalts are about 0 to 1 wt%. Transparent magnetic nanocomposites layer was directly deposited on glass substrate at room temperature by AD using composite magnetic-dielectric powder, as shown in FIG. 3. The typical layer thickness was about 1-10 μm with process conditions of 4 - 6 L/min of N_2 gas, a deposition time of about 3-5 min.

RESULTS AND DISCUSSION

Transparent thick nanocomposite cobalt/PZT films were obtained with AD for the first time and display very dense without any pores as well as any cracks. All specimens are room-temperature-deposited. Figures 4 and 5 show a TEM and an EDX picture of cobalt element in 0.1 wt% nanocobalt -containing PZT/cobalt nanocomposite layer where yellow parts indicate cobalt particles. The nanocobalt particles have good spatial distribution in PZT matrix, as shown in Fig.5. The size of nanocobalt particles are from 20 nm to less than 150 nm with some aggregation of magnetic particles. The magnetic properties and annealing behaviors of magnetic nanocomposite films will be subject of another report.

Figure 6 shows XRD patterns of room-temperature deposited 0.005, 0.02 and 0.05 w% cobalt-containing PZT films. All the diffraction peaks can be assigned to a randomly oriented perovskite structure. And no peak of cobalt is detected since too small amount is contained. Figure 7 shows optical photographs of 0.005, 0.02, 0.05 and 0.1 wt% nanocobalt -containing nanocomposite cobalt/PZT layer with a thickness of 1 μm. The colors of each specimens are very light brown to deep brown with the increase of cobalt wt% concentration. Figure 8 shows a plot of (a) transmittance spectra and (b) transmittance value of 0.005, 0.02, 0.05 and 0.1 wt% nanocobalt -containing nanocomposite cobalt/PZT layer. The optical properties of 0.005 wt% nanocobalt -containing nanocomposite cobalt/PZT layer are very similar to pure PZT films. As the wt% of nanocobalt increased, transmittance of nanocomposite layers at 633 nm gradually decreases with good linearity. Figure 9 shows a plot of (a) Faraday rotation hysteresis loop and (b) Faraday rotation angle of 0.005, 0.02, 0.05 and 0.1 wt% nanocobalt -containing nanocomposite cobalt/PZT layer at 532 nm. As the wt% of nanocobalt increased, magneto-optic effect measured in the transmission mode, especially Faraday rotation effect of nanocomposite layers gradually increases with good linearity.

From results of optical properties, we know that nanocoablt particles were spatially very well distributed in the PZT matrix and optical transmittance of the nanocomposite films could be precisely controlled by adjusting the concentration of nanocobalt in the host matrix. The optical transmittance of magnetic nanocomposite film is higher than that of conventional ferromagnetic metal films with continuous structure which show optical transmittance of approximately 23% at a 10 nm cobalt thin film deposited on a glass [7]. This is caused by reduction of the optical losses induced by the free electrons in nano-magnetic metal particles, which are much smaller than the wavelength of the incident light. Further, the Faraday rotation effect of 0.05 wt% nanocobalt -containing nanocomposite film is as strong as that of conventional 10 nm thick Co thin films (approximately 0.08 deg.).

CONCLUSION

Highly transparent magnetic nanocomposite thick films consisting of cobalt nanoparticles embedded in a host matrix were prepared by AD for the first time. The 0.1 wt% nanocobalt -containing nanocomposite film acquired the Faraday rotation angle of approximately 0.14 degree and the transmittance of more than 30%. The size control of the magnetic particle the host matrix may allows one to enhancement of faraday rotation effect beyond bulk magnetic metal. We are studying multiferroic behaviors of cobalt/ PZT nanocomposite systems. Transparent magnetic nanocomposite thick films fabricated by ADM would be good candidates for application to optical modulators and switches.

REFERENCES

[1] J. Homola, Sinclair S. Yee and Günter Gauglitz, Actuators B : Chemical, Vol. 54, pp. 3-15, 1999.
[2] C.Clavero, A. Cebollada, and G. Armelles, Phys. Rev. B, Vol. 72, pp.024441-1-6, 2005.
[3] A.L. Stepanov and R.I.Khaibullin, Rev. Adv. Matter. Sci., Vol. 7, pp. 108-125, 2004.
[4] M. Abe, M Gomi, and F. Yokoyama, J. Appl. Phys., Vol. 57, pp. 3909-3911, 1985.
[5] K. Baba, F. Takase, and M. Miyagi, Opt. Comm., Vol. 139, pp. 35-38, 1997.
[6] A. Horikawa, K. Yamaguchi, M. Inoue, T. Fujii, and K. I. Arai, Mat. Sci. Eng., Vol. A217/218, pp. 348-352, 1996.
[7] N. Richard et. al., J. Appl. Phys., Vol.88, pp.2541-2547, 2000.
[8] J. Akedo and M. Lebedev, Jpn. J. Appl. Phys., Vol.38, pp.5397–5401, 1999.
[9] J. Akedo and M. Lebedev, Appl. Phys. Lett., Vol.77, pp.1710–1712, 2000.
[10] M. Nakada, K. Ohashi, and J. Akedo, Vol.275, pp.e1275–e1280, 2005.
[11] J. Akedo, Material Science Form, Vol.449-452, pp. 43-48, 2004.
[12] J. Akedo, J. Am. Ceram. Soc., Vol. 89, pp. 1834-1839, 2006.
[13] Jae-Hyuk Park, Jun Akedo and Masafumi Nakada, Jpn. J. Appl. Phys., Vol.45, pp. 7512-7515, 2006.

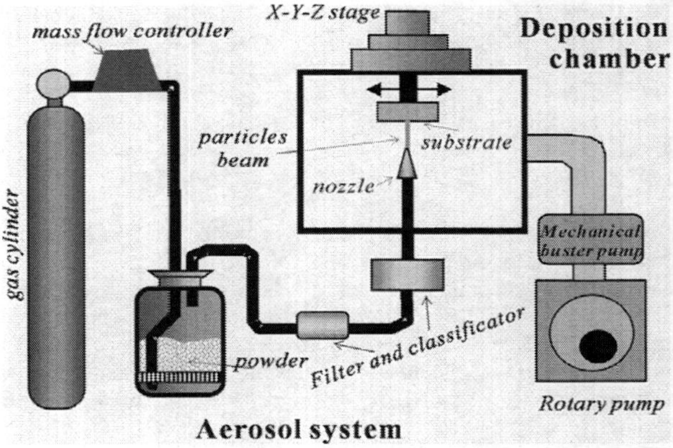

Fig. 1. An illustration of an AD system for preparing magnetic naoocomposite layer.

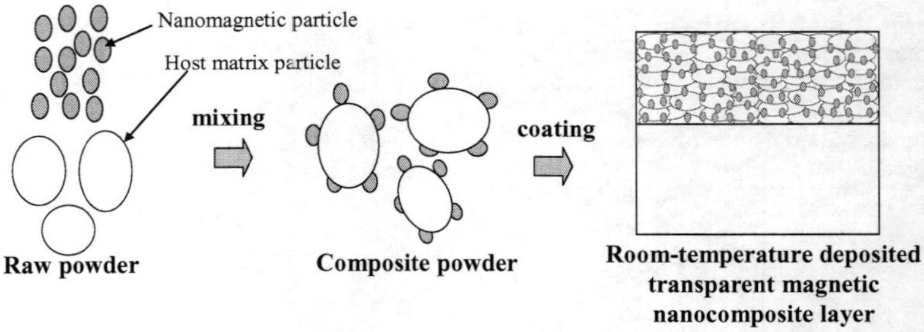

Fig. 2. A schematic of new concept for the preparation of nanocomposite systems with AD method.

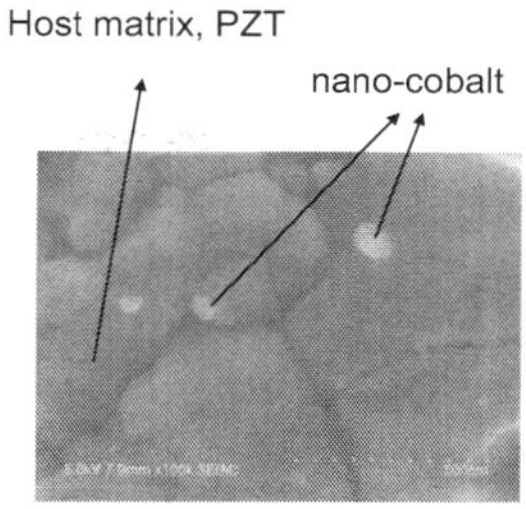

Fig. 3. FE-SEM picture of a magnetic nanocomposite powder of PZT powder and nano particles of cobalt (20- 50 nm)

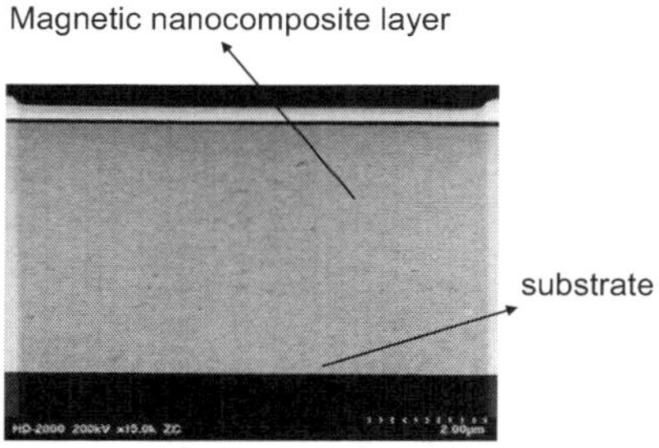

FIG. 4 is a TEM picture of 0.1 wt% nanocobalt -containing PZT/cobalt nanocomposite layer.

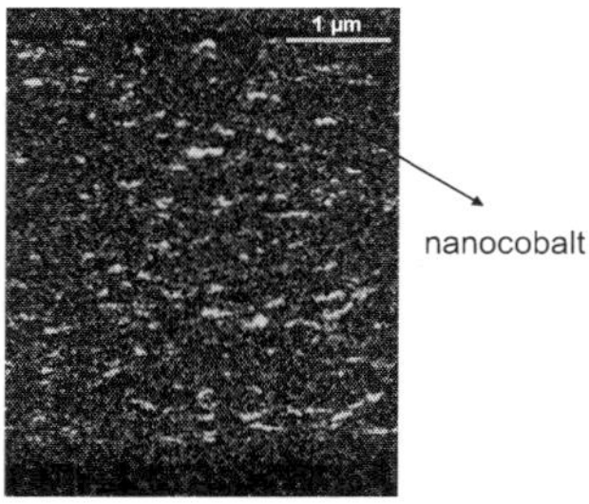

FIG. 5 is an EDX picture of cobalt element in 0.1 wt% nanocobalt -containing PZT/cobalt nanocomposite layer. Yellow parts allow cobalt particles.

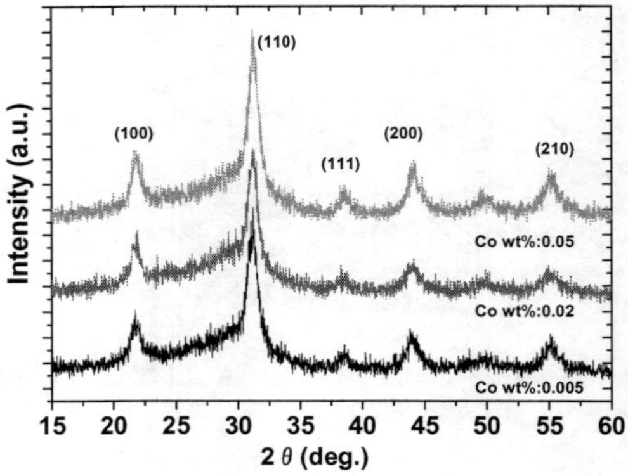

Fig. 6. XRD patterns of room-temperature deposited 0.005, 0.02 and 0.05 w% cobalt-containing PZT films

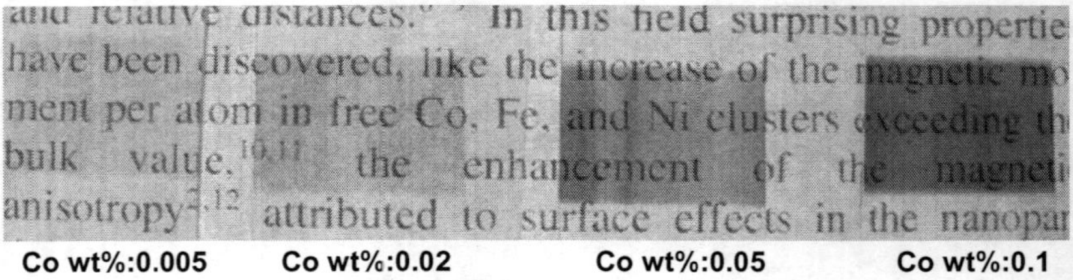

Fig. 7. Optical photographs of 0.005, 0.02, 0.05 and 0.1 wt% nanocobalt -containing nanocomposite cobalt/PZT layer with a thickness of approximately 1 μm.

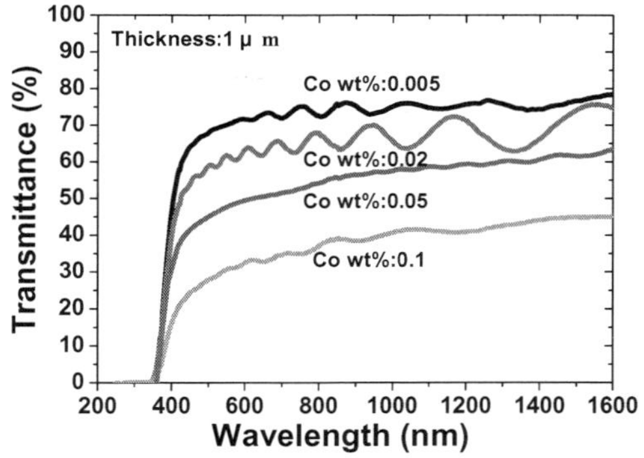

(a)

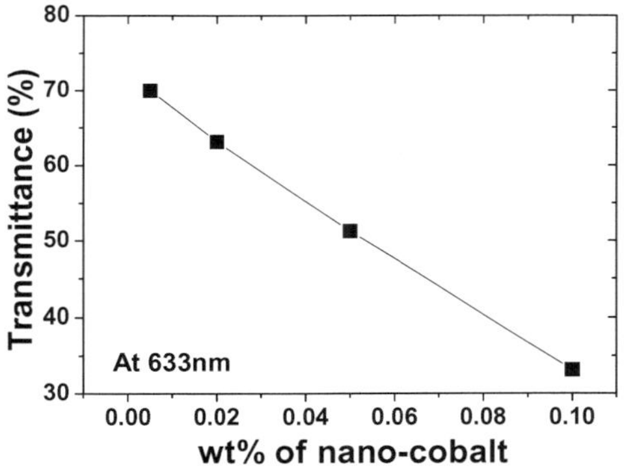

(b)

Fig. 8. A plots of (a) transmittance spectra and (b) transmittance value of 0.005, 0.02, 0.05 and 0.1 wt% nanocobalt -containing nanocomposite cobalt/PZT layer.

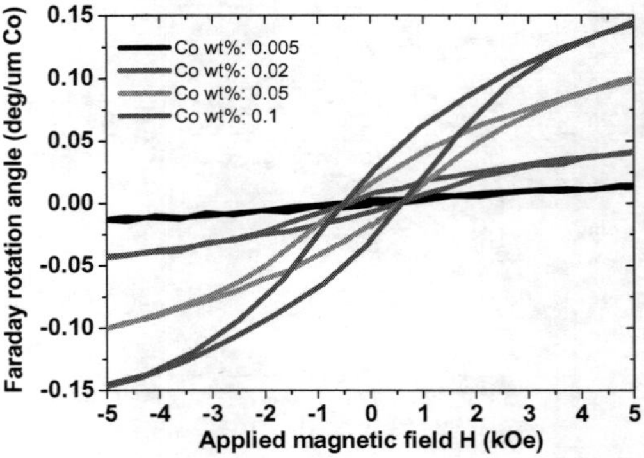

(a)

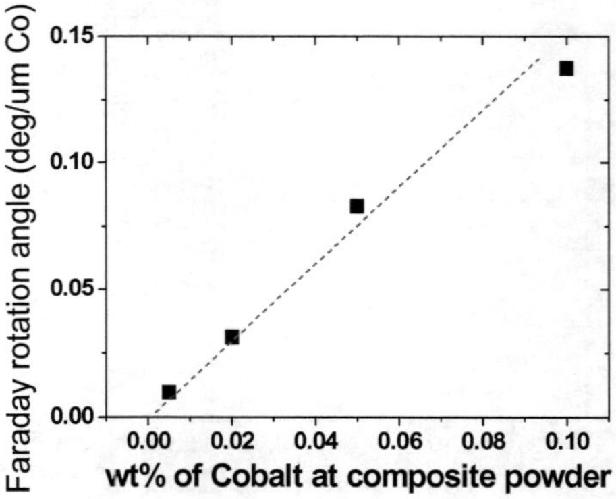

(b)

Fig. 9. A plot of (a) Faraday rotation hysteresis loop and (b) Faraday rotation angle of 0.005, 0.02, 0.05 and 0.1 wt% nanocobalt -containing nanocomposite cobalt/PZT layer at 532 nm.

INFLUENCE OF DISPERSANT ON RHEOLOGY OF ZIRCONIA-PARAFFIN FEEDSTOCKS AND MECHANICAL PROPERTIES OF MICRO PARTS FABRICATED VIA LPIM

Fatih A. Çetinel[†*], Marcus Müller [†], Joachim Rögner [‡], Werner Bauer [†] and Jürgen Hausselt [†]

[†] Karlsruhe Institute for Technology, Institute for Materials Research III, Hermann-von-Helmholtz-Platz 1, 76344 Eggenstein-Leopoldshafen, Germany

[‡] Karlsruhe Institute for Technology, Institute for Materials Science and Engineering I, Kaiserstrasse 12, 76131 Karlsruhe, Germany

ABSTRACT

The interest in miniaturization of components is rising in many sectors such as medical technology. Due to their specifically good mechanical, chemical, thermal, tribological and other physical properties ceramic micro components are gaining importance in microsystems technology (MST). Micro powder injection molding (μ-PIM) enables accurate manufacturing of complex-shaped micro parts. In particular, low-pressure injection molding (LPIM) is predestined for the fabrication of prototypes and small series production of ceramic micro components, as tooling costs are low and simple molds can be used. In this study the dispersant's influence on the rheological properties of zirconia-paraffin LPIM feedstocks and on the mechanical properties of sintered micro bending bars (200 x 200 x 1200) μm^3 was investigated. A considerable time dependency of the viscosity and the yield stress of the feedstocks were found, which affects further processing and reproducibility. Dispersant concentration has also an impact on the bending strength calculated from three-point bending tests. During thermal debinding the surface tension of the binder phase is influenced by the dispersant and, thus, leveling and healing of surface defects of the micro parts can take place. The thermal debinding step was investigated with real-time observations to assess the effect of various parameters on the defect healing process.

INTRODUCTION

Due to their unique properties, the interest in design and manufacturing of ceramic microcomponents is increasing in microsystems technology.[1,2] Plastic shaping techniques, in particular ceramic injection molding, enable near net shape fabrication of complex shaped ceramic components. Low-pressure injection molding (LPIM), sometimes also referred as hot molding, is predestined for the fabrication of prototypes and small series production of ceramic microcomponents, as simple molds with low tooling costs and low-viscosity feedstocks with favorable flow properties mostly consisting of paraffin and dispersants (surfactants) as the binder component are used.[1,3,4]

In order to obtain fairly isotropic behavior of the microparts, the particle size of the used powder should be at least about one order of magnitude smaller than the minimum internal dimension of the microcomponent.[5,6] For this, submicron or nano-powders have to be used to reproduce details down to micrometer range. Colloidal processing techniques have to be exploited to stabilize and to homogenize small particles in the dispersing medium.[7] For instance, the adsorption of carboxylic acids like stearic acid on zirconia enables the coupling of ZrO_2 particles to the nonpolar paraffin matrix, causes a decrease of attractive interparticle forces and enables well homogenized feedstocks with higher solids loadings compared to dispersant-free suspensions. As a result, in feedstock development special attention must be paid to the influence of the surfactant on feedstock rheology due to the fact that the binder composition and, thus, the flow behavior may have a remarkable effect on the properties of green and sintered microcomponents and also may impact the thermal debinding of green parts.[8-11]

In this study, the influence of dispersant concentration and storage time at elevated storage temperature on the rheological properties were investigated. Furthermore, the impact of dispersant on thermal debinding of green parts and mechanical properties of micro bending bars was evaluated. An increasing amount of dispersant results in a change of feedstock viscosity and yield point. Usually, high dispersant amounts result in enhanced flowability by decreasing the viscosity. However, this strongly depends on the applied shear rate and is discussed in this study. For the fabrication of microcomponents only a small amount of feedstock is necessary. In contrary, for the running of a LPIM machine feedstock volumes far above this amount have to be used. Therefore, feedstocks of proper amount have to be produced and stored for a certain time until used up completely. Initial flow properties can be regained by redispersing the feedstock after a short storage time.[12] The long-term stability of zirconia-paraffin feedstocks with varying amounts of dispersant, however, has not been studied yet. Depending on the used feedstock system, a change in rheological properties like viscosity or yield point of the suspensions with respect to the storage time may occur that would affect further processing and reproducibility. Real-time observations of the thermal debinding process of micro bending bars enable the evaluation of the influence of various parameters on the leveling and healing of surface defects, which represent the common flaw type for ceramic parts in the micrometer scale. The possibility to be able to control the defect leveling during thermal debinding opens up the chance to improve the mechanical properties of as-fired microcomponents during the debinding step cost and time-effectively, as surface treatments like grinding or polishing can be spared.

Previous studies showed that densities higher than 99.5 % of the theoretical density of sintered microcomponents could be achieved by using Hypermer LP1 as dispersant for low-pressure injection molding of TZ-3YS-E zirconia-paraffin feedstocks.[8,9] With other commercial dispersant systems such high sintered densities could not be achieved, which indicates that Hypermer LP1 results in a better microstructural homogeneity. The high sintered density, low porosity, low surface roughness and exceptional high strength of as-fired microcomponents fabricated via LPIM of zirconia-paraffin feedstocks illustrate the advanced feedstock quality, when Hypermer LP1 is used as dispersant.

EXPERIMENTAL

Feedstocks were prepared containing commercial zirconia powder with 3 mol% Y_2O_3 (TZ-3YS-E, Tosoh, Japan) with an average particle size of 0.39 µm, monomodal particle size distribution, virtually spherical particle shape and a specific surface area of 6.6 m^2/g. The binder formulation is based on paraffin (TerHell Paraffin 6403, Schümann Sasol, Germany) with a setting point of 64.5°C and a commercial dispersant (Hypermer LP1, Croda, UK). The powder was dried at 300°C for 2 h before homogenization and was kept at 120°C prior to adding to the molten binder. A laboratory vacuum dissolver (Dispermat, VMA-Getzmann, Germany) was used for plastification and homogenization of the powder-binder mixture at 90°C. The rheological properties of the feedstocks were analyzed with a rotational rheometer (MCR 300, Anton-Paar, Austria) by using a plate-plate measuring system (PP25, gap between plates: 0.5 mm). The yield stress was obtained by shear stress-controlled measurements and the viscosity was determined at shear rates of 10 s^{-1}, 100 s^{-1} and 1000 s^{-1} as reference values and at standard working temperature of 90°C. The solids loadings of the feedstocks were kept constant at 50 vol%, whereas the amount of the surfactant varied from 1.6 mg/m^2 to 3.2 mg/m^2 (Table I). The flow behavior of freshly prepared feedstocks and suspensions stored at 80°C (beaker covered by a petri dish to exclude possible vaporization of binder components) for 1, 2 and 7 days and stored at room temperature (set suspensions) for at least 50 days, respectively, was determined to evaluate the influence of storage time on rheology. Measurements of feedstocks aged at 80°C and room temperature were conducted after homogenization for 5-10 min. To ensure reproducibility each measurement was performed generally three times.

Table I. Composition and abbreviations of the investigated feedstocks.

Feedstock	Solids loading Φ [vol%]	Paraffin wax [vol%]	Surfactant [vol%]	Surfactant concentration [mg/m²]
Feed-1	50.0	46.6	3.4	1.6
Feed-2	50.0	46.1	3.9	1.8
Feed-3	50.0	45.7	4.3	2.0
Feed-4	50.0	44.9	5.1	2.4
Feed-5	50.0	44.0	6.0	2.8
Feed-6	50.0	43.1	6.9	3.2

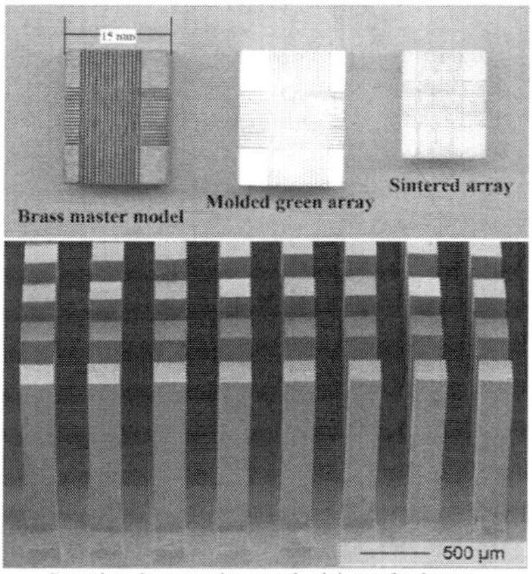

Figure 1. Illustrations of the brass master model, molded green array, sintered array (top) and scanning electron microscope image of sintered bending bars (bottom).

Bending bars were prepared by using a rapid prototyping process chain (RPPC).[1,13,14] A brass array of 20 mm in length, 15 mm in width and 8 mm in height with 15 x 15 columns of square cross section manufactured via micro milling was used as master model to cast silicone molds, which were used, in turn, for the replication of the test specimens via hot molding (Fig. 1). Feedstocks at standard working temperature were manually molded into the preheated silicone molds and evacuated under rigorous shaking by using a tempered jigging screen. This pressureless mold filling technique is possible due to the low viscosity of the applied feedstocks and, thus, was favored over a semi-automatic LPIM machine, as the processed feedstock amount is considerably lesser than using the latter. Thermal debinding of the column arrays was conducted at 500°C on porous alumina plates (Keralpor 99, Kerafol, Germany) by using a furnace with air circulation and drawing-off for the volatile pyrolysis products (HT6/28, Carbolite, UK). The pressureless sintering step was performed with a constant heating rate of 3 K/min up to 1450°C and a dwell time of 60 min in air atmosphere (VMK1800, Linn High Term, Germany). The columns exhibit dimensions of about (250 x 250 x 1500) μm^3 in the green state and approximately (200 x 200 x 1200) μm^3 after sintering (Fig. 1). For determination of the characteristic strength and Weibull modulus, a set of minimum 10 columns per array was characterized according to EN 843-5[15] using a micro three-point-bending device, which is described in detail elsewhere[16]. Real-time observations of the thermal debinding process were performed by using a video-based contact angle meter (OCA 10, DataPhysics Instruments, Germany) with a heating chamber up to 400°C.

RESULTS AND DISCUSSION

Influence of Surfactant Concentration on Rheological Properties

The yield point represents the rigidity of the particle networks resulting in a certain shear stress (yield stress), which is needed to overcome and to induce flow in the suspension.[12] Hence, sedimentation and phase separation in the suspension is prevented by the yield stress. Representative deformation curves of freshly prepared feedstocks with varying amounts of surfactant are illustrated in Fig. 2. With increasing surfactant content the yield point is decreasing remarkably, also reported in literature.[17,18] This observation can be explained by the fact that attractive interparticle forces among zirconia particles are lowered by the dispersant molecules due to steric or semi-steric stabilization.[19,20] According to the information of the manufacturer a relatively high molecular weight gives a steric stabilization barrier some 2 to 4 times thicker than conventional fatty acid based dispersants, e.g. stearic acid. However, no further information regarding the chemical structure of the surfactant is available. According to literature[21] Hypermer LP1 (formerly Hypermer KD4) is identical to the dispersant Solsperse 3000 (Lubrizol, formerly Zeneca, UK), which consists of a pentamer made up of one stearic acid unit and four units of 12-hydroxystearic acid ($C_{90}H_{172}O_{10}$) yielding a molecular weight of 1414 g/mol. Indeed, physical properties like density, viscosity, flash point and color of both dispersants are identical at first glance. Pursuant to available information we assume that Hypermer LP1 offers steric and not merely semisteric stabilization.

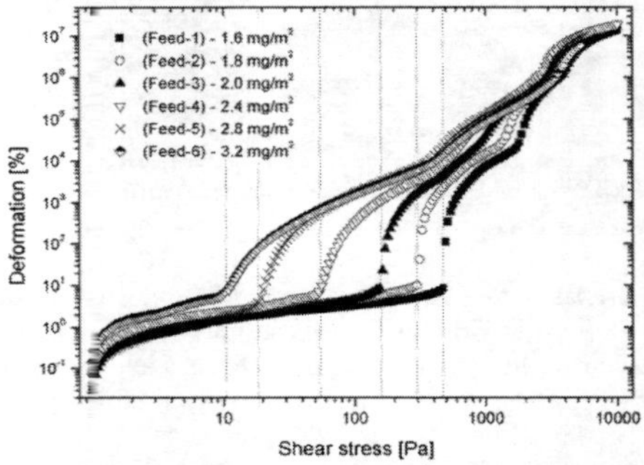

Figure 2. Representative deformation curves showing the yield stress dependency on the surfactant concentration of freshly prepared zirconia-paraffin feedstocks at 90°C.

In Fig. 3 the dependency of the viscosity of freshly prepared feedstocks on the surfactant content at different shear rates is illustrated. For shear rates at 10 s^{-1} the viscosity is decreasing almost exponentially. In contrary, a viscosity minimum for 100 s^{-1} and 1000 s^{-1} was found at 2.4 mg/m^2, respectively. Two effects are supposed to result in an increase of the viscosity after passing through a minimum. Firstly, entanglement of non-polar chains of the surfactant molecules may occur at high shear rates (>10 s^{-1}) for surplus amount of surfactant resulting in increased viscosity. Secondly, crosslinking of polar ends of the surfactant molecules may occur, which also would increase the viscosity.[22] Similar behavior was reported in literature.[20,21] Shear rates much higher than 10 s^{-1} have to be expected for molding by applying a significant injection pressure, e.g. by using a LPIM machine.[23]

For rapid prototyping processes applied in this study, where the feedstock is poured manually into soft silicone molds, shear rates below 1000 s^{-1} have to be expected.

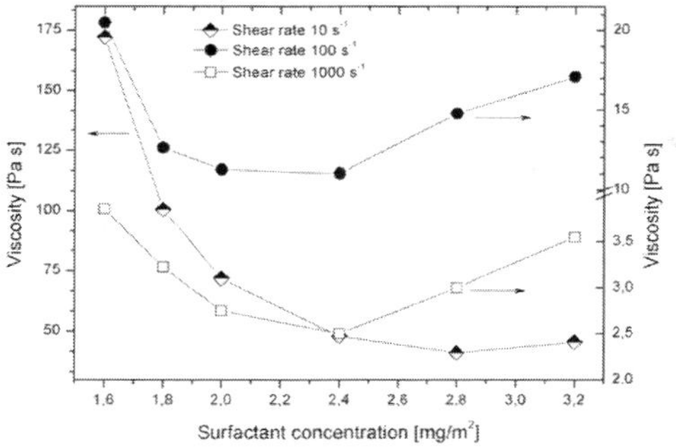

Figure 3. Feedstock viscosity versus surfactant concentration for different shear rates.

The power law according to Ostwald and de Waele was exploited to evaluate the flow behavior of the feedstocks in dependency on the surfactant concentration:

$$\eta = k\dot{\gamma}^{(n-1)} \tag{1}$$

where η stands for the viscosity, $\dot{\gamma}$ for the shear rate and n for the flow index. Flow indices n were calculated for the shear rate range between 10 s^{-1} – 1000 s^{-1}, as the power law generates accurate values only in middle range shear rates.[24] Values between 0.07 and 0.56 with correlation coefficients higher than 0.98 were calculated depending on the surfactant concentration revealing a pseudoplastical (shear thinning) behavior of the feedstocks, also reported elsewhere[25]. With increasing dispersant concentration the flow indices are also increasing, implying that the suspensions approximate to Newtonian behavior. For high amounts of surfactant (3.2 mg/m^2) the flow index reaches a plateau, which is in agreement with literature[21]. With increasing surfactant content the flow indices of the feedstocks also increase, enhancing the moldability of the LPIM feedstocks. Feedstock development for LPIM, however, requires a compromise between the moldability (low viscosity and yield stress, high flowability) and the resistance to deformation during thermal debinding (high viscosity and yield stress, low flowability).

Time Dependency of Rheological Properties

A time dependency of viscosity and yield point on storage time (stored at 80°C) of freshly prepared feedstocks was observed. An increase of viscosity and yield point was expected with time according to literature[26], as no powder pre-treatment like chemisorption of the surfactant onto the zirconia particles was applied to stabilize the powder against humidity or desorption of the merely physically adsorbed dispersant molecules. In contrary, a remarkable decrease of the yield point was found (Fig. 4). In the first 2 days of storage the yield point decreased gradually and remained almost constant. This implies that chemical or physical effects occur within the feedstock as a function of time. Even, re-melting of the feedstocks after aging of at least 50 days at room temperature caused no difference in the yield point, which is in agreement with a recent study[12]. A comparable behavior could be observed regarding the viscosity.

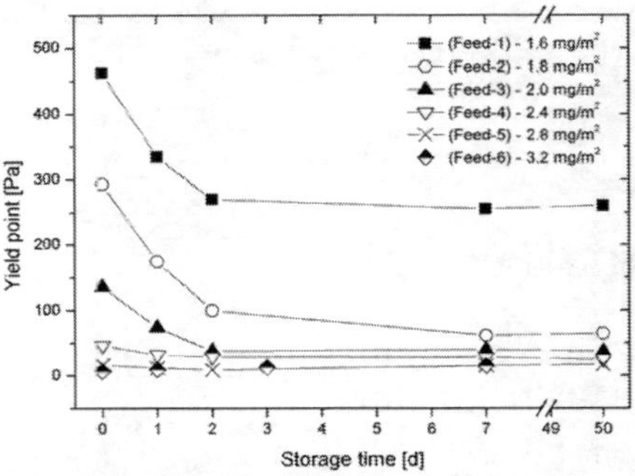

Figure 4. Yield point behavior of the investigated feedstocks versus storage time.

A decrease of the yield point is caused, when attractive interparticle forces among the ceramic particles are reduced. For the feedstocks investigated in this study the attractive interparticle forces are reduced as a function of time. Two reasons are assumed to have caused such an effect, in particular, (a) physical processes like degradation of the binder components or some rearrangement of dispersant molecules, or (b) chemical interaction among zirconia and paraffin, zirconia and dispersant, or paraffin and dispersant. It is supposed that a physical or chemical interaction among zirconia and dispersant takes place, as adding small amounts of dispersant (few drops) to the re-melted feedstock Feed-1 after at least 50 days significantly reduced the viscosity. No change in viscosity was observed for the pure binder components and for a mixture of dispersant and paraffin after storage at 80°C for 7 days. But a considerable change in viscosity and yield point takes place with feedstocks merely consisting of zirconia and paraffin. Further investigations revealed that degradation of binder components do not start before 150°C. It is unlikely that carboxylic groups at the polar end of the dispersant molecules react with the O-H group on the powder surface during storage at relatively low temperatures, as this would result in the formation of one water molecule per dispersant molecule[20] and, thus, would increase the viscosity of the suspension remarkably[22]. A physical process like rearrangement of surfactant molecules seems to be more likely. It is possible that the adsorbed molecules undergo some rearrangement with time and cause a significant decrease in viscosity and yield stress of the feedstocks with low amounts of dispersant (<2.4 mg/m^2), where the powder particles are assumed to be covered only partially by the surfactant molecules. It is supposed that an equilibrium condition of the surfactant concentration exists among already physically adsorbed dispersant molecules (static state) and not yet adsorbed molecules (mobile state), e.g. micelles. This equilibrium might be affected by storing the feedstocks at elevated temperature causing a shift of the equilibrium to an energetic more favorable static state and, hence, resulting in the observed time dependency of yield point and viscosity. For the feedstocks with surplus amount of Hypermer LP1 (>2.4 mg/m^2) no characteristic time dependency of the viscosity or yield point was observed, as the particles are fully covered with surfactant and, thus, fully stabilized due to the steric barrier induced by the dispersant molecules. It can be concluded that the surfactant used in this study has an exceedingly important time-dependent influence on the flow behavior of paraffin-based zirconia feedstocks.

Influence of Surfactant Concentration on Mechanical Properties

It could be revealed that the surfactant concentration impacts the strength of micro bending bars fabricated via hot molding of zirconia-paraffin feedstocks. A maximum in strength was found for arrays fabricated with feedstocks having a surfactant concentration of 2.4 mg/m^2, which correlated well with the evolution of the sintering density of the arrays and the feedstock viscosity (Fig. 5). In general, low densities result in poor mechanical strength. In turn, low densities are caused by flaws and defects introduced during processing. In this case, it is obvious that the feedstock with the lowest viscosity resulted in samples with the highest sintered density, as enhanced flow behavior of the feedstock enables a better filling of the cavities of the silicone molds and, hence, volume defects as pores caused by entrapped air could be avoided. Furthermore, it is assumed that a lower feedstock viscosity enhances the tendency to reduce the number and size of critical defects like pores and grooves at or near the surface by defect leveling during thermal debinding, yielding the observed improved mechanical properties.

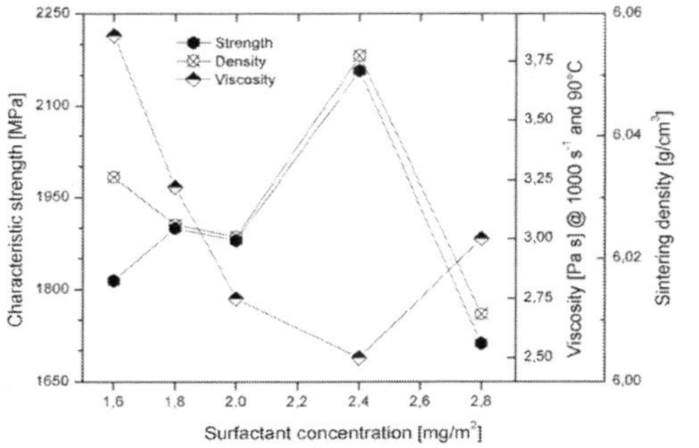

Figure 5. Actual characteristic three-point bending strength, density of sintered arrays and feedstock viscosity in dependency on surfactant concentration.

Leveling and Healing of Surface Defects

Edge rounding of micro bending bars as a result of leveling and healing of surface defects during thermal debinding was observed, also reported in previous studies[9,10]. Thereby, a liquid film at the surface of the column arrays was observed above the melting point of the paraffin wax also referred elsewhere[11]. It is assumed that the surface tension of this low-viscosity liquid might be able to induce leveling of critical surface defects like flaws, grooves and pores at or near the surface. But, this leveling may also result in remarkable dimensional change of the column arrays, viz. rounding of initially sharp edges (Fig. 6). Hence, the observed edge rounding represents an indicator for the occurrence of defect leveling processes. The development of well-defined understanding of the parameters and mechanisms influencing the edge rounding is object of current investigations.

A significant influence of the edge rounding as an indicator for the occurrence of defect leveling processes on the mechanical properties of micro bending bars was found, as illustrated in Fig. 7. Leveling and healing of critical surface defects during thermal debinding resulted in enhanced surface properties yielding an increase of the characteristic strength. Calculating the characteristic strength according to Weibull theory without considering the edge radius, strength values up to 2930 MPa were achieved for samples with distinctive edge rounding (28 μm). However, the strength increases remarkably up to 4110 MPa, when the radius is taken into account in the calculation. Thus, the moment of inertia changes from Eq. (2) for rectangular to Eq. (3) for rounded rectangle cross-section:

$$I_{y,rect} = \frac{bh^3}{12} \tag{2}$$

$$I_{y,round} = \frac{1}{12}\left(b_i r^3 + 3(h_i + r)^2 b_i r + r h_i^3 + b_i h_i^3\right) + \frac{1}{4}\pi r^4 + \frac{4}{3} r^3 h_i + \frac{1}{4}\pi h_i^2 r^2 \quad (3)$$

where b is the width and h the length of the bending bar cross-section; b_i and h_i stand for the inner width and inner length, when radius r is subtracted from b and h (Fig. 8). Therefore, depending on the degree of edge rounding, the actual strength values are effectively higher, when the edge rounding is considered in the calculations.

Figure 6. Micrograph of fracture surface of sintered bending bar without edge rounding (left) and with distinctive edge rounding of approximately 30 µm (right).

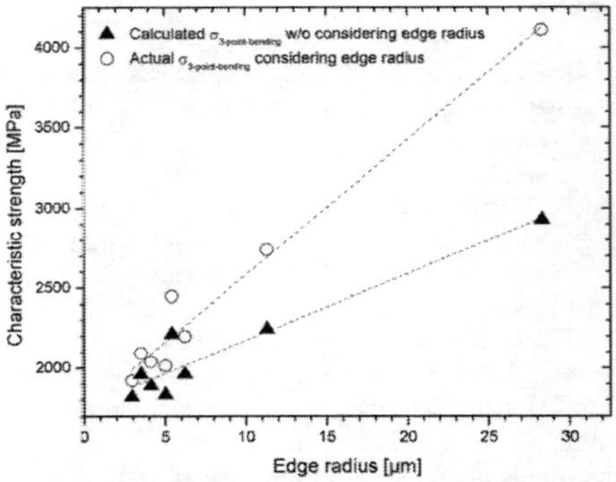

Figure 7. Influence of edge radius on the characteristic strength of hot molded zirconia micro bending bars with and without considering the rounding in the calculation of the strength values.

In Fig. 8 the ratio of the actual strength to the strength value without considering the radius versus edge rounding is illustrated. Two models were evaluated, which describe the evolution of the ratio of actual strength to the strength value without considering the edge radius. The first model is based on the presumption that the inner width b_i and length h_i are calculated merely by subtracting radius r from b and h. For increasing edge radius values, however, this causes a significant decrease of the cross-section area and a change from rectangular to circle shape, which results in a drastic increase of the calculated strength, as the moment of inertia also changes (Fig. 8, dashed line). Considering that the cross-section area stays constant and is decoupled from the change of the radius, a different behavior is predicted for the second model (Fig. 8, solid line). For the first model, the value of 2900 MPa (edge radius not considered in calculation yet) for the samples with mean 28 µm edge radius is increased by the factor 1.48, resulting in a predicted actual strength of 4290 MPa, whereas for the second model a factor of 1.44 is obtained yielding a predicted value of 4180 MPa. Comparing these results with the actual strength value of 4110 MPa, calculated applying Eq. (3), the second model predicts more precise than the first model. However, for edge roundings below 30 µm the difference between two models is negligible and gets more distinctive for edge roundings higher than 30 µm.

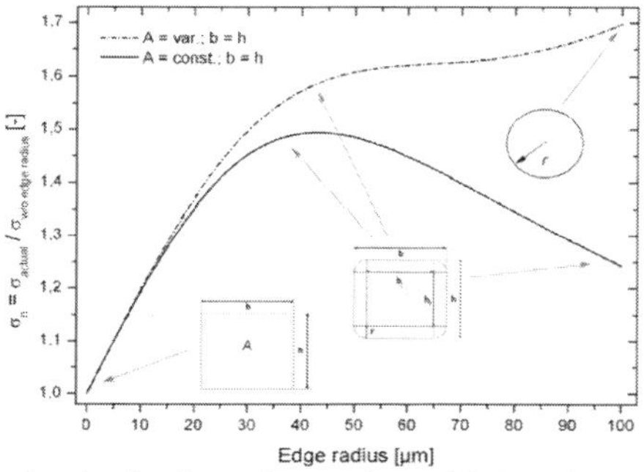

Figure 8. Models of calculated ratio of actual strength considering edge rounding and strength w/o considering edge rounding versus edge radius for constant and variable cross-section area

Comprehensive investigations on the defect leveling and healing mechanisms giving rise to the edge rounding are currently ongoing. Correlations among processing parameters, e.g. debinding conditions or composition, and edge rounding are investigated systematically to enable the reproducibility of the observed phenomenon. Some parameters influencing the observed behavior during thermal debinding were identified, viz. surfactant concentration of the feedstock and type of debinding support. In Fig. 9 images of the video-based real-time observations of the thermal debinding are illustrated (continuous heating ramp 0.5 K/min up to 300°C). Sample A-1, B-1 and C-1 were processed applying similar parameters in forming and debinding. The only difference is the surfactant concentration given as 2.4 mg/m^2 for A-1 and 3.2 mg/m^2 for B-1 and C-1 and, additionally, using non-absorbent glass support for A-1 and B-1 and porous absorbing alumina support for C-1 during debinding. At 90°C sample B-1 exhibits a pronounced binder film on the surface of the bending bars (Fig. 9-e), whereas for sample C-1 the binder film is less pronounced (Fig. 9-h). In comparison, for sample A-1 the binder film is not considerably apparent (Fig. 9-b). Moreover, grooves originating from the micro milling of the brass master model (Fig. 1) are still visible on the columns of sample A-1 (Fig. 9-c), whereas for C-1 the surface is obviously smoother (Fig. 9-i). For array B-1 it was not possible to maintain shape retention above 150°C due to the low yield point of the feedstock (Fig. 9-f), but a smoother surface is expected for the case, when the as-is state at 90°C (Fig. 9-e) could be preserved applying a dwell and a subsequent removal. of the binder.

It is obvious that a porous support enhances absorption of the binder by capillary flow or wicking, respectively.[27] In turn, this yields to the fact that the binder film, which is believed to cause the healing of surface defects, is absorbed by the porous support rapidly and is merely available for a certain time being not able to induce leveling of surface defects like grooves. For samples with low surfactant concentration and absorbing support the binder film was absorbed rapidly and a pronounced binder film could be observed only for a short time. Even dwell at adequate temperatures to maintain the binder film resulted in non-glossy surface of the columns after a short time subsequent to the melting of the binder. However, for high surfactant concentration and an absorbing support, as for C-1, a less pronounced glossy binder film could be maintained for a longer time in the upper part of the columns (Fig. 9-h). Using a non-absorbing support like glass caused a distinctive formation of a glossy binder film, which could be maintained easily. It seems that the interaction of various parameters impact the formation of edge rounding by healing and leveling of surface defects, e.g. surfactant concentration and type of debinding support. Though, complete thermal debinding on non-adsorbing support is not possible due to deformation of the columns (Fig. 9-f). Therefore, a "two-step-debinding process" was applied to sample B-1 consisting of (1) dwell of 5 h at 90°C using glass support and (2) heating up to 500°C with constant ramp of 0.5 K/min using the porous alumina support. After sintering, it could be revealed that bending bars with pronounced edge rounding and smooth surface were fabricated by using samples with high surfactant concentrations and the "two-step-debinding process" (Fig. 10). Three-point bending tests of the micro bending bars are currently conducted to enable correlations among the observed surface modification and the mechanical properties.

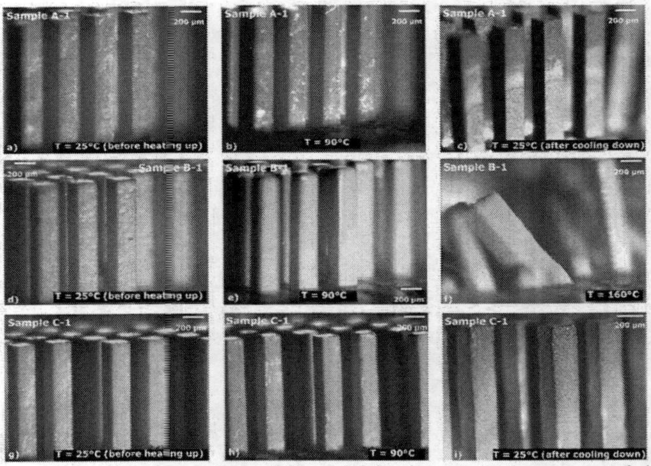

Figure 9. Images of real-time observations of thermal debinding; A-1: glass support, 2.4 mg/m^2; B-1: glass support, 3.2 mg/m^2; C-1: porous alumina support, 3.2 mg/m^2.

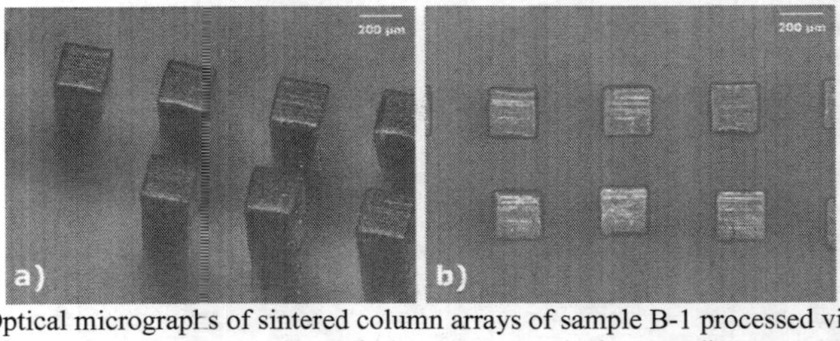

Figure 10. Optical micrographs of sintered column arrays of sample B-1 processed via "two-step-debinding process" exhibiting pronounced edge roundings.

CONCLUSIONS

The dispersant's influence on the overall flow behavior of zirconia-paraffin feedstocks was found to be very crucial. A significant decrease of both yield point and viscosity occurred with increasing dispersant concentration. For low shear rates (10 s^{-1}) the viscosity of the feedstocks decreased gradually with increasing dispersant concentration, whereas for high shear rates (100 s^{-1} and 1000 s^{-1}) a characteristic viscosity minimum was found due to entanglement of non-polar chains of the dispersant molecules or cross-linking of polar ends of the surfactant molecules for surplus amount of surfactant, respectively. An obvious time-dependency of both yield point and viscosity of feedstocks with low dispersant concentration (<2.4 mg/m^2) was observed, when stored at elevated temperature. It is supposed that physical effects like rearrangement of dispersant molecules or chemical interactions among zirconia and dispersant caused this time-dependency. Adsorbed Hypermer LP1 dispersant molecules are assumed to undergo some rearrangement with time at elevated temperature resulting in a decrease in viscosity and yield stress for low dispersant concentrations, where the powder particles are covered merely partially by surfactant molecules. An equilibrium condition of the dispersant concentration may exist between already physically adsorbed dispersant molecules (static state) and not yet adsorbed molecules (mobile state). By storing the feedstocks at elevated temperature, this equilibrium might be affected, causing a shift of the equilibrium to an energetic more favorable static state and, thus, resulting in the observed time dependency of the rheological properties. For feedstocks with sufficient or surplus amount of dispersant (>2.4 mg/m^2) no characteristic time dependency was observed, as the ceramic particles are fully covered with surfactant and, thus, fully stabilized due to the steric barrier induced by the dispersant molecules. This observation is crucial, as changes in rheological properties affect further processing and reproducibility.

The dispersant concentration was also found to impact the mechanical properties of micro bending bars fabricated via hot molding of zirconia-paraffin feedstocks. An enhanced flow behavior of the feedstock was obtained for optimum amount of surfactant that resulted in high sintering densities of micro bending bars due to reduced critical defects like pores, which, in turn, yielded enhanced mechanical properties of the microparts. Furthermore, a significant dependency of geometrical deformation (edge rounding) during thermal debinding on the three-point bending strength of micro bending bars was found. With increasing edge rounding of samples with rectangular cross-section the bending strength increased dramatically up to 4000 MPa. Mathematical models were evaluated successfully predicting the bending strength of samples with rounded rectangular cross-section. It could be revealed that the bending strength of bending bars with ideal rectangular cross-section may increase by factor up to 1.5 for edge roundings of approximately 30 μm. As yet, a well understanding of the parameters influencing the defect leveling and healing mechanisms giving rise to the observed edge roundings during thermal debinding was not available. Though, in this study some parameters could be identified thoroughly. It was found that inter alia dispersant concentration and type of debinding support influence the degree of edge rounding and deformation during thermal debinding remarkably. Low dispersant concentrations and absorbing debinding supports enhance shape retention and prevent edge rounding of bending bars, whereas high surfactant concentrations and non-absorbing supports enable edge rounding of bending bars to the point of total deformation. Bending strengths of micro bending bars fabricated by applying a "two-step-debinding process" are currently evaluated to enable reproducible correlations among the observed surface modification and the mechanical properties.

REFERENCES

[1]W. Bauer and R. Knitter, Development of a Rapid Prototyping Process Chain for the Production of Ceramic Microcomponents, J. Mater. Sci., **37**, 3127-3140 (2002).

[2]J. Wang, G. Liu, Y. Xiong, X. Huang, Y. Guo and Y. Tian, Fabrication of Ceramic Microcomponents and Microreactor for the Steam Reforming of Ethanol, Microsyst. Technol., **14**, 1245-1249 (2008)

[3]V. Piotter, W. Bauer, T. Benzler and A. Emde, Injection Molding of Components for Microsystems, Microsyst. Technol., **7**, 99-102 (2001).
[4]W. Bauer, J. Hausselt, T. Merz, M. Müller, G. Örlygsson and S. Rath, Microceramic Injection Molding, in: Microengineering of Metals and Ceramics (Part I), Weinheim, Wiley-VCH, 325-356 (2005).
[5]H.-J. Ritzhaupt-Kleissl, W. Bauer, E. Günther, J. Laubersheimer and J. Hausselt, Development of Ceramic Microstructures, Microsyst. Technol., **2**, 130-134 (1996).
[6]N.H. Loh, S.B. Tor, B.Y. Tay, Y. Murakoshi and R. Maeda, Fabrication of Micro Gear by Micro Powder Injection Molding, Microsyst. Technol., **14**, 43-50 (2007).
[7]J.A. Lewis, Colloidal Processing of Ceramics, J. Am. Ceram. Soc., **83**, 2341-2359 (2000).
[8]M. Müller, W. Bauer and H.-J. Ritzhaupt-Kleissl, Defect Levelling During Debinding for Improved Mechanical Properties of Ceramic Micro Parts, Proc. Third Inter. Conf. Shaping Adv. Ceram., 59-64 (2006)
[9]M. Müller, J. Rögner, B. Okolo, W. Bauer and H.-J. Ritzhaupt-Kleissl, Factors Influencing the Mechanical Properties of Moulded Zirconia Micro Parts, Proc. Tenth Inter. Conf. Exhib. Eur. Ceram. Soc., 1291-1296 (2007).
[10]D. Gronych, M. Auhorn, T. Beck, V. Schulze and D. Löhe, Influence of Surface Defects and Edge Geometry on the Bending Strength of Slip-Cast ZrO2 Micro-Specimens, Int. J. Mater. Res., **6**, 551-558 (2004).
[11] J.H. Song and J.R.G. Evans, Flocculation after Injection Molding in Ceramic Suspensions, J. Mater. Res., **9**, 2386-2397 (1994).
[12]A. Dakskobler and T. Kosmac, Rheological Properties of Re-melted Paraffin-Wax Suspensions Used for LPIM, J. Eur. Ceram. Soc., **29**, 1831-1836 (2009).
[13]W. Bauer, R. Knitter, R. Emde, G. Bartelt, D. Göhring and E. Hansjosten, Replication Techniques for Ceramic Microcomponents with High Aspect Ratios, Microsyst. Technol., **9**, 81-86 (2002).
[14]R. Knitter, W. Bauer, D. Göhring and J. Hausselt, Manufacturing of Ceramic Microcomponents by a Rapid Prototyping Process Chain, Adv. Eng. Mater., **3**, 49-54 (2001).
[15]DIN EN 843-5:2007-03: Hochleistungskeramik – Mechanische Eigenschaften monolithischer Keramik bei Raumtemperatur / Teil 5: Statistische Auswertung, Berlin, Beuth (2007).
[16]M. Auhorn, T. Beck, V. Schulze and D. Löhe, Quasi-static and Cyclic Testing of Specimens with High Aspect Ratios Produced by Micro-Casting and Micro-Powder-Injection-Moulding, Microsyst. Technol., **8**, 109-112 (2002).
[17]W.J. Tseng, D.M. Liu and C.K. Hsu, Influence of Stearic Acid on Suspension Structure and Green Microstructure of Injection-Molded Zirconia Ceramics, Ceram. Inter., **25**, 191-195 (1999).
[18]R. Lenk, and A.P. Krivoshchepov, Effect of Surface-Active Substances on the Rheological Properties of Silicon Carbide Suspensions in Paraffin, J. Am. Ceram. Soc., **83**, 273-276 (2000).
[19]D.M. Liu, Control of Yield Stress in Low-pressure Ceramic Injection Moldings, Ceram. Inter., **25**, 587-592 (1999)
[20]S. Zürcher and T. Graule, Influence of Dispersant Structure on the Rheological Properties of Highly-concentrated Zirconia Dispersions, J. Eur. Ceram. Soc., **25**, 863-873 (2005).
[21]V.M.B. Moloney, D. Parris and M.J. Edirisinghe, Rheology of Zirconia Suspensions in a Nonpolar Organic Medium, J. Am. Ceram. Soc., **78**, 3225-3232 (1995).
[22]S. Novak, K. Vidovic, M. Sajko and T. Kosmac, Surface Modification of Alumina Powder for LPIM, J. Eur. Ceram. Soc., **17**, 217-223 (1997).
[23]G. Bartelt, W. Bauer and J. Hausselt, Hot Molding of Zirconia Microparts, Progr. Colloid Polymer Sci., **121**, 19-22 (2002).
[24]T.G. Mezger, Das Rheologie Handbuch, Hannover, Vincentz Network (2006).
[25]D.M. Liu and W.J. Tseng, Rheology of Injection-Molded Zirconia-Wax Mixtures, J. Mater. Sci., **35**, 1009-1016 (2000).

[26]S. Novak, A. Dakskobler and V. Ribitsch, The Effect of Water on the Behaviour of Alumina-Paraffin Suspensions for Low-pressure Injection Moulding (LPIM), J. Eur. Ceram. Soc., **20,** 2175-2181 (2000).
[27]R.M. German, Theory of Thermal Debinding, Int. J. Powder Metall. Powder Tech., **23,** 237-245 (1987).

Advanced Composite Manufacturing

FIBER-REINFORCED CERAMIC MATRIX COMPOSITES PROCESSED BY A HYBRID PROCESS BASED ON CHEMICAL VAPOR INFILTRATION, SLURRY IMPREGNATION AND SPARK PLASMA SINTERING

Jerome Magnant[1]; René Pailler[1]; Yann Le Petitcorps[1]; Laurence Maillé[1]; Alain Guette[1]; Jimmy Marthe[1]; Eric Philippe[2]
[1] University of Bordeaux, Laboratory for Thermostructural Composites (LCTS), Pessac, France.
[2] Snecma Propulsion Solide, Le Haillan, France.

ABSTRACT

The fabrication of multidirectional continuous carbon and silicon carbide fiber reinforced Ceramic Matrix Composites by a new short time hybrid process was studied. This process is based, first, on the deposition of fiber interphase and coating by chemical vapor infiltration, next, on the introduction of Si_3N_4 powders into the fibrous preform by Slurry Impregnation and, finally, on the densification of the composite by liquid-phase Spark Plasma Sintering (SPS).
The homogeneous introduction of the mineral charges into the multidirectional fiber preforms was realized by slurry impregnation from highly concentrated (> 32 %vol.) and well dispersed aqueous colloid suspensions. The following densification of the composites by spark plasma sintering was possible with a 2 minutes holding period at 1650°C. The chemical degradation of the carbon fibers during the fabrication was prevented by adapting the sintering pressure cycle and the pressure media. The composites elaborated are dense. Our carbon fiber reinforced ceramic matrix composites have a damageable mechanical behaviour with a bending stress at failure around 250 MPa. Silicon carbide fiber reinforced ceramic matrix composites have a brittle mechanical behaviour with a bending stress at failure around 150 MPa. Microstructural analyses were conducted to explain the mechanical properties obtained. One main important result of this study is that spark plasma sintering can be used in some hybrid processes to densify multidirectional continuous fiber reinforced ceramic matrix composites.

INTRODUCTION

Non-oxide Ceramic Matrix Composites (CMC), i.e. mainly those consisting of a SiC or Si_3N_4 based matrix reinforced with either carbon or SiC-fibers, have the potential for being used up to about 1500°C, as structural materials, in different fields including advanced engines, gas turbines for power/steam co-generation, heat exchangers, as well as nuclear reactors of the future.[1-2]

CMC are processed according to (*i*) gas phase routes (CVI: Chemical Vapor Infiltration),[3] (*ii*) liquid phase routes either from polymers (PIP: Polymer Impregnation and Pyrolysis),[4-6] or molten elements reacting with mineral charges in the preforms or with the atmosphere (RMI: Reactive Melt Infiltration),[7] or finally (*iii*) the ceramic routes (SI-HP: Slurry Impregnation and Hot Processing).[8-10] Nevertheless, the time and cost production of CMC are some large obstacles that hamper their wider acceptance. Some hybrid processes, based on the combination of gas, liquid and ceramic routes, have been developed to optimize the densification of the fiber preforms, the microstructure of the composites and to reduce the time and cost production: CVI + (SI-MI or RMI),[11-15] $(PIP)_n$ + CVI,[16] CVI + (SI+HP) as the NITE process (for Nano Infiltration of Transient Eutectoid).[17-23] This last process was oriented to overcome the weakness of composites through other process routes for the improved matrix cracking stress and thermal conductivity.[18] The severe sintering conditions employed in NITE process require quasi-stoichiometric composition SiC fibers of third generation like Hi-Nicalon S, Sylramic or Tyranno SA fibers. $SiC_f/PyC_i/SiC_m$ composites with very low porosity, high mechanical and thermal properties as well as an excellent hermiticity with respect to gaseous cooling fluids such as helium were successfully elaborated at 1780°C-20 MPa by NITE process.

In this work, a new short time hybrid process was developed to (*i*) reduce the time and cost production, (*ii*) increase fiber preform densification and (*iii*) optimise the microstructure of the composite. This process is based on CVI fiber coating followed by Slurry Impregnation of ceramic powders into multidirectional continuous fiber preforms and then densification of the composite by Liquid Phase Spark Plasma Sintering (CVI + SI + LP-SPS).

Ex-PAN carbon fiber and SiC (Hi-Nicalon) fiber-reinforced Si_3N_4 matrix composites (C_f/Si_3N_{4m}, SiC_f/Si_3N_{4m}) were elaborated. Carbon fibers were used because of their unique high thermomechanical properties and their low cost, despite the fact that carbon undergoes oxidation at very low temperature (400-500°C). SiC (Hi-Nicalon) fibers of second generation are made of a mixture of SiC-nanocrystals (≈ 5 nm in mean size) and free carbon [C/Si (at) ratio ≈ 1.39] and were chosen for their excellent oxidation resistance and their cost which is about three times cheaper than quasi-stoichiometric SiC fibers of third generation. Hi-Nicalon SiC fibers creep at moderate temperature (1240°C) [24-25] but the soft sintering conditions achieve by SPS may limit their creep. Silicon nitride was chosen as constituent for the matrix because of its excellent thermomechanical properties, such as high strength, hardness, toughness, good shock resistance and good oxidation resistance.[26] First, the CVI technique was used in this process for the deposition of pyrocarbone interphase (PyC) and SiC coating because it yields deposits of relatively uniform composition, structure and thickness even with preforms of complex architecture as those employed in this study. Next, slurry impregnation was used to introduce the ceramic powders into the fiber preforms since this method is cost effective and can be easily employed with fiber preforms of complex shape. Then, liquid phase Spark Plasma Sintering was used to densify the composites since SPS technique allows shorter time densification than traditional techniques which, firstly, is cost beneficial and, secondly, may limit the chemical reaction of fibers with matrix phases during the fabrication. Furthermore, to our knowledge, no study reported the fabrication of continuous fiber reinforced ceramic matrix by Spark Plasma Sintering.

In this paper, the different steps of our hybrid process and the first mechanical characterisations of these materials will be presented and discussed.

EXPERIMENTAL

Fiber Preforms and Interphases

The main characteristics of Ex-PAN carbon fibers and SiC Hi-Nicalon fibers are given in Table I. As it is well known that fiber/matrix bonding is a key factor controlling performances of ceramic matrix composites, a pyrocarbon interphase (PyC_i) was deposited on the fiber surface from gaseous precursor in order to arrest and deflect the matrix microcracks and to assure a load transfer function.[27,28]

Table I. Main characteristics of the fibers used in this study and of Si_3N_4

	Density ($g.cm^{-3}$)	Tensile strength (MPa)	Young's modulus (GPa)	Failure strain (%)	CTE ($10^{-6}\,K^{-1}$)[a]	Thermal conduct. ($W.mK^{-1}$)[a]	Creep parameter ($T_{0.5}$)(°C)[b]
Ex-PAN C-fibers	1.77	2100-2400	230-297	1.05-0.71	-0.1 - 10	17	/
SiC Hi-Nicalon fibers	2.74	2800	270	1.0	3.3-3.5	5.0-7.8	≈ 1240
Si_3N_4	3.21	800-1600[c]	260-330[c]	≈ 0.2	2.8-3.5	15-40	[c]

[a] At 25°C; [b] Temperature at which m = 0.5 in BSR-test; [c] Depending on processing conditions.

To minimize possible chemical reactions of carbon fibers with the matrix constituents during the fabrication, a $SiC_{(c)}$ coating was deposited by CVI on the PyC interphase (PyC_i). The different combinations of PyC interphase, SiC coating and fibers used in this study are: {C_f/ PyC_i}; {C_f/ PyC_i + SiC_c}; {SiC_f (Hi-Nicalon) / PyC_i + SiC_c}.

Mineral Powders

Sub-micronic silicon nitride powders + 7 wt% Y_2O_3 + 3 wt% Al_2O_3 sintering-aid powders were used (Table II). Yttria and alumina sintering-aids react in temperature with the silica layer covering the surface of silicon nitride powders to form an amorphous liquid phase that firstly promotes the densification of Si_3N_4,[26] and secondly, may accommodate the stress in the material due to external pressure and, as a result, limit possible fiber damage.

Sintering Conditions for the Unreinforced Matrix

Silicon nitride+yttria+alumina powders were mixed by wet ball-milling in polyethylene pots with ethanol and Al_2O_3 balls during 24 hours. Then, mixed powders were dried and milled by dry ball-milling in polyethylene pots during 5 hours to eliminate the agglomerates. The powders were then fitted in a cylindrical carbon die with an inner diameter of 20 mm for Spark Plasma Sintering (Dr Sinter 2080, SPS Syntex Inc.). The sintering temperature varied from 1500°C to 1700°C, the heating-rate was set to 200°C/min and the holding period varied from 1 to 20 minutes. First, the powders were degassed under vacuum at a temperature of 700°C, next a mechanical pressure of 30 MPa was applied and finally a nitrogen atmosphere (P_{N2} = 0.02 MPa) was introduced in the chamber to avoid Si_3N_4 decomposition at high temperature. The heating process was controlled using an optical pyrometer focused on the surface of the carbon die. The linear shrinkage of the unreinforced matrix during sintering was recorded by an extensometer.

The bulk density of the samples was measured by Archimedes method. The microstructures were characterised by scanning electron microscopy (Quanta 400 FEG, FEI) and energy-dispersive X-ray spectroscopy (EDAX) on polished surfaces. Crystalline phases were determined from X-ray diffraction patterns obtained in a θ-θ Bragg-Brentano configuration (D8 advance, Bruker).

Table II. Some characteristics of the powders used.

	Si_3N_4	Y_2O_3	Al_2O_3
Provider	HC Starck	HC Starck	Alfa Aesar
grade	M11	C	-
SSA ($m^2.g^{-1}$)	12-15	10-16	10

Preparation of the Mineral Suspensions

Electrokinetic behaviour of Si_3N_4 powders in aqueous suspension is governed by the ionisation of the surface groups according to the pH and was extensively studied.[29-33] By acoustophorometric measurement, Jung *and al.* shown that Si_3N_4 M11 Starck powder has a maximum absolute zeta potential value in the pH range [10-11.5].[32] Hruschka *and al.* studied the evolution of zeta potential versus pH for Al_2O_3 and Y_2O_3 aqueous suspensions and showed that in the pH range [10.5-11.5], Al_2O_3 and Y_2O_3 particles are respectively charged with the same sign as Si_3N_4 particles.[33] As a result, mixed aqueous suspensions of Si_3N_4 + 7 wt.% Y_2O_3 + 3wt.% Al_2O_3 particles were prepared in the pH range [10.5-11.5], being award that above pH=11 the decomposition of Y_2O_3 particles becomes important.[33] To adjust the pH, tetramethylammonium hydroxide (TEAH, 35 wt% in water, Sigma-Aldrich Chemical Co.) was employed.

The rheological behaviour of the suspensions was investigated with a cone-plate rheometer (Rotovisco RT 20, Haake) by ascending shear rates ramp from 0 to 400 s^{-1} in 5 min to determine the exact pH value for which the viscosity of the suspension is the lowest. To avoid undesired influence from mechanical histories, fresh samples were homogenized by shearing at an identical shear rate of 100 s^{-1} for 1 min and left standing for an additional 1 min prior to measurement.

Introduction of the Mineral Powders into the Fiber Preforms

Different techniques can be used to introduce ceramic particles inside fiber preforms from colloidal suspensions: Electrophorese Impregnation (EPI),[34] submicronic powder aspiration[35] and slurry impregnation under vacuum. This last method was chosen in this work for its ability to be easily transposed at plant-level to impregnate preforms with complex shape. The impregnated preforms were then dried during 48h at 25°C. Some samples were cut, polished and characterised by optical microscopy.

Fiber Preforms Densification by Spark Plasma Sintering

Silicon nitride + sintering-aid powder charged preforms were stacked in a graphite die for SPS. The Spark Plasma Sintering conditions are a compromise between minimizing fibers degradation and achieving full densification. The selection of the sintering conditions was based on the densification of monolithic Si_3N_4. The sintering temperature was 1650°C and hold 2 min at this temperature, while the pressure was varied from 30 to 150 MPa. Fibers degradation can be the result of:

(i) chemical reactions of the carbon fibers or pyrocarbone interphase with the matrix phases. Luthra and Park shown that Al_2C_3 and Y_2O_3 are stable with carbon provided they are no cracks for CO gas to escape.[36] However, silica undergoes carbothermic reduction at 1650°C with $p_{CO} < 3$ bar (reaction 1). The silicon nitride can react as well with the carbon according to the reaction 2.

$$SiO_{2\,(s)} + 3\,C_{(s)} = SiC_{(s)} + 2\,CO_{(g)} \quad (1) \quad \Delta rG(1950°K) \approx -53\ kJ.mol^{-1}$$

$$Si_3N_{4\,(s)} + 3\,C_{(s)} = 3\,SiC_{(s)} + 2\,N_{2\,(g)} \quad (2) \quad \Delta rG(1950°K) \approx -117\ kJ.mol^{-1}$$

Silicon carbide can react in a same way with silica or alumina to form $CO_{(g)}$.

The fast kinetic of densification obtain with Spark Plasma Sintering may limit these reactions. Furthermore the sintering was conducted with the sintering chamber under N_2 atmosphere to limit possible decomposition of Si_3N_4.

(ii) physical degradations due to the external pressure applied.

Then, to minimize fiber degradation the pressure was applied only when the grain boundary phases began to soften but soon enough (1300°C) to limit gas evolution and thus to inhibit any possible reaction of Si_3N_4 or secondary grain boundary phases with the pyrocarbone interphase or the fibers.[36] Furthermore, to obtain a system as closed as possible to inhibit gas evolution, the preforms were embedded in a powder-bed of graphite before leading on the die. For comparison, some preforms were embedded in a powder-bed of BN+Si_3N_4 and sintered with the pressure applied at 1500°C instead of 1300°C.

The sintered composites were subsequently cut and ground into 45x12x2.5 mm^3 specimens for three-point-bending tests, with a crosshead speed of 0.5 mm/min and a span of 30 mm (Model 5569, Instrom Corp). The polished cross sections of the composites before bending tests were characterized by scanning electron microscopy and energy-dispersive X-ray spectroscopy on polished surfaces.

RESULTS AND DISCUSSION

Sintering Conditions for the Unreinforced Matrix

Figure 1 shows the shrinkage and shrinkage rate curves of Si_3N_4 + 7 wt% Y_2O_3 + 3 wt% Al_2O_3 during sintering up to 1650°C under 30 MPa. The shrinkage rate curve clearly shows two maximum around 1260°C and 1500°C.

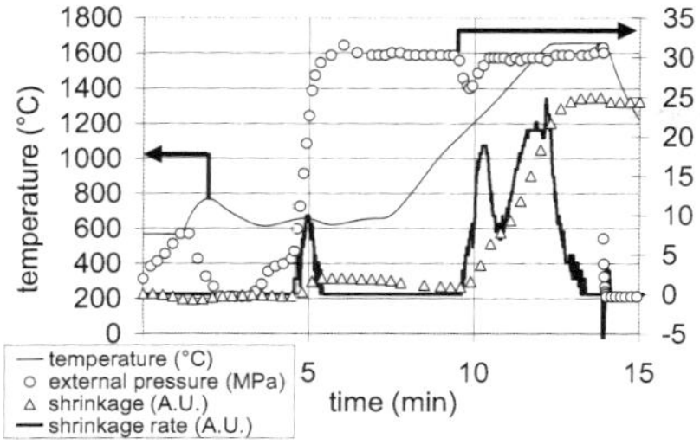

Figure 1. Temperature, pressure applied and shrinkage, shrinkage rate of the sample during the Spark Plasma Sintering cycle.

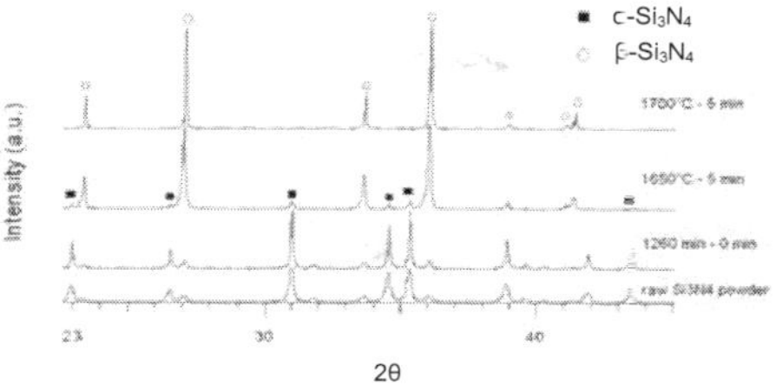

Figure 2. XRD patterns of the Si_3N_4 + sintering-aids samples as a function of the sintering temperature.

Figure 2 and Figure 3 reveal the phases composition and the microstructures of Si_3N_4 + 7 wt% Y_2O_3 + 3 wt% Al_2O_3 samples sintered at 1260°C with no holding period and at 1700°C after 5 min holding period. Sample sintered at 1260°C is still very porous (65% of theoretical density) whereas sample sintered at 1700°C during 5 min is fully dense (99% of theoretical density) and is constituted of only β-Si_3N_4 grains embedded in an amorphous glassy phase formed from liquid phase solidification. In the 1700°C sintered sample, the liquid phase forms a continuous network which is not the case in the 1260°C sintered sample. The first important increase in density observed around 1260°C can be attributed to particles rearrangement due to liquid phase formation and the second one around 1500°C to a reconstructive α to β phase transition and densification. The isothermal kinetic of densification in the temperature range [1550°C-1700°C] was reported on Figure 4.

(a) (b)

Figure 3. Si_3N_4 + 7 wt% Y_2O_3 + 3 wt% Al_2O_3 unreinforced matrix (a) secondary electron image of sample sintered at 1260°C with no holding period and (b) back-scattered image of fully dense sample sintered at 1700°C during 5 min.

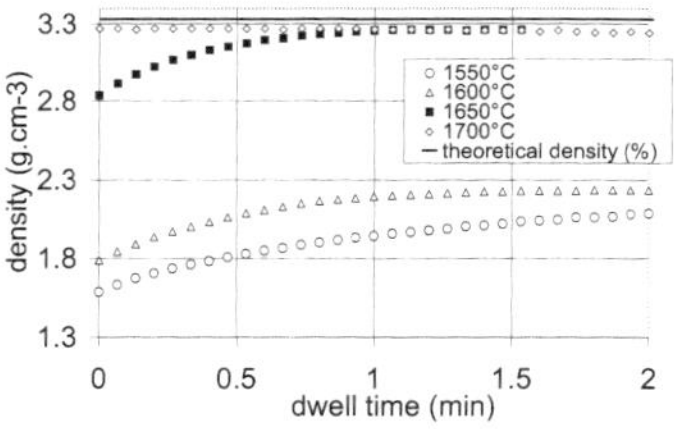

Figure 4. Evolution of shrinkage of Si_3N_4 + 7 wt% Y_2O_3 + 3 wt% Al_2O_3 samples during isothermal sintering at different temperatures.

A sintering temperature of 1650°C during 1 min allows a relative density of 99 % whereas a sintering temperature bellow 1650°C does not permit to obtain relative density upper to 68 % even for long sintering period (Figure 4). The XRD phase composition of samples sintered 5 min at 1650°C and 1700°C are presented Figure 2. Sample sintered at 1650°C is constituted of α and β-Si_3N_4 grains whereas sample sintered at 1700°C is only constituted of β grains. As mechanical properties are changing according to the α/β phase ratio,[37] one has to choose the couple of parameters (sintering temperature - sintering time) which corresponds to the required properties. In this study, as fiber-degradation needs to be avoided, a sintering temperature of 1650°C applied during 2 minutes was chosen.

Preparation of the Mineral Suspensions

The minimum in the viscosity of the suspension, which reveals the better conditions for the dispersion of Si_3N_4, occurs for pH=10.5+/-0.1 (Figure 5). As a consequence, the sintering-aids, i.e. Y_2O_3 and Al_2O_3 powders, were added to the 33 vol.% Si_3N_4 suspensions prepared at pH=10.5. The sintering-aid powders in the Si_3N_4 suspension do not destabilize it (Figure 6). As a result Si_3N_4+sintering aids suspensions for infiltration of the fibrous perform were prepared at pH=10.5.

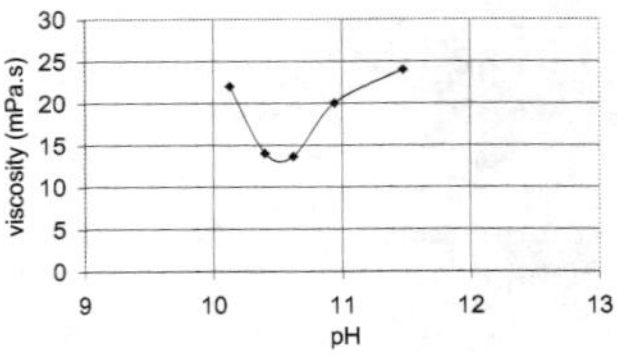

Figure 5. Viscosity vs. pH for the 33 vol.% Si_3N_4 suspension at a shear rate of 150 s^{-1}.

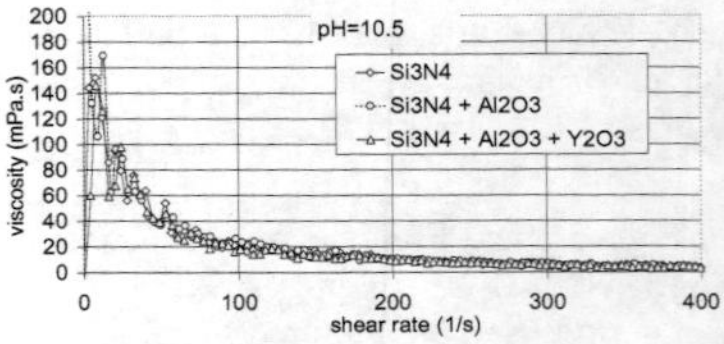

Figure 6. Influence of the sintering-aids on the rheological behaviour of the 33 vol.% Si_3N_4-based suspension.

Introduction of the Mineral Powders into the Fiber Preforms

After the slurry impregnation step, the inter-tows porosity of the fiber preforms is filled homogenously by the mineral charges (Figure 7).

(a) (b)

Figure 7. Optical micrographs of fiber preforms (a) before impregnation by the mineral powders and (b) after impregnation. Silicon nitride + sintering aid powders correspond to white contrast between the tows (b).

Fiber Preforms Densification by Spark Plasma Sintering

The state of densification of the composites as a function of the applied pressure is presented in the Table III.

Table III. State of densification of the matrix inside the composites as a function of the mechanical pressure applied during Spark Plasma Sintering.

	30 MPa	50 MPa	75 MPa	150 MPa
C_f/ [PyC_{i1}] *	incomplete	incomplete	complete	complete
C_f/ [PyC_{i1}+SiC_{c1}] *	incomplete	incomplete	incomplete	complete
SiC_f/ [PyC_{i2}+SiC_{c2}] *	incomplete	complete	complete	complete

* The thicknesses of PyC_i interphase and SiC_c coating are different in C_f and in SiC_f composites.
PyC thicknesses of C_f / [PyC_{i1}] and of C_f / [PyC_{i1}+SiC_{c1}] are the same.

When a SiC coating is deposited around the carbon fibers, the pressure required to obtain a full dense matrix is more important than without SiC coating. This lies in the fact that the SiC coating withstands the transmission of the pressure to the matrix. The composite with the SiC coating is fully densify at 150 MPa, but a maximum pressure of 75 MPa is available for large mechanical testing samples due to SPS machine limitations. Hence mechanical tests were carried out on samples sintered at this pressure.

Microstructural Analyses

The microstructures of the composites were studied in term of fiber degradation, densification of the inter and intra-tows matrix, microcracks inside the matrix, decohesion between the fibers and the matrix and will be successively discussed.

SEM micrographs and EDX analyses (Figure 8, 9 and 10) reveal the influence of the pressure cycle and of the pressure media on the fiber degradation. When pressure is applied at 1500°C, the carbon fibers embedded in a powder-bed of BN+Si_3N_4 are degraded which is not the case when the pressure is applied as soon as 1300°C and the preform embedded in a powder-bed of graphite. The degradation of the carbon fibers can be the result of the chemical reactions between the carbon and the matrix phases, as confirm by EDX analysis (Figure 10). When a $CO_{(g)}$ atmosphere surrounded the preform and the sintering pressure applied soon enough to limit gas evolution inside the system, fiber degradation can be avoided.

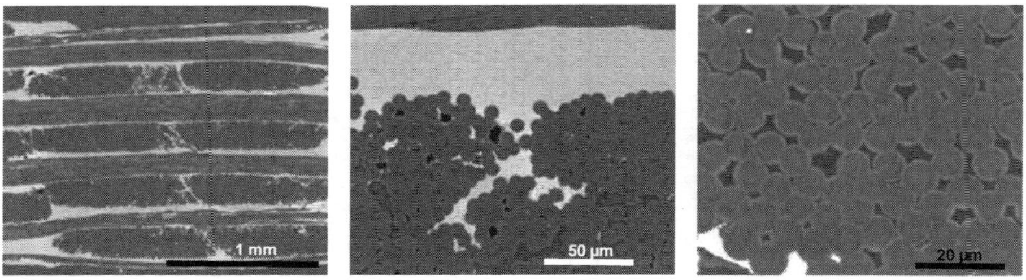

Figure 8. SEM micrographs of C_f / PyC_i / Si_3N_{4m} Spark Plasma Sintered composites embedded in a powder-bed of graphite with a 75 MPa mechanical pressure applied at 1300°C.

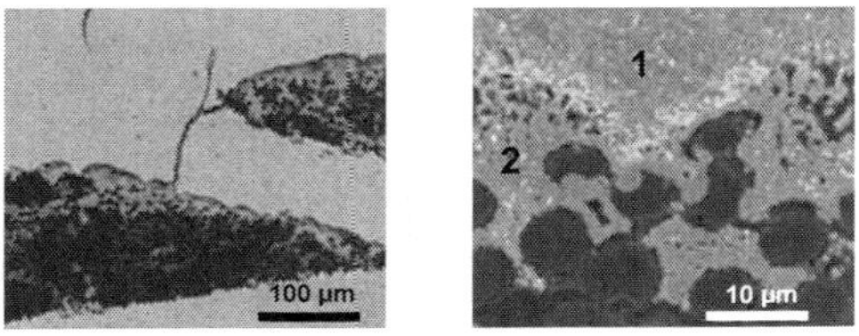

Figure 9. SEM micrographs of C_f / PyC_i / Si_3N_{4m} Spark Plasma Sintered composites embedded in a powder-bed of Si_3N_4+BN, with a 75 MPa external pressure applied at 1500°C, showing fiber degradation.

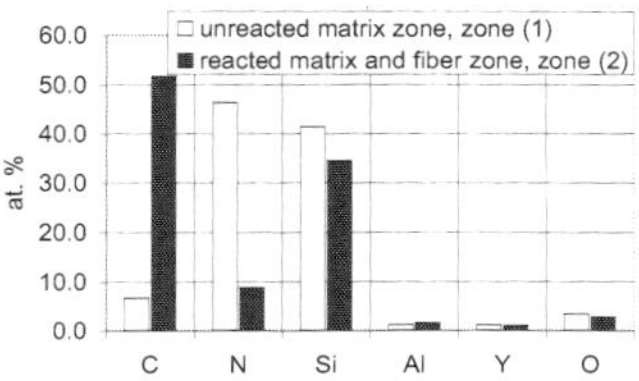

Figure 10. EDX analysis of zones marked 1 and 2 on Figure 9.

The SiC (Hi-Nicalon) fibers are chemically degraded after sintering of the composite (Figure 11). Indeed, the elements Al, Y and O were detected at the periphery of the SiC+C fibers (Figure 12). These elements have diffused inside the fiber from the sintering aids liquid phase or from its vapour. Any chemical degradation was observed in the carbon fiber composites.

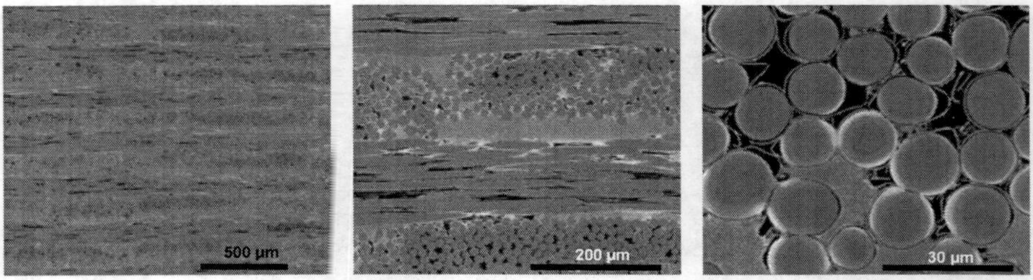

Figure 11. SEM micrographs of SiC_f Hi-Nicalon / PyC_{i2} + SiC_{c2} / Si_3N_{4m} Spark Plasma Sintered composites embedded in a powder-bed of graphite with a 75 MPa mechanical pressure applied at 1300°C.

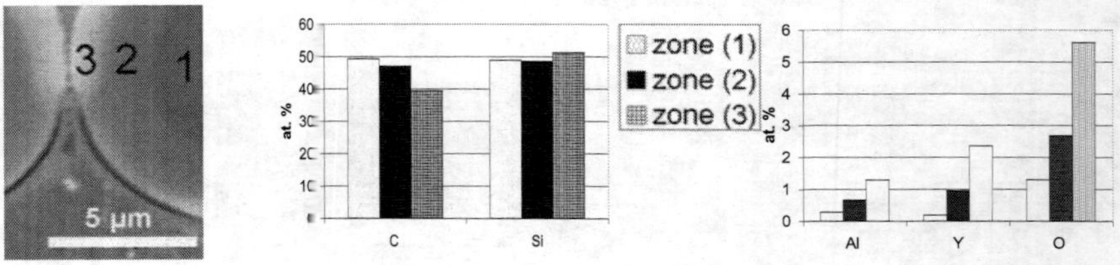

Figure 12. EDX analysis of SiC-Hi-Nicalon fibers in the SiC_f / PyC_i + SiC_c / Si_3N_{4m} Spark Plasma Sintered Composites embedded in a powder-bed of graphite with a 75 MPa mechanical pressure applied at 1300°C.

The inter-tows matrix of all composites is fully dense except in some located areas between axial tows for C-fibers composites (Figure 8, 11 and 13). The fiber architecture of the carbon fiber preform which is not exactly the same that the SiC fiber preform is at the origin of this partial densification.

In C_f/Si_3N_4 composites, the fibers are bonded inside the tows by the pyrocarbone and eventually by the SiC coating whereas in SiC_f/Si_3N_{4m} composite the thickness of the pyrocarbone interphase and of the SiC coating is too low to bond correctly the fibers. Carbon fiber composites without SiC coating have some intra-tow areas densified by the Si_3N_4-based matrix. For carbon fiber composites with the SiC coating, the accessibility to the intra-tow porosity is blocked by the SiC layer prior to powder impregnation.

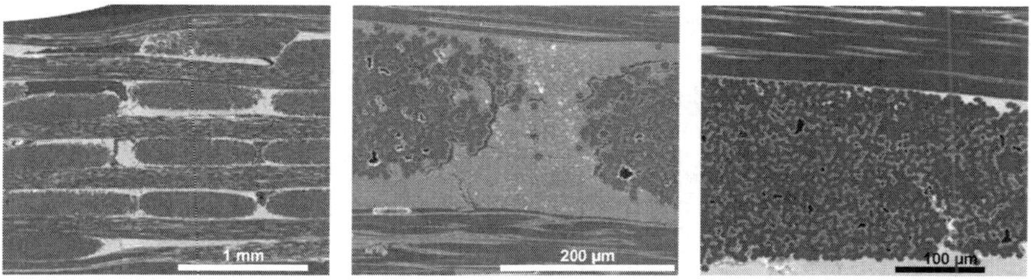

Figure 13. SEM micrographs of C_f / PyC_i + SiC_c / Si_3N_{4m} Spark Plasma Sintered composites embedded in a powder-bed of graphite with a 75 MPa mechanical pressure applied at 1300°C.

All C_f/Si_3N_{4m} composites exhibit microcracks after fabrication but not SiC_f/Si_3N_{4m} composites (Figure 8, 11 and 13). These microcracks are the result of the mismatch between the coefficient of thermal expansion of carbon-fibers and Si_3N_4-matrix. Sabouret *and al.* have computed around 450 MPa the internal stress inside the matrix in unidirectional C fiber-reinforced Si_3N_4 matrix composite (with $E_{C\text{-fiber}}$=250 GPa).[38] In the multidirectional C-fiber arrangement composites studied, the repartition of the thermal stress inside the matrix is more complicated than in unidirectional composites but are higher than the intrinsic stress at failure of the Si_3N_4 matrix.

No important decohesion between the fibers and the interphase/coating is observed in carbon fiber composite. In contrast, SiC fiber composites show important heterogeneous decohesions (Figure 11). SiC (Hi-Nicalon) fibers are known to creep at moderate temperature (around 1240°C).[25] Hence, these decohesions can be attributed to different strains between the SiC fiber and the SiC coating during sintering under pressure at 1650°C. As a consequence, the soft sintering conditions in SPS do not permit to limit enough the fiber strain. In carbon fiber composite the strong thickness of PyC interphase can accommodate some strain differences between the constituents of the composite which is not the case in SiC fiber composite due to lower PyC thickness. Furthermore carbon fibers show no important strain. Perspectives to limit the decohesions between the SiC fibers and the SiC coating are (*i*) to stabilize the fiber microstructure by a heat treatment prior to the SiC deposit,[25] (*ii*) to deposit a thicker PyC interphase between the fiber and the SiC coating, (*iii*) to formulate a Si_3N_4 based matrix sinterable at lower temperature. These perspectives will be addressed in a further paper.

Some decohesions or cracks between the PyC interphase or the SiC coating and the matrix are observed only in C_f/Si_3N_{4m} composite. These decohesions are the result of thermal stresses produced by the mismatch between the CTE of the C-fiber and the matrix.

Mechanical Behavior

The stress-displacement curves obtain during bending mechanical tests are presented in Figure 14. SiC-fibers composites elaborated have a brittle behavior with a stress at failure relatively low ($\approx$ 150 MPa). This relatively low stress at failure can be attributed to the chemical degradation of the SiC-fibers during the fabrication, the absence of intra-tows matrix and to some decohesions between interphases and fiber. The brittle behavior is more probably due to the chemical degradation of the SiC-fibers. Others possibilities to explain the brittle behavior are a too thin PyC interphase and/or inhomogeneous bond areas between the fibers and the SiC coating. C-fibers composites exhibit damageable behavior with the maximum stress at failure around 250 MPa. The presence of the SiC coating around the fibers enhances the mechanical properties. Hypotheses formulated to explain this point are that (*i*) the strength of the interface bond between the SiC coating and the matrix phases (more precisely the intergranular SiO_2-Y_2O_3-Al_2O_3 glassy phase) is stronger than the bond between the PyC interphase and the amorphous matrix phase, (*ii*) the SiC coating prevented the fibers from chemical degradation during the fabrication. However, as fiber degradations were not observed, the first hypothesis is the more probable. Characterisations are in progress to better understand the influence of the SiC coating on the mechanical properties of carbon fiber composites. For comparison, a raw preform (i.e. without matrix and not sintered), with a PyC + SiC fiber coating, was mechanically tested (Figure 14). Even if the SiC coating contribute to consolidate the fiber preform, the influence of the matrix on the mechanical properties is obvious.

The mechanical behavior of the C_f/Si_3N_{4m} composites elaborated by our hybrid process is promising even if their stress at failure around 250 MPa is lower than the 20-years under-development C_f/SiC_m composites fabricated by CVI (≈ 350 MPa).[39] Experiments should be conducted with different fiber architectures and different thickness of PyC interphases and SiC coating to find the more appropriate ones to this hybrid process.

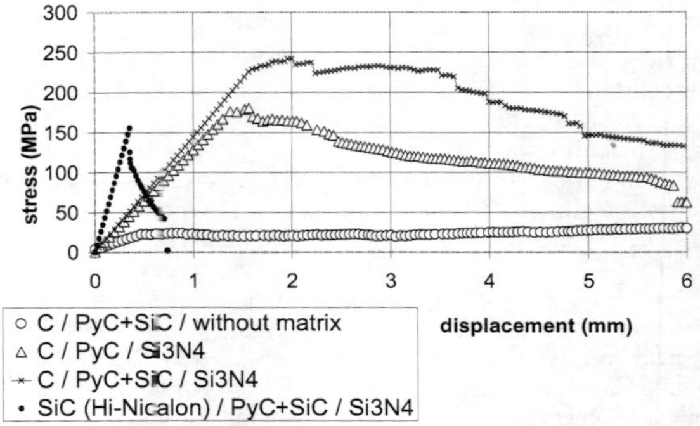

Figure 14. Stress/displacement curves from bending test for composites embedded in a powder-bed of graphite and sintered with a 75 MPa mechanical pressure applied at 1300°C.

CONCLUSION

The hybrid process developed here allows short-time fabrication of ceramic matrix composites. Firstly, a mechanical fuse pyrocarbon layer and a SiC coating were deposited around the fiber in the preform by Chemical Vapor Infiltration, secondly, silicon nitride and sintering-aid powders were introduced in the preform by slurry impregnation and, thirdly, mineral-powder-charged-fiber preforms were consolidated by Spark Plasma Sintering. SiC Hi-Nicalon fiber-reinforced Si_3N_4 matrix composites have a brittle-type mechanical behavior with a relatively low bending stress at failure (≈ 150 MPa) due to fiber chemical degradation and the lack of bond between fibers inside the tows or to some decohesions between interphases and fiber. C_f/Si_3N_4 composites can be elaborated with no chemical degradation of the fibers if they are embedded in a powder-bed of graphite during sintering and if the external sintering pressure is applied soon enough in temperature to inhibit gas evolution and thus fiber degradation. The value of the pressure required to fully sinter the matrix depends on the presence of a SiC coating because this one withstands the application of the pressure inside the preform. The carbon fiber composites show a mechanical damageable behavior and a maximum bending stress at failure around 250 MPa. As a consequence Spark Plasma Sintering can be used in some hybrid processes to densify multidirectional continuous fiber reinforced ceramic matrix composites.

ACKNOWLEDGEMENTS

This work was supported by DGA (French Direction Générale pour l'Armement) and Snecma Propulsion Solide through a grant. The authors acknowledge S. Bertrand (Snecma Propulsion Solide) for material supply; G. Chevallier, C. Estournes and G. Raimbeaux (PNF^2, University of Toulouse, France) for Spark Plasma Sintering tests.

REFERENCES

1. R. Naslain, SiC-matrix composites: Nonbrittle ceramics for thermo-structural application, *Int. J. Appl. Ceram. Technol.*, **2** , 75-84 (2005).
2. R.H. Jones, D. Steiner, H.L. Heinisch, G.A. Newsome, and H.M. Kerch, Radiation resistant ceramic matrix composites, *J. Nucl. Mater.*, **245**, 87-107 (1997).
3. R. Naslain, F. Langlais, G. Vignoles, R. Pailler, The CVI-process: State of the art and perspective, *Ceram. Eng. Sci. Proc.*, **27**, 373-86 (2006).
4. L.V. Interrante, C. W. Whitmarsh, and W. Sherwood, Fabrication of SiC matrix composites using a liquid polycarbosilane as the matrix source, *Ceram. Trans.*, **58**, 111-18 (1995).
5. M. Takeda, Y. Kagawa, S. Mitsuno, Y. Imai, and H. Ichikawa, Strength of a Hi-Nicalon/silicon-carbide-matrix composite fabricated by the multiple polymer infiltration-pyrolysis process, *J. Am. Ceram. Soc.*, **82**, 1579-81 (1999).
6. Z.S. Rak, Novel CMC material based on a Si_3N_4 matrix, *Adv. Eng. Mater.*, **2**, 503-07 (2000).
7. W.B. Hillig, Making ceramic composites by melt infiltration, *Am. Ceram. Soc. Bull.*, **73**, 56–62 (1994).
8. Y. Katoh, S.-M. Dong, and A. Kohyama, A novel processing technique of silicon carbide-based ceramic composites for high temperature applications, *Ceram. Trans.*, **144**, 77–86 (2002).
9. G. Grenet, L. Plunkett, J.B. Veyret, and E. Bullock, Carbon fibre-reinforced silicon nitride composites by slurry infiltration, *In: Proceeding HTCMC II, Manufacturing and materials Development*, 125-30 (1995)
10. J.-B. Veyret, P. Tambuyser, C. Olivier, E. Bullock, and M.H. Vidal-Setif, Hi-Nicalon reinforced silicon nitride matrix composites, *J. Mater. Sci.*, **32**, 3457-62 (1997).
11. Y. Xu, L. Cheng, and L. Zhang, Carbon/silicon carbide composites prepared by chemical vapour infiltration combined with silicon melt infiltration, *Carbon*, **37**, 1179–87 (1999).
12. W. B. Hillig, Melt Infiltration Approach to Ceramic Matrix Composites, *J. Am. Ceram. Soc.*, **71**, C96–9 (1988).
13. K.L. Luthra, R.N. Singh, and M.K. Brun, Toughened Silcomp composites, Process and preliminary properties, *Am. Ceram. Soc. Bull.*, **72**, 79-85 (1993).
14. R. Kochendörfer, Low Cost Processing for C/C-SiC Composites by Means of Liquid Silicon Infiltration, *Key Engin. Mater.*, **164-165**, 451-56 (1999).
15. W. Krenkel, Cost Effective Processing of CMC Composites by Melt Infiltration (Lsi-Process*)*, *Ceram. Engin. Sci. Proc.*, **22**, 443 – 54 (2001).
16. K. Suzuki, S. Kume, K. Nakano, and T.W. Chou, Fabrication and Characterization of 3D C/SiC Composites via Slurry and PCVI Joint Process, *Key Engin. Mater.*, **164-165**, 113-16 (1999).
17. A. Kohyama, S. Dong, and Y. Katoh, Development of SiC/SiC composites by nano-infiltration and transient eutectoid (NITE) process, *Ceram. Engin. Sci. Proc.*, **23**, 311-18 (2002).
18. Y. Katoh, A. Kohyama, S. Dong, T. Hinoki, and JJ Kai, Microstructure and properties of liquid phase sintered SiC/SiC composites, *Ceram. Engin. Sci. Proc.*, **23**, 363-70 (2002).
19. D. Shaoming, Y. Katoh, and A. Kohyama, Preparation of SiC/SiC Composites by Hot Pressing Using Tyranno-SA Fiber as Reinforcement, *J. Am. Ceram. Soc.*, **86,** 26–32 (2003).
20. K. Shimoda, J.S. Park, T. Hinoki, and A. Kohyama, Influence of pyrolytic carbon interface thickness on microstructure and mechanical properties of SiC/SiC composites by NITE process, *Compos. Technol.*, **68**, 98–105 (2008).

[21] J S. Park, A. Kohyama, T. Hinoki, K. Shimoda, and Y H. Park, Efforts on large scale production of NITE-SiC/SiC composites, *Journal of Nuclear Materials*, **367-370**, 719-24 (2007).

[22] S. Dong, Y. Katoh, and A. Kohyama, Preparation of SiC/SiC Composites by Hot Pressing, Using Tyranno-SA Fiber as Reinforcement, *J. Am. Ceram. Soc.,* **86**, 26-32 (2003).

[23] M. Suzuki; M. Sato, N. Miyamoto, and A. Kohyama, Preparation and properties of dense SiC/SiC composites, *Ceram. Engin. Sci. Proc.*, **26**, 319-26 (2005).

[24] R. Bodet, X. Bourrat, J. Lamon, and R. Naslain, Tensile creep behaviour of a silicon carbide-based fibre with a low oxygen content, *J. Mater. Sci.*, **30**, 661-77 (1995).

[25] G. Chollon, R. Pailler, R. Naslain, and P. Olry, Structure, composition and mechanical behavior at high temperature of the oxygen-free hi-nicalon fiber, *Ceram. Trans,* **58**, 299-04 (1995).

[26] F. L. Riley, Silicon Nitride and Related Materials, *J. Am. Ceram. Soc.*, **83**, 245–65 (2000).

[27] R. Naslain, The design of the fibre-matrix interfacial zone in ceramic matrix composites, *Composites Part A,* **29**, 1145–55 (1998).

[28] J.F. Després, and M. Monthioux, Mechanical properties of C/SiC composites as explained from their interfacial features, *J. Europ. Ceram. Soc.*, **15**, 209-24 (1995).

[29] MP. Albano, and L.B. Garrido, Processing of concentrated aqueous silicon nitride slips by slip casting, *J Am. Ceram. Soc.*, **81**, 837-44 (1998).

[30] L. Bergstrom, and RJ. Pugh, Interfacial characterization of silicon nitride powders, *J. Am. Ceram. Soc.*, **72**, 103-9 (1989).

[31] J. Zhang, F. Ye, D. Jiang, and M. Iwasa, Dispersion of Si_3N_4 powders in aqueous media, *Colloids and Surfaces A: Physicochem. Eng. Aspects*, **259**, 117-23 (2005).

[32] YS. Jung, U. Paik, C. Pagnoux, and Y. Jung, Consolidation of aqueous concentrated silicon nitride suspension by direct coagulation casting, *Mater. Sci. Eng. A.*, **342**, 93-100 (2003).

[33] MKM. Hruschka, W. Si, S.Tosatti, TJ. Graule, and LJ. Gauckler, Processing of β-silicon nitride from water-based α-silicon nitride, alumina, and yttria powder suspensions, *J. Am. Ceram. Soc.*, **82**, 2039-43 (1999).

[34] R. Boccaccini, C. Kaya, and K. K. Chawla, Use of electrophoretic deposition in the processing of fibre reinforced ceramic and glass matrix composites: a review, *Composites Part A: Appl. Sci. Manufacturing*, **32**, 997-06 (2001).

[35] C. Bernard, C. Robin-Brosse, Method of densification of a porous substrate by a matrix containing carbon, *European Patent EP0495700.*

[36] K. L. Luthra, and H. D. Park, Chemical Considerations in Carbon Fiber/Oxide-Matrix Composites, *J. Am. Ceram. Soc.*, **75**, 1889–98 (1992).

[37] H.A. Kawaoka, T.A. Kusunose, Y.Hb Choa, T.A. Sekino, and K.A. Niihara, Precise property control in silicon nitride ceramics by α/β phase ratio control, *J. Ceram. Proc. Res.*, **2**, 51-53 (2001).

[38] E. Sabouret, Composites à matrice Nitrure de Silicium et fibres de carbone : Elaboration, comportement mécanique, *PhD Thesis, University of Limoges-France* (1996).

[39] N. Eberling-Fux, Matériaux composites à matrice nanostructurée, *PhD Thesis, University of Bordeaux1-France* (2003).

MANUFACTURING OF THE CMC NOSE CAP FOR THE EXPERT SPACECRAFT

Christian Zuber, Thomas Reimer, Kornelia Stubicar, Bernhard Heidenreich, Hermann Hald
DLR – German Aerospace Center
Institute of Structures and Design
Pfaffenwaldring 38-40
D-70569 Stuttgart

ABSTRACT

DLR is responsible for the manufacturing of the CMC nose cap for the experimental re-entry spacecraft EXPERT. The goal of the EXPERT project led by the European Space Agency ESA is to collect aerothermodynamic data during atmospheric re-entry. The challenge was to manufacture a large (diameter: 716 mm, height: 413.5 mm) nose cap structure with integrated load introduction elements via the liquid silicon infiltration process (LSI) developed by DLR [1]. Two major requirements were the stagnation point temperature of 2000 °C and the aerodynamic pressure of 3 bar. The LSI process is divided into three main steps - the CFRP manufacturing, the pyrolysis and the siliconizing. To reduce internal stresses, which can cause delaminations on multi curved parts during pyrolysis, the CFRP lay-up was modified. The size of the prepreg patches was adapted, the fiber orientation was varied and the fiber length of the fabric layers was shortened with a cutting tool. To prevent distortions of the nose cap during the pyrolysis, a new C/C framing was used. To transmit the loads caused by the stagnation point pressure from the nose cap into the substructure, 16 load introduction elements were joined by the so-called in situ joining into the nose cap. The launch of the nose cap as a part of the EXPERT space-capsule is scheduled for 2010. In this paper the manufacturing process of the nose cap is described in detail. Furthermore results from the load introduction element tests and from the full-scale tests are presented.

INTRODUCTION

The Instrumented Nose Assembly (NAP) is a sub-system of the EXPERT vehicle [2][3][4], consisting of the CMC nose cap and embedded payloads to measure important aerothermodynamic data like temperature, heat flux and spectroscopic data. EXPERT will be launched on a sub-orbital trajectory with a peak velocity of 5 km/s (Figure 1). The launch will be carried out from a submarine in the Pacific ocean using a converted Russian military system, the Volna rocket. The maximum altitude will be approximately 120 km with a total traveled downrange distance of 1600 km, with 520 km after entry interface at 100 km altitude. The landing will take place in Eastern Russia on the peninsula of Kamchatka.

During re-entry the peak heat flux will reach up to 1.5 MW/m^2 in the stagnation area assuming a partial catalytic behaviour of the Ceramic Matrix Composite (CMC) material of the nose cap. Since EXPERT is a ballistic vehicle with a relatively high ballistic coefficient the total pressure in the stagnation area reaches a value of up to 3 bar, the deceleration peak load factor is 15 g. The aerothermal environment for the EXPERT vehicle during re-entry is typical for a ballistic flight from low earth orbit with a relatively small capsule. The ballistic factor is for most of the time during the peak load phase slightly below 1000 kg/m^2. The aerothermodynamic environment was looked at more closely in [5]. EXPERT in general accommodates fourteen experiments related to the topic of re-entry aerothermodynamics.

Figure 1: EXPERT vehicle

LIQUID SILICON INFILTRATION (LSI) PROZESS

The manufacturing of C/C-SiC composites by the Liquid Silicon Infiltration (LSI) process can be split into three major steps (Figure 2). Basically the process starts with the fabrication of carbon fiber reinforced plastic (CFRP) composites. It's essential to use phenolic precursor matrices of high carbon yield therefore. After curing, the composites are tempered for 4 h at 240 °C to complete the polymerization of the matrix as well as to initiate first matrix cracks. Subsequently, these CFRP composites are pyrolysed in a two step process. Firstly the CFRP composites are pyrolysed under inert atmosphere (e.g. nitrogen) at a temperature of 900 °C to convert the polymer matrix to amorphous carbon. Afterwards the C/C composites are heat-treated under vacuum at a temperature of 1650 °C. The pyrolysis of the polymer matrix leads to a macroscopic shrinkage and forms a microscopic network of cracks within the resulting C/C composites. The siliconizing is the third and final processing step. During the siliconizing the capillary effect of the translaminar crack system and the low viscosity of the molten silicon enable a quick filling of the micro cracks with molten silicon. The exothermic reaction between the carbon matrix and the molten silicon results in carbide crystals which encapsulate the carbon fibers. The siliconizing takes place under vacuum at a temperature of 1650 °C. The C/C-SiC composites in principle comprise three phases. These are the carbon phase consisting of carbon fibers and residual carbon matrix, silicon carbide and a small share of unreacted silicon.

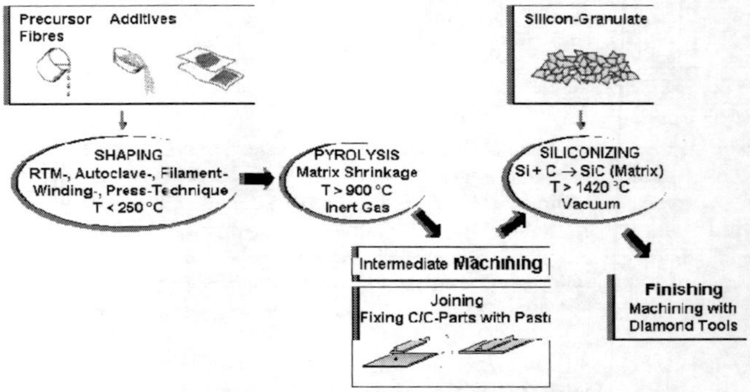

Figure 2: Liquid Silicon Infiltration (LSI) process

MANUFACTURING OF THE CFRP NOSE CAP

In total 5 nose caps were produced during the EXPERT project. Two of them were taken through the complete LSI process and were completed as qualification and flight unit. In Table 1 an overview of all manufactured nose caps is given.

Table 1: Overview manufactured nose caps

nose cap	characteristic	condition	use
# 1 (PH1596)	405 pre-cut parts	delaminations at stagnation point after pyrolysis	Siliconized (shell only), pressure test
# 2 (PH1641)	405 pre-cut parts	malfunction of autoclave control	Siliconization demonstrator with profiles
# 3 (PH1668)	525 pre-cut parts	significantly reduced delaminations	backup
# 4 (PH1746)	1428 pre-cut parts	acceptable condition	QM (qualification model)
# 5 (PH1800)	1428 pre-cut parts	as #4	FM (flight model)

Based on the outer shape requirement of the final nose cap, given by ESA, an autoclave mould for the near net shape manufacture of the C/C-SiC structure was designed. During the later pyrolysis the CFRP part shrinks up to 10 % in thickness direction, this shrinkage had to be taken into account when the mould was designed. The mould itself is an iron cast part which was machined on the inner surface. To allow an online monitoring of the autoclave process, the mould was equipped with 6 thermocouples. The positioning of the thermocouples is shown in Figure 3 and Figure 4.

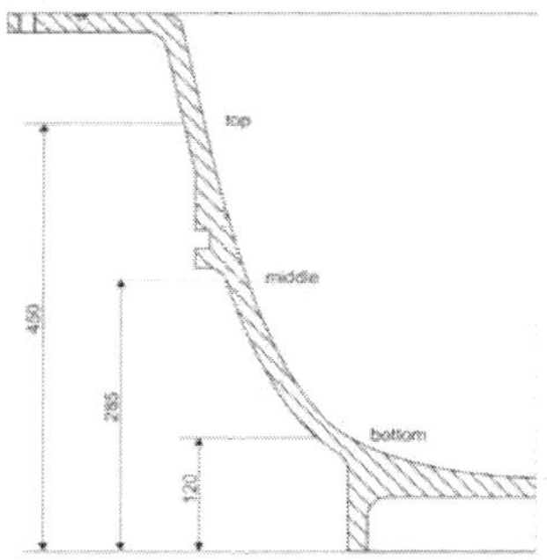

Figure 3: Position of thermocouples (vertical)

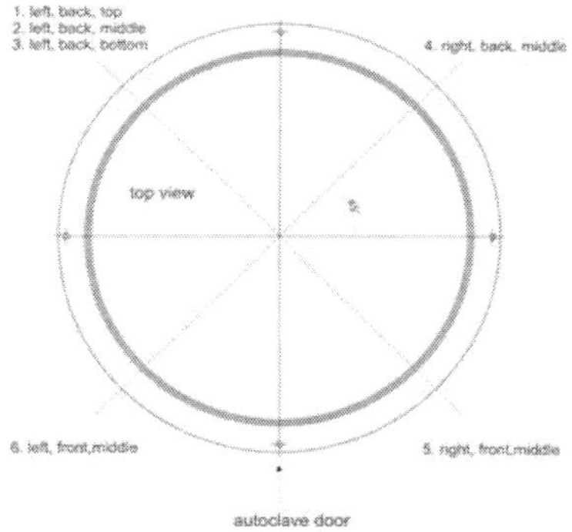

Figure 4: Position of thermocouples (horizontal)

Two laser line generators were used to allow a reproducible and accurate lay-up of the 1428 pre-cut parts for each of the nose caps #4 and #5. The laser line generators project two fine bright lines onto the inner surface of the mould which enclose an angle of 51.4° (Figure 5). The line generators were mounted on an aluminum ring, which was fixed to the mould. On the ring they can be rotated around the rotation axis of the mould in 51.4°-steps. The aluminum itself can be rotated around the symmetry axis of the mold in 3.67°-steps.

Figure 5: CFRP lay-up

LAY-UP CFRP NOSE CAP

The nose cap was built up out of 27 layers prepreg. Every layer consisted of 7 sectors, each with 51.4°, which again were subdivided in 7 to 9 pre cut parts. The configuration for one of these sectors is shown in Figure 6 and Figure 7. After depositing a complete prepreg layer, the aluminum ring was rotated by 3.67°. By this rotation a radial overlap between the edgings of the pre-cut parts was avoided.

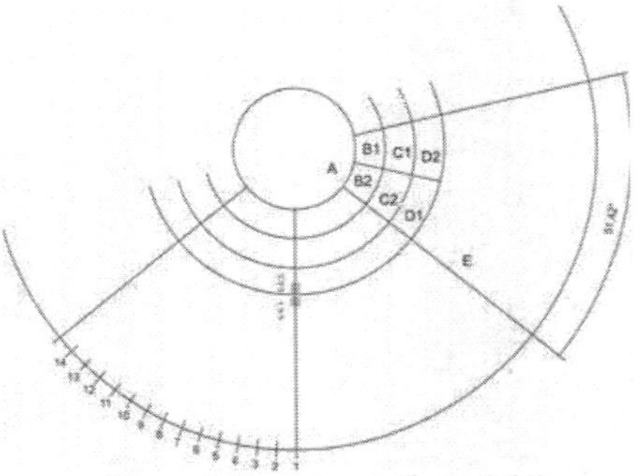

Figure 6: Pre-cut parts (layer 1-21)

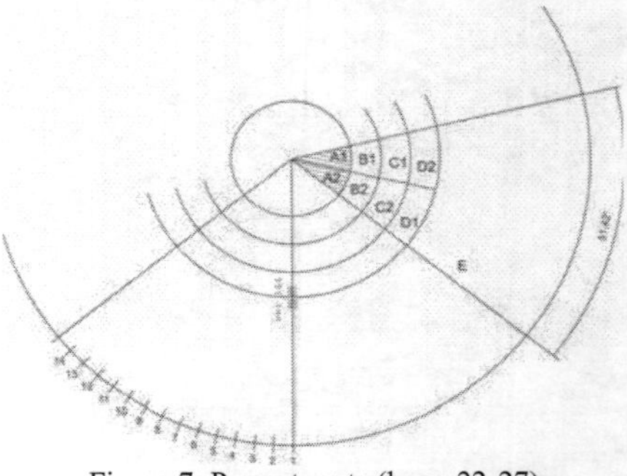

Figure 7: Pre-cut parts (layer 22-27)

For the layers 1 to 21 the pre-cut parts A, B1, B2, C1, C2, D1, D2 and E (lower half of part E) were cut with a die-cutting tool in 0° and 90° direction of the fabric (Figure 8). The resulting maximum fiber length was 30 mm. The idea of cutting the fabric was to "soften" it, to avoid delaminations of the laminate during later pyrolysis. These delaminations are caused by tensions as a result of the matrix shrinkage during pyrolysis in interaction with the stiff fabric which doesn't shrink.

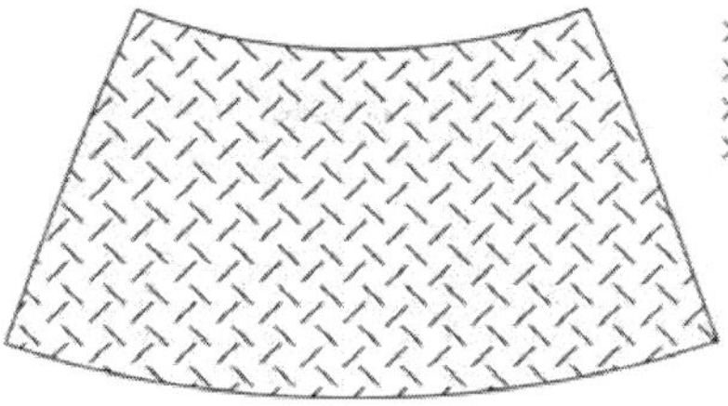

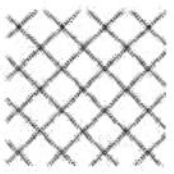

Fiber orientation ± 45°

Figure 8: Cutting pattern in 0° and 90° direction

For the layers 22 to 27 the fiber orientation of the pre-cut parts was changed from ±45° to 0°/90°. The 0° direction of a single prepreg pre-cut part is shown in picture 5 (area A in the middle). Additionally the circular part A in the stagnation point was separated into 14 sectors. The pre-cut parts for Layer 22-27 weren't cut with the die cutting tool (for details Table 2). The intention was support the delamination reduction by introducing a stiff inner structure made up of layer 22 to 27. The shrinkage of the six inner layers in circumferential direction is due to the 0°/90° fibre orientation extremely limited. In contrast, the outer layers can easily shrink onto the inner layers during pyrolysis, reducing inner stresses during pyrolysis.
The last prepreg layer, number 27, was covered with a release film. The release film was covered with two layers of breather material. They were necessary to suck the air out of the prepreg lay up. In addition the breather material absorbed the surplus of the phenolic precursor. To enable sufficient air suction, four glass fabric bands were placed between the nose cape center and the border of the mold. Finally the vacuum bag was placed on the set-up. After finishing the autoclave cycle (t= 480 min, T_{max}= 195°C), the nose cap was tempered for 4 h at 240 °C to complete the polymerization of the polymer matrix.

Table 2: Cut of pre-cut parts for NC4 (1428 pre-cut parts)

Outer layers (1-21)	Fiber orientation	Cutting (in 0° and 90° direction of the fabric)
A (circular)	-	yes
B1, B2	± 45°	yes
C1, C2	± 45°	yes
D1, D2	± 45°	yes
E	± 45°	yes (lower half of the pre-cut part)
Inner layers (22-27)		
A1, A2	0°/90°	no
B1, B2	0°/90°	no
C1, C2	0°/90°	no
D1, D2	0°/90°	no
E1	0°/90°	no

PYROLYSIS OF THE CFRP NOSE CAP

To convert the CFRP into a porous C/C structure, the CFRP green body was pyrolysed in a two step pyrolysis process.

Figure 9: C/C kiln furniture

To avoid distortion and additional inner stresses during pyrolysis, commonly massive graphite moulds are used as kiln furniture. However, a massive graphite mould couldn't be used in this case for two reasons – firstly the weight, which would exceed the maximum load of the available furnace and secondly the high procurement price. Therefore a kind of C/C jig made out of plate material was used for the first time (Figure 9). The ribs of the jig define the reference shape for the final nose cap. To press the CFRP nose cap against the ribs, the cap was filled up with graphite granulate.

MANUFACTURING OF LOAD INTRODUCTION ELEMENTS

Each nose cap is equipped with 16 hat profiles as load introduction elements on the inside. The CFRP green bodies were made via warm pressing technique. The warm pressing technique allowed a quick and simple manufacturing of the raw profiles. Therefore 23 rectangular prepreg layers were stacked onto the positive aluminum mould. After closing the mould and forming the profile the precursor was cured during the 290 min lasting process ($T_{max.}$= 210 °C ,p= 8 bar). Figure 10 shows the female mould with the CFRP profile after polymerization as well as some surplus of the phenolic precursor, as it typically flows out of the mould during the warm pressing cycle. After finishing the subsequent tempering cycle with a maximum temperature of 250 °C, the profiles were pyrolysed with the same two step pyrolysis processes used for the nose cap. Therefor the CFRP green bodies were placed in a graphite mould (Figure 11). The shape of the graphite mould corresponds to the finale shape of the hat profiles. As shown in Figure 11 the graphite mould doesn't fit to the profile in the initial CFRP condition. During pyrolysis shrinkage in thickness direction and a change of the angles of the CFRP green body occurs, which is supported by the graphite mold. After pyrolysis the mould fits exactly to the C/C part. In the C/C condition two hat profiles, as shown in Figure 12, were machined out of one C/C raw profile. In the run-up to the hardware manufacturing, stress peaks in the joining area between the hat profiles and the nose cap were detected via FEM analysis. To reduce these stress peaks, intermediate elements (referred to as C/C shoes) were joined to the hat profiles via in-situ joining (picture 10).

Figure 10: CFRP hat profile after polymerization

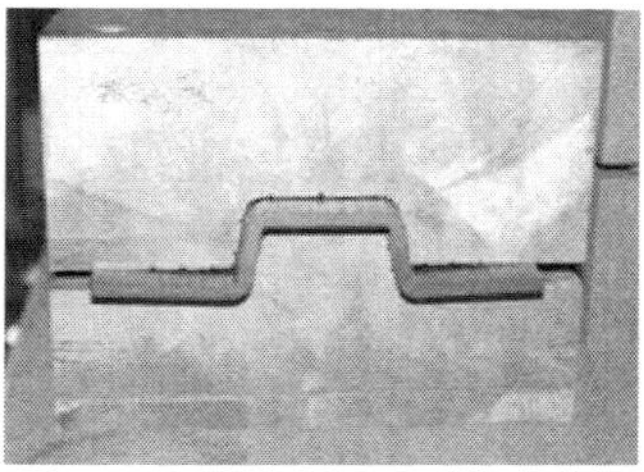

Figure 11: CFRP hat profile inside graphite kiln furniture

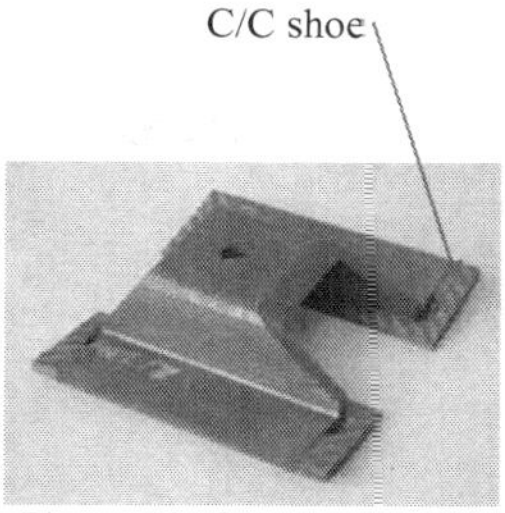

Figure 12: C/C hat profile prepared for joining

IN-SITU JOINIG OF NOSE CAP AND LOAD INTRODUCTION ELEMENTS

After pyrolysis the nose cape was calibrated with a 3-D measuring machine. The deformations caused by the pyrolysis were rated to be in an acceptable range. Based on the 3D measuring results planar faces were machined as pockets into the inside of the nose cap. The hat profiles were joined into the pockets with a joining paste, developed for the LSI process. The paste is based on a phenolic resin with a high residual carbon content of approx. 60 mass-% as well as a high amount of carbon powder. The high carbon content is needed to ensure a stable joint during the subsequent Siliconization, were the polymer content of the joining past firstly is converted into carbon and subsequently transformed to SiC.

To assure a homogeneous pressure in the entire joining area between the hat profiles and the nose cap, aluminum pushers were used, which fit to the contour of the nose cap respectively the shape of the hat profiles. The needed force during the curing of the joining paste was achieved by using heavy screw clamps which allow an operation by torque wrench. The torque wrench guarantees a reproducible joining force for all hat profiles during the curing. The joining past was cured in a tempering furnace at $T_{max.}$= 135° for 2 hours (Figure 13).

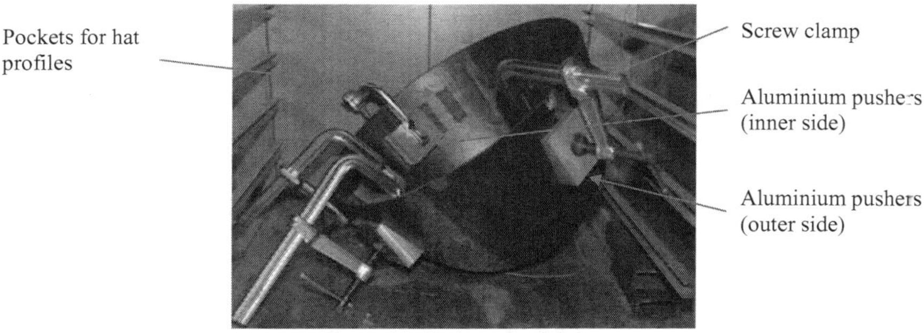

Figure 13: Curing after joining

SILICONIZING OF THE NOSE CAP

After the joining of the nose cap and the hat profiles, the nose cap was siliconized at a temperature of 1650°C in vacuum. Usually the offered silicon amount is calculated using the open porosity of

the C/C preform measured via the Archimedes principle. Because of the large dimensions of the part, a porosity measurement wasn't possible. Therefore a silicon amount of 57 mass-% of the C/C preform was offered, as it is common for the siliconization of C/C plates manufactured by autoclave technique. The silicon granulate was applied from the bottom (2/3 of total amount) and the top (1/3 of total amount) of the nose cap. Silicon carbide buffer plates were used to regulate the silicon infiltration during siliconizing. The whole build-up was placed in a graphite crucible which was also coated with boron nitride (Figure 14). The boron nitride coating was applied to prevent the graphite from reacting with the liquid silicon during infiltration.
After siliconizing the nose cap was inspected via computer tomography (CT) to detect inhomogeneously siliconized areas, e.g. areas which were not infiltrated with silicon as a result of delaminations. The hat profiles were considered to be particularly critical, because the silicon needed for a proper infiltration of the hat profile has to flow through the joining layer between them and the nose cap. It could be shown via CT that the silconization of the nose cap in general and the hat profiles in particular was successful – no inhomogeneously siliconized areas could be detected.
After the CT inspection the nose cap was desiliconized to remove residual free silicon which hasn't reacted to silicon carbide. The desiliconizing was done at 1650 °C in vacuum. Therefore the C/C-SiC part was partly embedded into graphite granulate which reacts with the residual silicon to silicon carbide.

Figure 14: Siliconiziing of nose cap #5

FINAL MACHING AND COATING

The final manufacturing steps towards a flight-ready component comprised mainly the machining of the siliconized nose cap and a coating process. In detail the following tasks had to be carried out:

- Drilling of the holes for sensor heads
- Machining of planar faces on the inside for sensor heads
- Drilling of holes in the hat profiles
- Drilling of holes on the inside for thermocouple placement
- Cutting of the nose cap to the required height
- Reduction of surface roughness (residua of siliconizing) with diamond coated abrasive pads

- Cleaning of with osmosis water with subsequent drying at 110°C in tempering furnace
- CVD-SiC coating with a resulting coating thickness of 90 – 120 µm

The finished nose cap is shown in Figure 15 and Figure 16.

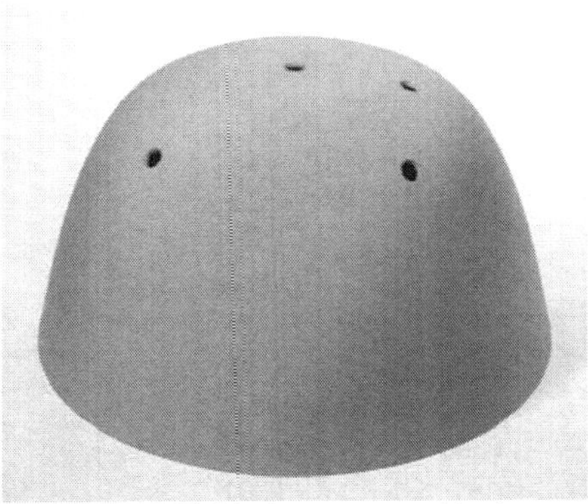

Figure 15: Finished nose cap outside view

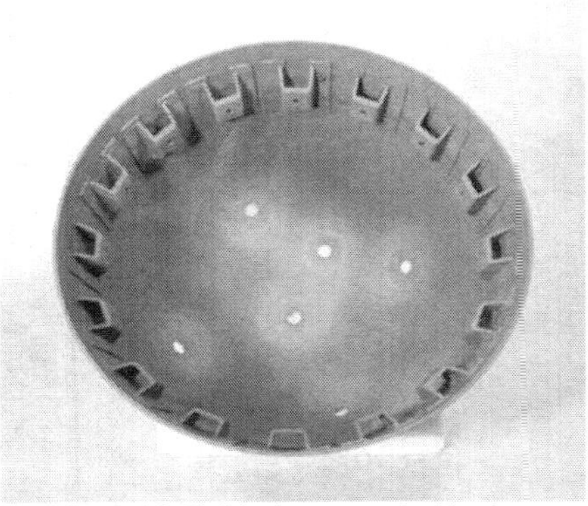

Figure 16: Finished nose cap inside view

DESIGN VERIFICATION

The design of the nose cap was verified via several major elements:

- Finite element analysis
- Pressure test of the nose shell
- Mechanical test of the load introduction elements
- Material characterization in arc jet tests
- Payload subsystem arc jet tests

Of these, the numerical analysis is not described in this paper because the focus is on the hardware and the related hardware tests. A description of some aspects of the numerical design work can be found in [6]. Also not described are the dedicated arc jet tests with material samples and the embedded elements of the sensor subsystems in the nose cap. More details on these topics can be found in [7].

PRESSURE TEST OF THE NOSE CAP

The dynamic pressure acting on the stagnation region of the nose cap is 150 kPa. This equals a total pressure of roughly 270 kPa. A pressure test with the nose shell only was carried out in order to verify that kind of load. The test was set up as a pressure difference test in the autoclave of the Institute. The nose cap was positioned on an aluminium plate as shown in Figure 17. The edge of the nose cap was sealed against the plate. Thereby a sealed volume was created enclosed by the nose cap and the plate. A vacuum line was connected to the plate and to a vacuum pump on the far end outside of the autoclave. The nose cap on the plate was placed in the autoclave and the pressure inside of the nose cap was decreased via the vacuum pump. At the same time the pressure in the autoclave was raised so that a pressure difference was created between the outside and inside of the nose cap. The targeted pressure difference was 300 kPa. Figure 18 shows that a maximum of 315 kPa pressure difference was achieved. The nose cap was equipped with strain gauges during the test. The maximum recorded levels were -22MPa compression stress. The bending strength of the material is in the range of 180 MPa, tension strength is 120 MPa and compression strength is -200 MPa.

Figure 17: Nose cap pressure test

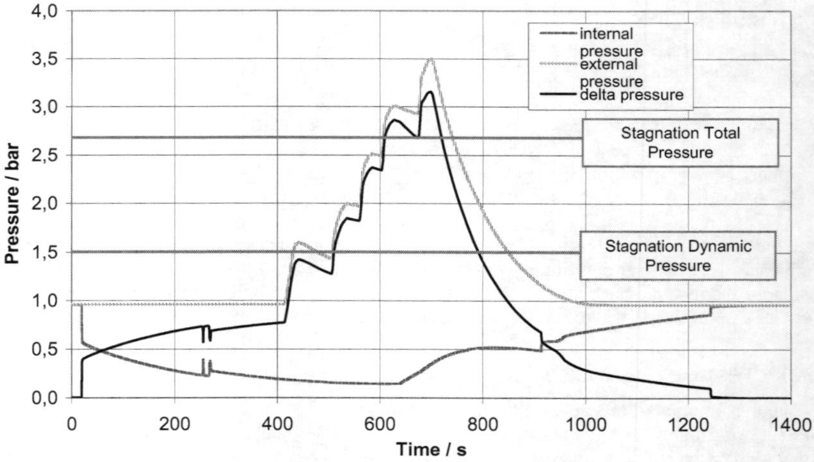

Figure 18: Nose cap pressure test recorded data

LOAD INTRODUCTION ELEMENTS

The load introduction elements on the inside of the nose cap have to transmit the aerodynamic loads from the nose cap to the structure of the capsule. An analysis of the transient deceleration loads on the vehicle resulted in a value of approximately 60 kN that act as a mechanical load on the nose cap and have to be sustained by the load introductions. For each individual load introduction chain, that results in 3.75 kN. Since that load is due to the aerodynamic pressure it is a compression load that acts on the load transfer elements. A test was devised to verify the capability of the nose cap to handle that situation. The test was set up as a uni-axial compression test of structural samples that represent a nose cap cut-out with the corresponding hat profiles and other related components.

Figure 19 shows a sample as prepared for the testing and Figure 20 shows the test machine with a sample fixed to the lower sample holder. A number of 32 tests were carried out in total. The test parameters included mainly variations in the lay-up of the CMC components and in the materials used as shown in Table 3.

Table 3: Load introduction test parameters

Test Series	Samples	Material	Fiber orientation hat profile	Fiber orientation plate	Reinforcement
a	6	XB	0°/90°	45°	yes
b	6	XB	45°	45°	no
c	6	XB	45°	45°	yes
d	6	XB-T800	45°	45°	yes
e	6	XB	0°/90°	0°/90°	yes
f	2	XB	45°	0°/90°	yes

The results of the tests were very favorable. A summary of all tests is presented in Figure 21. The average failure load of the samples was 12.3 kN. Two failure modes could be observed. Either there was a debonding between carrier plate and profile or there was a cracking failure in the top face of the profile. Both failure types were fairly well distributed among all the tested samples.

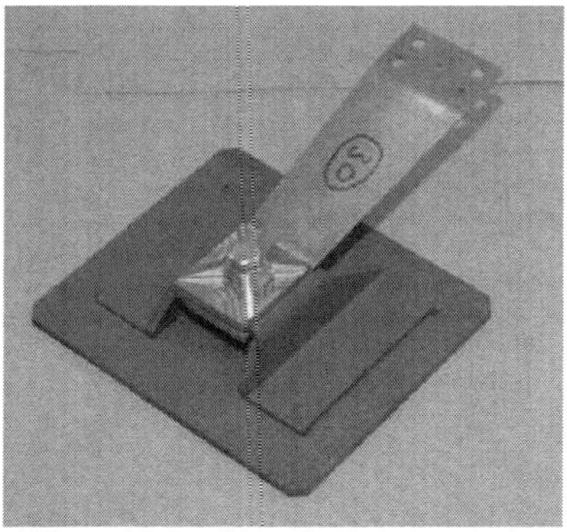

Figure 19: Load introduction test sample

Figure 20: Load introduction test set-up

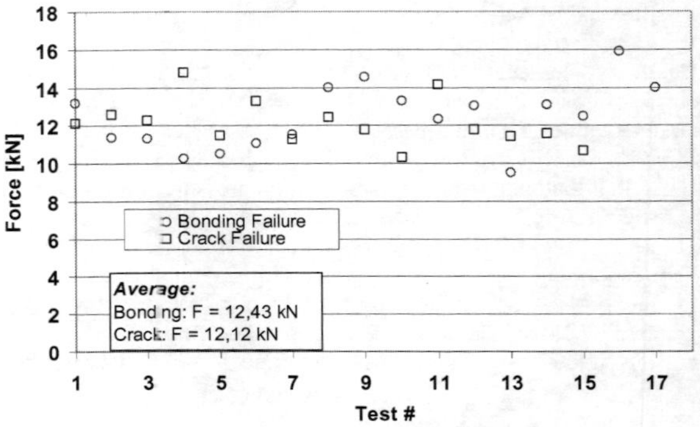

Figure 21: Load introduction test results

CONCLUSIONS

With the EXPERT nose cap, a large and complex CMC structure was successfully developed. Initial lay-up problems could be solved by the variation of the lay-up. The nose cape was siliconized successfully. It could be shown that a siliconization over the joining boundaries between the load introduction elements and nose cap is possible. During the whole manufacturing process advanced NDE equipment as e.g. computer tomography was utilized.

It becomes obvious that a further understanding of the mechanisms leading to the delaminations is required for future projects. In this context a comprehensive characterization of the remedial effects resulting from the lay-up variations should be done.

REFERENCES

[1] W. Krenkel, "Entwicklung eines kostengünstigen Verfahrens zur Herstellung von Bauteilen aus keramischen Verbundwerkstoffen", Dissertation Universität Stuttgart, DLR-Forschungsbericht 2000-04, Stuttgart, Germany, 2000

[2] F. Massobrio, R. Viotto, M. Serpico, A. Sansone, M. Caporicci, J.-M. Muylaert, „EXPERT:An atmospheric re-entry test-bed", Acta Astronautica, Volume 60, Issue 12, June 2007, p974-985

[3] G. Reibaldi, J. Gavira, F. Ratti, S. de Mey, J-M. Muylaert, R. Provera, F. Massobrio, "EXPERT: The ESA experimental re-entry test-bed programme", 59th International Astronautical Congress, 29 Sep.-03. Oct. 2008, Glasgow, Scotland, IAC-08.D2.6.3

[4] J. Gavira, "EXPERT - A European Aerothermodynamics In-Flight Testbed", 6th European Workshop on Thermal Protection Systems and Hot Structures, 1-3 April 2009, Stuttgart, Germany

[5] T. Reimer, T. Laux, "Thermal and Mechanical Design of the EXPERT C/C-SiC Nose", 5th European Workshop on Thermal Protection Systems and Hot Structures, Noordwijk, The Netherlands,17-19 May 2006 (ESA SP-631, August 2006)

[6] K. Stubicar, T. Reimer, "Thermo-Mechanical Design of the EXPERT Nose and Testing of the Load Introductions", 6th European Workshop on Thermal Protection Systems and Hot Structures, 1-3 April 2009, Stuttgart, Germany

[7] K. Stubicar, T. Reimer, G. Koppenwallner et al., *Testing of EXPERT Nose Material and Sensor Heads in the DLR ARCJET Facilities*, 6th Symposium on Aerothermodynamics for Space Vehicles, Versailles, France, 3-6 November 2008

THE NATURE OF SILICON CARBIDE PHASES DEVELOPED FROM DIFFERENT CARBONACEOUS SOURCES AND ITS IMPACT ON THE MICROSTRUCTURE OF C_f/C-SiC COMPOSITES

Andrew Leatherbarrow and Houzheng Wu
Department of Materials, Loughborough University
Loughborough, Leicestershire, England, LE11 3TU

ABSTRACT

A variety of carbonaceous matters have been explored and used as sources of manufacturing carbon fibre reinforced silicon carbide ceramics (C_f/C-SiC). The nature of the produced silicon carbide could be influenced by the format of the carbon. In this research, the phase composition and microstructure were examined and understood for silicon carbide phases generated through liquid silicon infiltration (LSI) of C_f/C preforms. The preforms included carbon produced through a carbonization of methane via chemical vapour infiltration (CVI), pyrolyzation of impregnated epoxy and phenolic, and CVI carbon being further graphitised. Optical microscopy, transmission electron microscopy and X-ray diffraction were used to characterise the microstructure and phase composition. Six possible distinct regions were identified inside the developed composites: pyrolytic carbon enfolding carbon fibre filaments or bundles; dispersed nano-SiC particles inside pyrolytic carbon medium; nano-crystalline SiC layer; dispersed nano-SiC particles inside course SiC grain region; course SiC layer; un-reacted re-solidified pure silicon with or without interspersed SiC islands. A hierarchy order composed of these regions existed in all composites regardless of the carbonaceous source. However, the size and morphology of each region varied and were attributed to differences in the format of pyrolytic carbon. Combined TEM and XRD analysis indicated that the generated SiC was the same regardless of the carbon format and that the most likely SiC polytype was face centered cubic (FCC) β 3C-SiC.

INTRODUCTION

Damage tolerant ceramic matrix composites comprising silicon carbide matrices and carbon fibre reinforcements known as C_f/SiC or C_f/C-SiC composites if residual carbon is accounted as part of the matrix, have so far been developed as weight saving and high temperature resistant applications [1-3]. Such composites are manufactured by either liquid [4] or gas phase routes as already reviewed elsewhere [5, 6], or alternate derivatives [7-11] that operate on the principle of impregnation and densification of a porous carbon fibre based carbon (C_f/C) preform. In recent years, liquid phase routes such as liquid silicon infiltration (LSI) have gained popularity because of its rapidity, economy and the satisfactory achievements in engineering performance, in comparison to traditional gas phase routes [4]. During LSI, silicon is heated beyond its melting point (1410 °C) and infiltrated into the C_f/C preform. The silicon reacts directly with any existing carbon with a consequence of converting them into a mixture of SiC and residual free un-reacted silicon/carbon, if not all SiC. The microstructure generated by LSI has been widely documented [12, 13] and comprises carbon fibres as reinforcements, carbon, SiC and silicon (C/SiC/Si) as the matrix, apart from typical microstructural features such as interfaces and voids.

The mechanisms governing the microstructural evolution during LSI have been investigated [14-19] although only limited and in some cases conflicting experimental support has been available. Schule-Fischedick *et al* [14] documented the presence of a fine SiC layer and attributed its formation to silicon vapor that readily accessed this open porosity at temperatures prior to reaching the melting point of the silicon. Verala-Feria *et al* [17] observed a similar fine SiC layer, but also that was not present throughout the entire structure; they attributed this to the effect of LSI temperatures and possible higher localized reaction temperatures.

Meanwhile, diffusion of carbon through the evolving SiC is the widely accepted mechanism for the creation of the bulk SiC layer. As the layer develops thicker, a barrier is essentially formed between the carbon matter and molten silicon, which increasingly inhibits further diffusion and conversion reactions. Consequently, a region of un-reacted re-solidified silicon exists beyond the SiC. However, in this region, SiC islands can be observed, the formation mechanism of which is still not agreed upon. It was speculated the mechanism as a combination of solution and precipitation [14]. Alternatively, Favre *et al* [15] hypothesized alternating periods of crystallization and breaking away of SiC from the SiC region. Zollfrank *et al* [16] believed the mechanism of diffusion, dissolution and re-crystallization.

One of the factors that may greatly influence the conversion is the carbonaceous material for producing the carbon constituent; with the difference in the molecular structure of the typically used ones, where heat treatment could lead the pyrolized carbon to a state from amorphous to highly crystallized one [20]. This is in agreement with Zollfrank *et al* [16] who stated the microstructure was dependent on the initial morphology of the carbon as well as the manufacture parameters: reaction time and temperature, where short reaction times and low temperatures favoured nano-sized SiC grains and extended reaction times and high temperatures favoured formation of a coarse grained SiC. In the investigation by Favre *et al* [15], comparisons were made between glassy carbon that was heat-treated to 2500 °C and polycrystalline graphite. The determining mechanism and microstructure after LSI were found to be similar, which leads to a speculation of how different carbon formats would influence the microstructure of the composites.

The most common carbonaceous sources used to generate carbon constituent inside carbon fibre preform include methane through gaseous deposition of carbon via chemical vapour infiltration (CVI), or polymeric resins through impregnation and pyrolysis (PIP). In this paper, we will investigate the development of SiC phases when liquid silicon is infiltrated into carbon regions that are developed from carbonaceous materials in different formats, including crystallinity, size and orientation. These include carbon derived from PIP of phenolic resin, epoxy resin, carbon deposition of methane by CVI and highly graphitilized CVI carbon.

EXPERIMENTAL PROCEDURES

(a) Materials Preparation

Different types of carbonaceous sources, i.e. epoxy, phenolic, carbon deposited from vapour, were selected in this study to generate C_f/C preforms. These preforms were achieved through different heat treatment routes after the carbonaceous materials were impregnated into carbon fibres via different ways; each of which is outlined as following:

(i) PyC-Epoxy C_f/C preform was achieved from a direct pyrolyzation of a single woven AS4-3k PAN carbon fibre pre-preg layer in an inert atmosphere. The resin in this pre-preg is of an industry standard epoxy (e.g. Hexply 8552, Hexcel Ltd).

(ii) PyC-Phenolic A pyrolysed single woven AS4-3k PAN carbon fibre pre-preg was stacked in a steel die and impregnated with phenolic resin (Hexcel Ltd). The preform was thermally cured under pressure before being post cured to ensure complete polymerization to generate carbon fibre reinforced plastic (CRFP) preforms. The CFRP preforms were subsequently pyrolysed at elevated temperatures to convert the polymer matrix into pyrolytic carbon.

(iii) CVI- Carbon These preforms were prepared through gaseous deposition of carbon derived from the cracking of methane at elevated temperature.

(iv) CVI-Graphite The C_f/C preformed derived from CVI process was further heat treated at a temperature typically larger than 2000 °C in an inert atmosphere to have a higher degree of graphitisation.

All C_f/C preforms were subjected to the same siliconisation. This process was performed using silicon granulate of commercial grade 99.9 % purity that was placed in direct contact with each preform prior to heating, in an alumina tube furnace (Lenton, High Wycome, UK). Samples were heated at 5 °C/min in an inert atmosphere up to 1500 °C and held for a period of 3 minutes.

(b) Microstructure Characterisation

All specimens were polished to a 1-µm diamond grits finish using a polishing wheel (Struers, Tegra-pol). Optical microscopy (Reichert-Jung/MEF-3) was used to characterize the microstructure of each sample after LSI using a combination of polarised and differential interference contrast (DIC) light where appropriate to aid distinction between the interfaces and regions. A focus ion beam (FIB) microscope (FEI Nova 600 Nanolab) was used to remove cross-sections across the C/SiC/Si interfaces measuring approximately 20-µm wide by 200 nm thick for TEM (JEOL 2000FX) analysis. Selected area diffraction (SAD) patterns were generated using a dual tilt arm at a camera length of 100 cm at 200 kV. SAD patterns were taken at the carbon region to characterize the crystallinity of the carbon, and at the SiC regions to examine their crystallographic structure. The following equation, derived from the Bragg diffraction condition was used to calculate the d-spacings of crystal the planes (Å): -

$$d = \frac{\lambda L}{R} \qquad (1)$$

where, λ is the wavelength of electrons 2.5 pm, L the camera length (cm) and R the distances of diffraction spots to the central spot or radius of the diffraction rings (mm). The measured d-spacings from the SAD were matched to values that could be acquired from the international powder diffraction database, and correlated to the d-spacings measured from XRD spectra to ascertain the most likely polytype of the SiC. Being aware of the limitations of that XRD alone in detecting compound typically under 5 % of the bulk size of the sample, instead of being implemented as the sole method to prove the type of SiC, it was used to support TEM investigations in the micro-scale regions. All X-ray diffraction was performed on Philips PW17-30 using CuK_α incident radiation through a 2θ range from 0° to 90° with a step rate of 1°/min.

RESULTS AND DISCUSSION

The aim of the current investigation was to characterise the phase composition and the structure of SiC phases generated by the reaction of pyrolytic carbon from different carbonaceous sources with impregnated liquid silicon.

(a) Microstructure of C_f/C-SiC generated from different carbonaceous materials

Optical micrographs in fig. 1 show that the hierarchy order of the microstructure maintained the same as carbon fibre embraced by pyrolytic carbon, and then possible silicon carbide; silicon regions are always surrounded by silicon carbide. It seems that the carbon format had no direct influence on the overall hierarchy of the microstructure of the composites. It is believed that silicon carbide was generated when the silicon melt impinged the pyrolytic carbon. The estimated thickness of the SiC layer, based on optical microscopy images, ranged between 1-30 µm among these samples and only extended further in specific regions of where the carboneous matter was enclosed. The thinnest SiC layer was observed from the CVI-Carbon between 5-10 µm, increasing to 10-20 µm for both pyrolytic carbon samples and further still to 10-30 µm from CVI-Graphite. Differences in the thicknesses could be explained at least in part by the diffusion mechanism governing the formation of the SiC. For sustained reactions and growth of the SiC, the carbon from pyrolytic carbon region must either diffuse to the silicon region or vice versa. Considering the work by Hon *et al* [21-22] who found that the lattice self-diffusion coefficient rate of carbon was up to 100 times higher than silicon. It is more likely the case that the carbon diffused outwards through the evolving SiC layer. However, matter transportation through non-atomic diffusion ways could boost the development speed of silicon carbide. For instance, connected micro-channels or pores in the pyrolytic carbon could lead to a quick sucking-in of silicon melt or vapour into the carbon matrix and then a reaction between carbon and silicon to convert the region into silicon carbide completely. The residual un-reacted silicon inside the silicon carbide region shown in CVI-Graphite (fig. 1(b)) has demonstrated such a possibility. It would not be an unreasonable assumption that the carbon region would be left with more porosity in CVI-Graphite after being treated at much higher temperature than the other samples, due to the loss of more volatile species and any further shrinkage as a consequence of high degree of graphitization (see the section (b) on the carbon structure). In comparison to CVI-Carbon, pyrolytic carbon from polymer resins could be more porous due to more volatile species could be generated at the heat treatment temperatures, which likely lead to thicker SiC layers as observed in fig. 1(c) and (d).

Figure 1. Optical micrographs of (a) CVI-Carbon, (b) CVI-Graphite, (c) PyC-Phenolic and (d) PyC-Epoxy samples, showing that the hierarchal order is maintained irrespective of the carbon format, but different SiC thicknesses were generated.

(b) Structure of the pyrolytic carbon

As shown in fig. 2, TEM SAD patterns of the carbon materials illustrate the differences in the carbon format for representative carbon region of different samples. The {002} reflections of carbon from CVI-Carbon sample give a well defined circles in the SAD pattern (fig. 2(a)); those from other three samples present in short arch, whilst the lengths of the arches in PyC-Phenolic and PyC-Epoxy are clearly longer than these in CVI-Graphite. The length of the arch is likely related to the size and the orientation of the graphite units. The results in fig. 2 show that the CVI-Graphite contains larger graphite units and/or smaller orientation angle among the units than all others; the pyrolized carbon from two types of resins has clearly larger graphite units and/or smaller orientation angle than CVI-Carbon. Indeed, the ring in the diffraction pattern indicates that over the length scale defined by the SAD aperture the carbon is isotropic in CVI-Carbon. In fig. 2, the diffusion of the {002} reflections has only a marginal difference among CVI-Carbon, PyC-Phenolic and PyC-Epoxy, apart from CVI-Graphite where the reflections have less diffusion. A similar level of diffusion of the reflections provides a clear indication that the graphite units have a fairly close $d_{(002)}$-spacing distribution and/or similar thickness along the normal direction of the basal plane of the graphite units, and the reduced diffusion in CVI-Graphite indicates that the $d_{(002)}$–spacing is slightly more consistent and/or the thickness along normal direction of the basal plane is larger than that other three samples. It is also noted in fig. 2 that the {110} reflections have much longer arches than the {002} ones for all of the samples. It is believed that the prismatic planes of the graphite units could be heavily distorted, whilst the graphene layers have been more defined and stacked together.

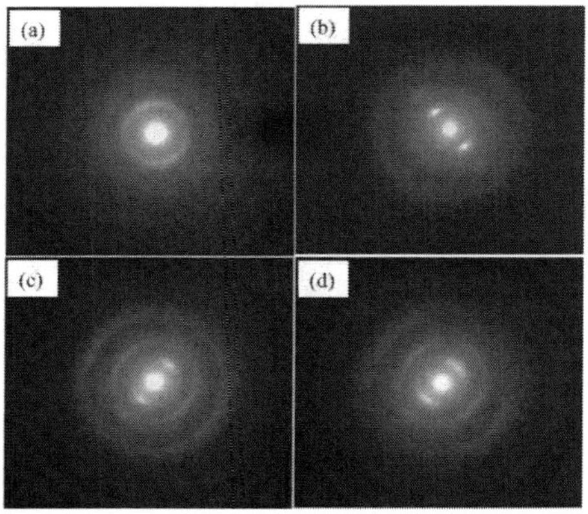

Figure 2. SAD patterns from carbon regions in sample of (a) CVI- Carbon, (b) CVI- graphite, (c) PyC- Phenolic and (d) PyC-Epoxy.

The extracted information from the SAD patterns shows the state of the carbon generated from different carbonaceous sources and manufacture routes introduced differences in size and orientation of the graphite unit, but not significantly in the degree of graphitization. This could be defined by the temperature regime used in this study; except the CVI-Graphite, the other three samples were pyrolized under 2000 °C, which is a temperature regime that only limited amount of graphitisation could be completed, as indicated in Oberlin's study on the pyrolization of different carbonaceous materials [20].

(b) Structure of the silicon carbide

Fig. 3 shows the sliced TEM sample from PyC-Phenolic using the FIB microscope. A silicon carbide zone was developed between the pyrolytic carbon and silicon, and all interfaces between the carbon fibre and pyrolytic carbon, pyrolytic carbon and silicon carbide, silicon carbide and silicon are exposed in this view. Under diffraction contrast-imaging condition, there is no particular contrast displayed in the carbon and silicon regions. More details on the SiC are specified and discussed as following.

(i) Silicon carbide region

The course SiC layer was of variable thickness ranging typically between 5-30 μm and comprised of multi-crystals that are in accordance with previous investigations. Apart from strain contours appeared in all SiC grains, significant stacking faults were also developed across the grains, as shown in fig. 3(a), and highlighted in fig. 3(b). The appearance of stacking faults implies that the SiC region was under strain after LSI processing. This strain could come from (a) thermal residual stress developed during cooling down from manufacture temperatures, due to the difference of thermal expansion coefficient among silicon carbide, silicon and carbon; (b) resistance to the growth of silicon carbide crystallites during LSI process, due to the space competition of reactants and products. These phenomena were noticed in reaction bonded silicon carbide ceramics [23, 24].

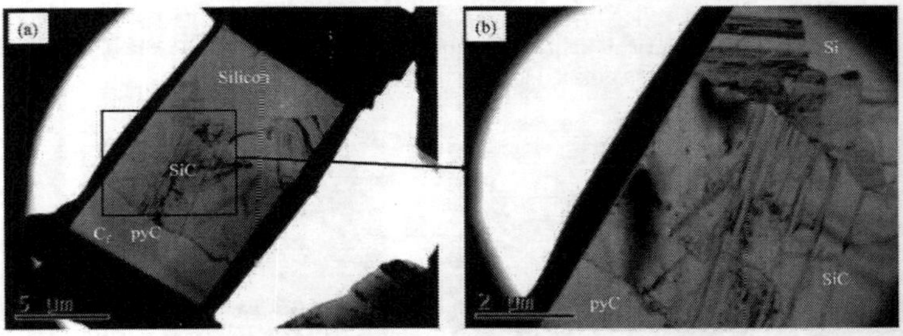

Figure 3. TEM micrograph of (a) the general microstructure of PyC-Phenonic after LSI, (b) detailed features of the evolved SiC grains.

(ii) Structure of the interface region between SiC and C, Si

Fig. 4 shows a typical interfacial region that is composed of fine particles with a size of around 50 nm; a similar structure was shown in all studied samples. Using SAD pattern, shown as an inset in fig. 4(b), from this interface region, we confirmed these fine particles are all silicon carbide ((more details and discussion in (iii)). The fine particles around the interface were originally assumed to have originated from a layer of fine particles between the pyrolytic carbon region and large sized SiC region, as reported by other researchers [14]. By carefully examining both bright and dark field images in fig. 4, we believe that, apart from a possible layer composed of the fine SiC particles alone, there may exist two dispersoid regions on each side of the interface layer, where the dispersants are all SiC within the same size range as the interface layer. As seen in fig. 4, near the carbon region, the continuous phase for the dispersoid should be pyrolytic carbon, and near the SiC region, the continuous phase silicon carbide. The first one was also noticed in previous investigations [14, 15, 25], but the later one has not been previously documented.

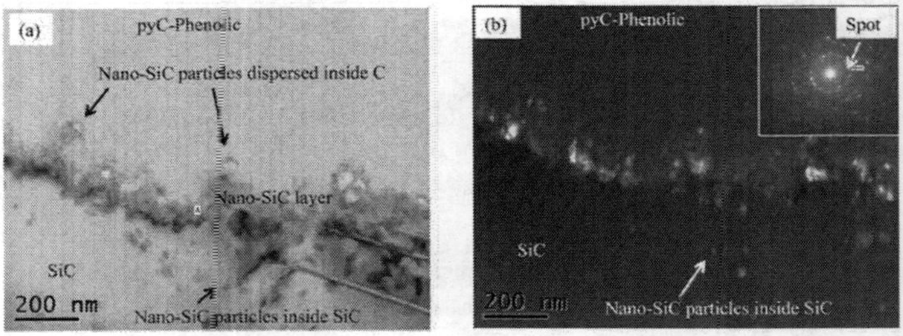

Figure 4. TEM bright field micrograph (a) and supporting darkfield micrograph (b) showing the nano-crystalline SiC layer with evidence of dispersed nano-sized SiC inside the pyC- carbon and coarse SiC.

The dispersoids within carbon matrix were speculated to have developed from silicon vapour that infiltrated prior to reaching the melting point of the silicon [14], but the development of a dispersoids within the silicon carbide matrix has no understanding yet, inside the knowledge of the authors. It might be an artificial effect due to the large inclination of the interface layer in this observation. Obviously, further effort is needed to clarify the structure near the interface; we believe this information could help us further understand the mechanisms governing the development of SiC phase during LSI. Among the PyC-Phenolic, PyC-Epoxy and CVI-Carbon samples, the structure of the fine SiC between the course SiC and pyrolytic carbon was fairly similar, with a scale of about tens to hundreds nanometer across the dispersoid band. However, in the CVI-Graphite sample, this scale was extended up to several microns, as shown in fig. 5. This observation seems to support the previously proposed mechanism that silicon vapour infiltrated the porous carbon and then developed these fine SiC grains prior to reaching the melting point of the silicon, because the graphitized carbon should be the most porous one among these samples, readily granting access of the vapour into and throughout the architecture that could account for the increased visibility of the nano-SiC particles.

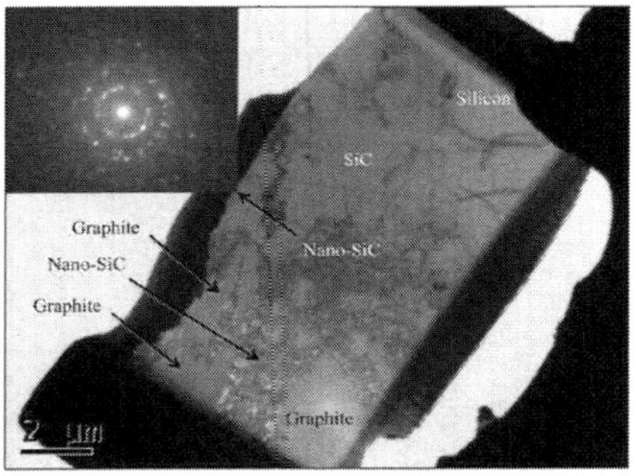

Figure 5. TEM micrograph from the CVI-Graphite samples, illustrating the extended thickness of the nano-SiC region and the inset showing a SAD pattern of the identified fine grain SiC region.

Fig. 6 illustrates the morphology of the microstructure around the SiC-Si interface. Unlike the interface of SiC-C, there were no fine particles or dispersoids in or around the interface. The silicon itself contains a few dislocation lines, which seem to have developed from the interface between SiC and Si; away from the interface, there was no observed evidence of any other crystal defects. In this

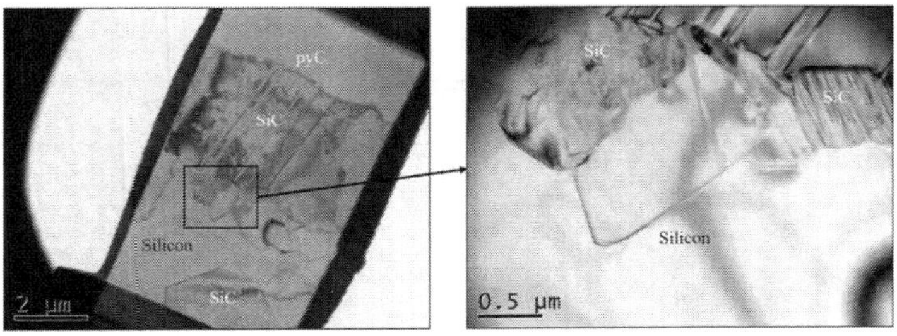

Figure 6. TEM bright field micrograph from PyC-Phenolic sample; enlarged image illustrating the interface in greater detail.

region, we noticed that faceted SiC grains of around a micron in size that were interlocked with the silicon region. Following the diffusion-controlled theory for the development of the SiC phase [21], these relatively small particles could be the front of the SiC production during the LSI process. Again, the grains in the frontier of the reaction were heavily strained with a large density of stacking faults inside, as shown in fig 6. It is impossible however, to speculate how these SiC grains were generated and is therefore clear that more research is needed on the details of the structure in this region.

(iii) Phase composition of the SiC

Reflection spots/rings in SAD patterns (fig. 7) were used to calculate the d-spacing's of the coarse SiC region and nano-crysalline SiC respectively. The coarse SiC was highly crystalline; the calculated d-spacing from the diffraction pattern in fig. 7(a) were 2.54 Å for $(\bar{1}\bar{1}1)$ crystal place, 2.19 Å for (200) and 1.56 Å for $(\bar{2}20)$, and in fig. 7(b) are 2.52, 1.54, 1.31 for (111), (220) and (113) respectively. Comparing these measurements with those shown in table I, where the d-spacing's of the corresponding crystal plane measured by XRD data, we are ensured that the course and the fine SiC belongs to the same polytype in all of the four samples. These measurements match the international powder diffraction data sheet (00-029-1129) [26], and indicate the polytype should be FCC β 3C-SiC, which is in accordance with prior investigations [14-15, 19, 27].

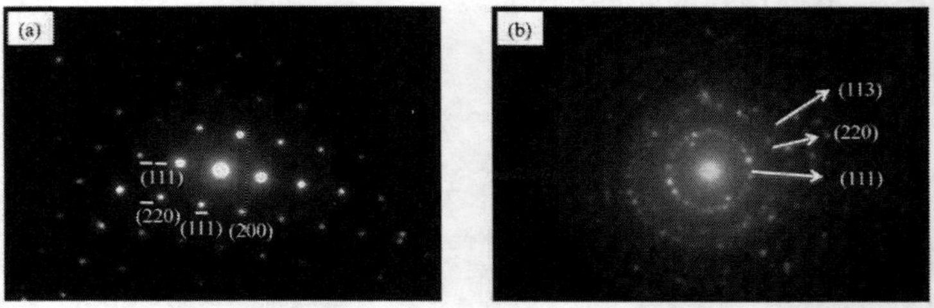

Figure 7. TEM SAD pattern from PyC-Phenolic of (a) coarse SiC region with a [110] zone axis and (b) nano-crystalline SiC.

Table I. d-spacing's measured by X-ray diffraction.

Sample	Measured 2θ	h	k	l	Measured d (Å)
CVI-Carbon	35.64	1	1	1	2.52
	42.28	2	0	0	2.14
	59.96	2	2	0	1.54
CVI-Graphite	35.66	1	1	1	2.52
	41.40	2	0	0	2.18
	59.96	2	2	0	1.54
PyC-Phenolic	35.28	1	1	1	2.54
	41.26	2	0	0	2.19
	59.20	2	2	0	1.56
PyC-Epoxy	35.66	1	1	1	2.52
	41.42	2	0	0	2.18
	59.98	2	2	0	1.54

(c) Silicon region

Among all the samples, un-reacted silicon was always left with a surrounding of the silicon carbide, as shown in fig. 1. Understandably, these silicon residuals were re-crystallized from the infiltrated silicon melt that failed to react with the carbon. However, in amongst the silicon were interspersed SiC islands (fig. 1) that were evident in all samples and in most cases situated in close proximity to the SiC-Si interface. Through grinding/polishing of the surface, it was proved that most of these islands were not part of the general architecture of SiC regions, but isolated islands.

As discussed before, the diffusion rate of carbon inside the SiC should be the controlling factor for the development of SiC by the reaction of diffused carbon within the silicon melt. If this mechanism is genuine, it would imply that the developed products through a diffusion-controlled reaction should set in between the silicon carbide and silicon. Most of the faceted SiC peninsulas shown in fig. 1 and fig. 6 should be a consequence of this process, but not for the isolated SiC islands. These SiC islands could be explained using a process of dissolving-precipitation. According to the Si-C phase diagram [28], carbon can dissolve in silicon melt with a limit of about 5 atm% at 1500 °C, and its dissolubility increases with the increase of the temperature. As the melt is cooled down, SiC crystals should precipitate and set inside the silicon crystal. Further experimental study is nevertheless needed in order to validate this speculation. Additionally, Favre *et al* [15] speculated that these islands might have formed from broken off carbon, as this most likely occurred during the violent molten silicon infiltration reaction. It is believed both dissolving-precipitation and violent reaction could have lead to the creation of these SiC islands.

SUMMARY

C_f/C preforms comprising different formats of pyrolytic carbon as the matrix were manufactured from different carbonaceous sources and processing means, and C_f/C-SiC composites were thereafter developed though liquid silicon infiltrating these preforms under same conditions. Microstructure examination revealed that the hierarchy order of the evolved SiC was not dependent on the carbon format.

TEM examination confirmed the presence of a fine SiC layer, with SiC particle size of about 50 nm, between pyrolytic carbon and course SiC region. The thickness of the fine SiC layer is in a range of 10 to 100 nm in CVI-Carbon, PyC-Phenoic and PyC-Epoxy samples, and up to a micron in CVI-Graphite sample. Apart from that, dispersoids may have been developed inside carbon and SiC regions: one with fine SiC particles dispersed in carbon, and one in the course SiC with both near the nano-SiC layer. Combined TEM and XRD investigations indicated that the generated SiC in the nano-SiC and coarse SiC regions belonged to the same polytype of SiC regardless of the carbon format and that the most likely type was FCC β 3C-SiC.

TEM results showed significant stacking faults existing in developed SiC grains, but few defects were noticed in the silicon. The appearance of isolated SiC islands inside the un-reacted silicon region were speculated to have formed due to dissolving-precipitation and/or violent reactions of molten silicon with carbon.

ACKNOWLEDGEMENTS

The authors express their gratitude to the EPSRC and Technology Strategy Board* (TSB) for their continued financial support toward this research and members of the Loughborough University ceramic group.

FOOTNOTES

* The Technology Strategy Board is a business-led executive non-departmental public body, established by the government. Its role is to promote and support research into, and development

and exploitation of, technology and innovation for the benefit of UK business, in order to increase economic growth and improve the quality of life. It is sponsored by the Department for Business, Innovation and Skills (BIS). For more information please visit www.innovateuk.org.

REFERENCES

[1]R. Naslain: Design, preparation and properties of non-oxide CMCs for application in engines and nuclear reactors: an overview, *Composites Science and Technology* **64** [2], 155-170 (2004).
[2]R. Naslain, SiC-Matrix Composites: Nonbrittle ceramics for thermo structural application, *Int. J. Appl. Ceram. Tech.* **2** [2], 75–84 (2005).
[3]W. Krenkel & F. Berndt, C/C–SiC composites for space applications and advanced friction systems, *Mat. Sci. and Eng.* **412** [1-2], 177–181 (2005).
[4]W. Krenkel, Cost effective processing of CMC composites by melt infiltration (LSI), *25th Annual Conference on Composites, Adv. Cer., Mats, and Structures: A: Cer. Eng. and Sci. Pro.* **22** [3] 443-454 (2001)
[5]Y. Zhang, Y. Xu, J. Lou, L. Zhang & L. Cheng, Braking Behavior of C/SiC Composites Prepared by Chemical Vapor Infiltration, *Int. J. Appl. Ceram. Tech.* **2** [2], 114-121 (2005).
[6]D. L. Zhao, H.F. Yin, F. Luo and W.C. Zhou, Microstructure and mechanical properties of 3D textile C/SiC composites fabricated by Chemical vapour infiltration, *Advanced materials research* **11-12**, 81-84 (2006).
[7]J.M. Lin, C.M. Ma, N.H. Tai, W.J. Wu & C.Y. Chen, Preparation and properties of SiC modified carbon-carbon composites by carbothermal reduction reaction, *J. Mat. Sci. Letts.* **18** [16], 1353-1355 (1999).
[8]J. Li, J. Tian & L. Dong, Synthesis of SiC precursors by a two-step sol-gel process and their conversion to SiC powders, *J. Eu. Ceram. Soc.* **20** [11], 1853-1857 (2000).
[9]E. Muller, R. Dittrich & K. Moritz, Studies on a Novel Route to C/SiC, *Adv. Eng. Mat.* **6** [7], 568-572 (2004).
[10]A.H. Wu, W. B. Cao, C. C. Ge, J. F. Li & A. Kawasaki, Fabrication and characteristics of plasma facing SiC-C functionally graded composite material, *Mat. Chem. & Phy.* **91** [2-3], 545-550 (2005).
[11]Y. Ding, S. Dong, Z. Huang and D. Jiang, Fabrication of short C fiber-reinforced SiC composites by spark plasma sintering, *Ceram. Int.* **33** [1], 101-105 (2007).
[12]W. Krenkel, Carbon fibre reinforced Silicon Carbide composites (C/SiC, C/C-SiC), Handbook of Ceramic Composites, Springer, 117-148 (2005).
[13]S. W. Fan, Y. D. Xu, L. T. Zhang, L. F. Cheng, L. Yu, Y. D. Yuan, F. K. Zhang, G. L. Tian, Z. J. Chen and J. J. Lou, Microstructure and properties of 3D needle-punched carbon/silicon carbide brake materials, Elsevier, *Composites Science and Technology* **67** [11-12], 2390-2398 (2007).
[14]J. Schule-Fischedick, A. Zern, J. Mayer, M. Ruhle, M. Friess, W. Krenkel & R. Kochendorfer, The morphology of silicon carbide in C/C–SiC composites, *Mat. Sci. & Eng.* **332** [1-2], 146-152 (2002).
[15]A. Favre, H. Fuzellier and J. Suptil, An original way to investigate the siliconizing of carbon materials, *Ceram. Int.* **29** [3], 235-243 (2003).
[16]C. Zollfrank and H. Sieber, Microstructure evolution and reaction mechanism of biomorphous SiSiC ceramics, *J. Am. Ceram. Soc.* **88** [1], 51-58 (2005).
[17]F. M. Vaarela-Feria, J. Ramirez-Rico, A. R. de Arellano-Lopez, J. Martinez-Fernandez & M. Singh, Reaction-formation mechanisms and microstructure evolution of biomorphic SiC, *J. Mat. Sci.* **43**, 933-941 (2008).
[18]J. C Margiotta, D. Zhang & D. C. Nagle, Microstructural evolution during silicon carbide (SiC) formation by liquid silicon infiltration using optical microscopy, *Int. J. of Refractory Metals and Hard Materials* **28**, 191-197 (2010)

[19]S. Xu, G. Qiao, D. Li, H. Yang, Y. Liu & T. Lu, Reaction forming of silicon carbide ceramics using phenolic resin derived porous carbon preform, *J. Eu. Ceram. Soc.* **29**, 2395-2402 (2009).
[20]A. Oberlin, Carbonization and Graphilization, *Carbon* **22** [6], 521-541 (1984)
[21]M. H. Hon & R. F. Davie, Self diffusion of ^{14}C in polycrystalline β-SiC, *J. Mat. Sci.* **14**, 2411-2421 (1979).
[22]M. H. Hon & R. F. Davie, Self diffusion of ^{30}Si in polycrystalline β-SiC, *J. Mat. Sci.* **15**, 2073-2080 (1980).
[23]J. N. Ness & T. F. Page, Microstructural evolution in reaction-bonded silicon carbide, *J. Mat. Sci.* **21**, 1377-1397 (1986).
[24]O. P. Chakrabarti, P. K. Das & J. Mukerji, Growth of SiC particles in reaction sintered SiC, *Materials Chemistry and Physics* **67**, 199-202 (2001).
[25]S. Xu, G. Qiao, D. Li, H. Yang, Y. Lui & T. Lu, Reaction forming of silicon carbide ceramic using phenolic resin derived porous perform, *J. Eu. Ceram. Soc.* **29**, 2395-2402 (2009).
[26]FCC 3C-SiC (00-029-1129), International centre for data diffraction (2009).
[27]H Zhou & R. N. Singh, Kinetics Model for the Growth of Silicon Carbide by the Reaction of Liquid Silicon with Carbon, *J. Am. Ceram. Soc* **79** [9], 2456 – 2462 (2005).
[28]R.W. Olesinski & G.J. Abbaschian, The C-Si (Carbon-Silicon) system, *J. Phase Equilibria* **5** [5], 486-489 (1984).

SHAPING RADIATION CURABLE COLLOIDAL DISPERSIONS – FROM POLYMER/CERAMIC FIBERS AND MICROSPHERES TO GRADIENT POROSITY CERAMIC BULK MATERIALS

Yoram de Hazan[†], Maciek Wozniak, Judit Heinecke[§], Gregor Müller, Veronika Märkl[§], Thomas Graule[§]

Laboratory for High Performance Ceramics, Empa, Swiss Federal Laboratories for Materials Testing and Research, Überlandstrasse 129, CH-8600 Dübendorf, Switzerland

[§]and TU Freiberg, Institute of Ceramic, Glass and Construction Materials, D-09596 Freiberg, Germany

[†]Author to whom correspondence should be addressed. e-mail: Yoram.dehazan@empa.ch

ABSTRACT

In this work we present novel technologies for shaping of radiation curable ceramic colloidal dispersions into polymer/ceramic and ceramic fibers, microspheres and cellular articles. High loading, low viscosity ceramic nanoparticle dispersions in UV curable monomer mixtures are achieved by proper choice of monomers or use of special comb-polyelectrolyte surfactants. Novel ceramic/polymer nanocomposite fibers are produced by continuous "on the fly" curing of the UV curable colloidal ceramic dispersions structured through a micro-extrusion die. The method is advantageous to obtain multifunctional composite fibers, small diameter ceramic fibers and robocasting applications. The wide choice of monomers enables composites with tailored properties required for woven products and the low processing temperature benefits applications in biomedicine. UV curable colloidal dispersions containing pore formers are found advantageous for the production of inorganic microspheres and cellular articles. Cellular ceramic articles with high porosity and well controlled cell size and shape are produced.

INTRODUCTION

Ceramic and ceramic/polymer composite fibers find diverse applications in filtration, catalysis, reinforcement, textiles, sensing and detection, biomedicine, renewable energy and smart materials. These versatile applications often require new and general approaches to produce fibers with increased microstructural control, functionality and material combinations.

Ceramic and ceramic/polymer composite fibers are commonly produced by extrusion/spinning technologies such as melt spinning,[1] wet and dry spinning,[2] and electrospinning.[3] Another class of advanced extrusion based processes are free-form or direct-write fabrication of colloidal inks (sometimes referred to as robocasting) where fibers are assembled into predetermined 3d shapes.[4-8]

With the exception of templating techniques used for producing microtubes,[9] the shaping of elongated objects invariably uses extrusion or drawing methods frequently aided by tensioning/elongation of the extruded filament. The various extrusion methods differ mainly in their feedstock composition and method of shape stabilization (hardening) of the extruded filament. Along with the method of elongation, it is the method in which the viscosity of the feedstock is increased after the elongation step until shape stabilization is attained, which essentially distinguishes between different extrusion/spinning processes. This determines also the useful viscosity range in the elongation step. Melt spinning uses the viscosity dependence of polymer melts on shear rate and temperature, wet and dry spinning use solvent exchange and evaporation, respectively, to achieve shape stabilization of polymer based solutions. Direct write techniques frequently use coagulation of oppositely charged colloids,[4-5] coagulation solutions,[5] freezing,[6] sol-gel transition[7] and most recently UV curing of hydrogels.[8]

In this paper we present a new, general and versatile spinning process capable of producing continuous ceramic/polymer nanocomposite elongated articles such as fibers, tubes or tapes. The new method is not limited to, but advantageously uses feedstock consisting of monomer based, radiation curable dispersions of inorganic colloidal particles. Such dispersions are frequently used in gel casting and stereolithography applications.[10-13] The use of low viscosity, radiation curable colloidal dispersions for the production of inorganic microspheres and cellular articles is also presented.

EXPERIMENTAL

Materials and Dispersion Preparation

Model nanoparticles used in this study were SiO_2 (Aerosil OX50, Evonic Degussa, Germany, amorphous, 40 nm average particle size), Al_2O_3 (TMDAR, Taimei Chemical Co., LTD, Japan, α-Al_2O_3, 150 nm average particle size), ZnO (Sigma Aldrich, Switzerland, <100 nm), Hydroxyapatite (Sigma Aldrich, Switzerland, <200 nm) and TiO_2 (Aeroxide P25, Evonic Degussa, Germany, 80% anatase/20% rutile, 25 nm). A coarser Al_2O_3 powder, Ceralox HPA-0.5 (Condea Vista, USA) having an average particle size of 400 nm was also investigated. The SiO_2 and Hydroxyapatite powders were used without surfactants. Al_2O_3 particles were pre-stabilized with 4.5% and ZnO and TiO_2 with 10% MelPers4343 per particle weight, respectively. Melpers4343 is a commercial comb-polyelectrolyte surfactant from BASF, Germany.[13] The UV curable monomers used in this study are 2-Hydroxy ethyl acrylate (2-HEA), 4-hydroxy butyl acrylate (4-HBA) from BASF, Germany and Polyethylene glycol 200 diacrylate (PEGDA) from Rahn, Switzerland. The UV curable mixture contained either 2-HEA/PEGDA (for cellular materials and microspheres, viscosity measurements, unless noted) or 4-HBA/PEGDA (for fibers) in a ratio of 14:1. The monomer mixture is used without any additional solvents/diluents. Genocure LTM, a liquid photoinitiator blend with a broad absorption peak around 380 nm from Rahn, Switzerland was used as the photoinitiator. Dispersions of SiO_2 (30 vol%) and Hydroxyapatite (33 vol%) in UV curable monomer mixtures were prepared by the procedure described by Wozniak et al.[12] Dispersions of Al_2O_3 (41.5 vol%), ZnO (37 vol%),, Al_2O_3/ZnO 1:1 (42 vol%) and TiO_2 (17 vol%) in UV curable monomer mixtures were prepared as described by de Hazan et al.[13] The viscosities of the SiO_2 and Al_2O_3 dispersions, measured with a rotational viscosimeter (Rheolab MC 120, Physica Messtechnik GmbH) are 0.8 Pas (23 °C, 100 s^{-1}) and 0.3 Pas (4-HBA/PEGDA, 23 °C, 500 s^{-1}), respectively. The photoinitiator in an amount of 3-5% (per monomer mixture) was added to the dispersions shortly before use and mixed thoroughly.

Pore forming agents were polyethylene (PE) microspheres (Sumitomo, Japan) with an average diameter of 12 micrometers and a broad size distribution (3-25 μm), and polystyrene (PS) microspheres (Microbeads, Norway, 40 μm) with a narrow size distribution. Mazolla oil (Coop, Switzerland) was used as an emulsifying agent for the preparation of microspheres.

Microshaping Procedures, Debinding and Sintering

The extrusion system consists of 8 ml stainless steel syringe (Hamilton, USA) equipped with an exit die. The syringe which contains the UV curable dispersion is driven by a precision syringe pump (PHD 2000, Harvard Scientific, USA). Spinning dies were either 9.5 mm long luer-lock needles with diameters of 110, 160 or 500 μm (Hamilton, USA) or home made brass disc with 50 μm exit hole drilled in its center. The dispersions were extruded through the various dies into a glass beaker with a diameter of 10 cm filled with DI water to a level of 15 cm. The exit die was positioned 1-3 cm inside the DI water and was shielded from UV radiation by an aluminum tape. The average linear velocity of the dispersion stream passing the die was varied in the range of 1-20 cm s^{-1}. One UV lamp producing 120 mW cm^{-2} in the UVA (Fe bulb, 100 W, Dr. Hönle AG, Germany) positioned outside the beaker, illuminated the bottom 10 cm of the beaker and cured the dispersion stream shortly after exiting the die. The continuous fibers such produced were separated from the water, washed with water and dried. All materials (e.g., dispersions, solvents) and all manipulations (e.g., curing and washing) were carried out at 25 °C.

The die delivering the dispersion for co-extrusion or hydrodynamic tensioning consisted of a 160 µm needle (310 µm OD) concentrically positioned within a 500 µm tube delivering a second dispersion, water or glycerol in the annular gap. The streams were combined 3-5 mm before exiting the die. The water and glycerol average linear velocity in the annular gap was varied between 0 to 150 cm s^{-1} (0-0.13 cm^3 s^{-1}), corresponding to a Reynolds number in the range of 0-300 and 0-0.3, respectively. Mechanical elongation was achieved with a cylindrical fiber spooling device positioned at the bottom of the beaker. Elongated fiber was collected when the spooling device was rotating at a linear velocity up to 2-3 the extrusion velocity.

Curable pastes for cellular materials were prepared by mixing the pore formers into 41.5% or 31% alumina dispersions in an amount equivalent to 74 and 83% porosity, respectively. The porosity is estimated by the following equation: porosity = 100 Vpf/(Vpf+Vpa), where Vpf and Vpa are the volumes of pore formers and inorganic particles, respectively. The dispersion was either extruded to fiber through a 500 µm die as described above or cast in 2 cm diameter circular molds to a thickness of 2-3 mm and then cured for several minutes using the UV lamp described above.

Al_2O_3/polyacrylate microspheres were prepared by emulsifying 41.5% Al_2O_3 dispersion with Mazolla oil followed by UV curing of the dispersion droplets in the emulsion obtained. The solid composite microspheres were separated from the emulsion, washed with isopropyl alcohol and water and dried before debinding and sintering procedures.

Debinding of Al_2O_3 and SiO_2 composite fibers, cellular materials and microspheres was carried out in air at 650 °C for 2 h. The pore formers are removed during the debinding step. SiO_2 fibers were sintered at 1250°C for 1 h and quenched rapidly to 800°C. Al_2O_3 fibers, microspheres and cellular articles were sintered at 1500 °C for 2 h followed by 15 h cooling to room temperature.

Material Characterization

Ceramic and ceramic/polymer composite fibers were characterized by optical microscopy (Leica Wild M3Z equipped with micrometer) and SEM (VEGA Plus 5136 MM, Tescan instruments, Czech republic). Hg porosimetry (Pascal 140/440, Porotec GmbH, Germany) up to a pressure of 2000 bar was used for the determination of apparent and skeletal density, open and closed porosity and pore size distribution in the range of 4 nm -100 µm.

RESULTS AND DISCUSSION

Feedstock Development

Before discussing the details of the new spinning technology it is interesting to compare the viscosity of different feedstock used in common spinning processes (Fig. 1). Melt spinning typically uses melts with viscosity $>10^3$ Pas whereas electrospinning uses solutions with viscosities in the range of 0.1-10 Pas. Wet and dry spinning feedstock shows intermediate viscosity range, in the order of 5-500 Pas. In comparison, the new method uses homogeneous dispersions of inorganic colloidal particles in UV curable resin which can be produced at high solid loadings in a wide viscosity range (Fig. 1).

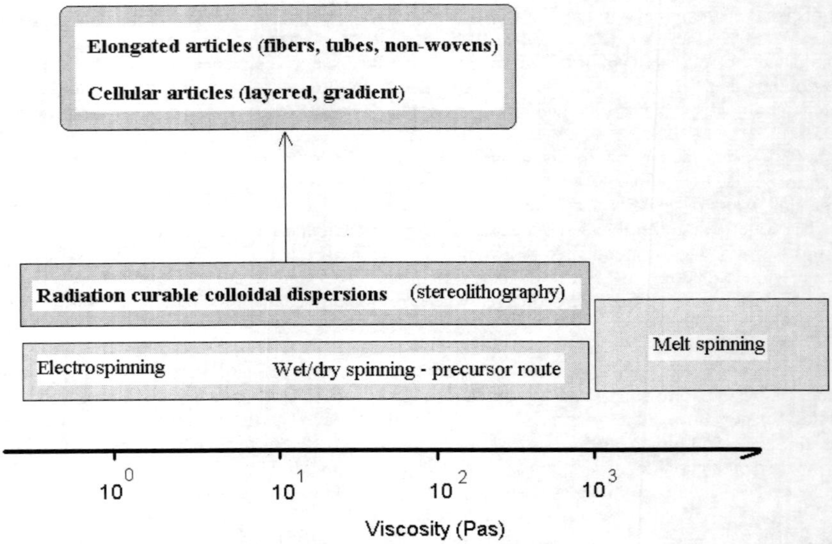

Fig. 1 – Classification of extrusion techniques by viscosity of feedstock.

Due to the ability to solidify most curable liquids rapidly by UV radiation, the new technology has no lower limit for feedstock viscosity. High viscosity/high loaded radiation curable dispersions can be readily obtained by using nanosized particles or high viscosity resins. At the same time, highly loaded monomer based dispersions with viscosities in the range of electrospinning feedstock can similarly be produced, without the necessity of solvents or diluents.[10-13] For example, Fig. 2 shows the viscosity of an Al_2O_3 dispersion (average particle size of 150 nm) in an 2-HEA/PEGDA (14:1) monomer mixture. The particles are stabilized by comb-polyelectrolyte surfactants.[13] Fig. 2 shows that at a solid loading of 41 vol% Al_2O_3 (150 nm), the viscosity is 0.18 Pas (23°C, 500 s^{-1}). At 48 vol% the viscosity increases sharply to 0.73 Pas, but is still well below the maximum viscosity range required for stereolithography.[10] For comparison, Fig. 2 shows a 48 vol% dispersion made with a coarser powder (400 nm). While a higher viscosity 4-HBA/PEGDA (14:1) monomer mixture is used, the viscosity is only about 0.36 Pas due to the larger particles.

The UV curable dispersions described are used here for the fabrication of composite and ceramic fibers, and cellular ceramics (Fig. 1).

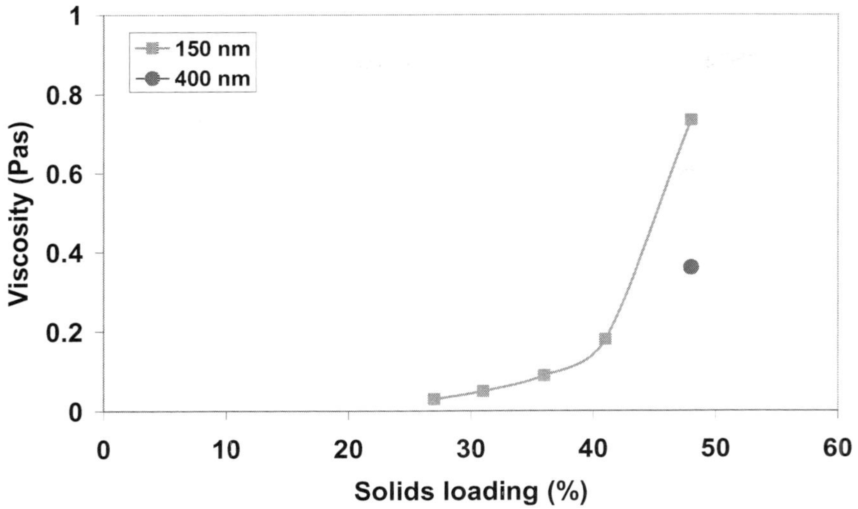

Fig. 2 – Viscosity of dispersions containing Al_2O_3 in UV curable monomer mixtures (at 500 s^{-1} and 23°C). The monomer mixtures were 2-HEA/PEGDA (14:1) and 4-HBA/PEGDA (14:1) for the 150 nm and 400 nm powders, respectively.

Fabrication of Fibers

Fig. 3 presents the general scheme for fiber production and several of its variations[14]. In the general scheme (Fig. 3a), a curable dispersion is shaped into a thin filament through an extrusion/spinning die and hardened to a solid fiber shortly after exiting the die by radiation (e.g. UV) induced photopolymerization of the monomer phase. Such 'on the fly' curing scheme have recently been introduced for non-filled, organic polymer materials using microfluidic devices[15,16] or direct write fabrication.[8] Further tensioning/elongation of the relatively low viscosity extruded filament can be achieved by external fields produced by mechanical or electrohydrodynamical means (Fig. 3b) or hydrodynamically using sheathing fluids (Fig. 3c).

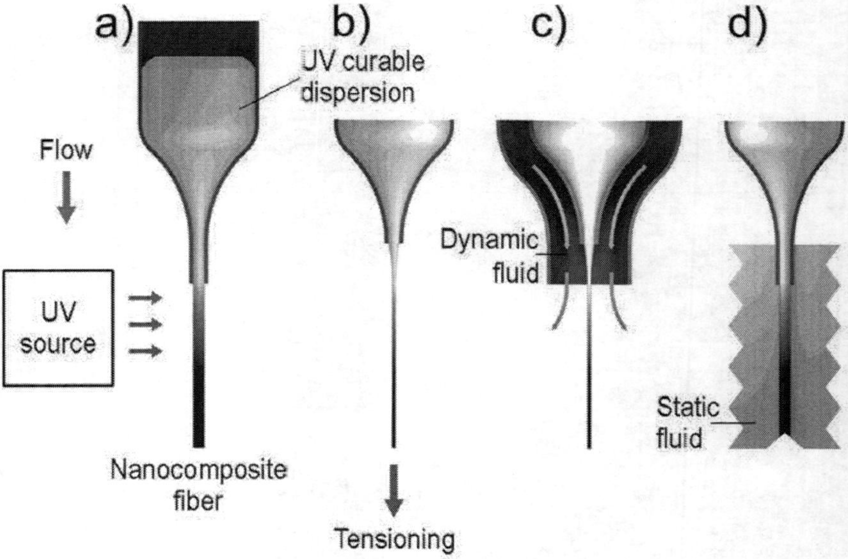

Fig. 3 - Fabrication schemes for ceramic/polymer composite fibers. a) general scheme, b) tensioning by external devices or fields, c) hydrodynamic tensioning, d) stabilization of flow using static fluids.

Dispersions exhibiting high surface tension and low viscosity are frequently prone to flow instabilities when forced at low flow rates through small orifices directly to air. These instabilities can be minimized or avoided altogether if extrusion is carried out into higher viscosity static or dynamic liquids such as water (Fig. 3d).

The curing of the dispersions can be done thermally using appropriate thermal initiators such as peroxides or Azo compounds. At the same time, UV curing enables operating the new spinning process at low temperatures with very short and controllable curing times. These may be advantageous for biological applications (e.g., immobilization of enzymes[15]). In addition, UV curing implies lithographic techniques can also be employed to further produce complex microstructures by selective curing.[17,18] The product of this spinning process is a new ceramic/polymer nanocomposite fiber. The process is especially advantageous (but not limited) for nanocomposites with particle concentration above the percolation threshold; facilitating functional properties such as conductance and low sintering temperatures for ceramics.

The method described above can in principle be employed to produce an unlimited variety of inorganic/polymer nanocomposites. It is demonstrated here for several model nanoparticles and resins. Fig. 4 shows several examples of fibers produced by the methods described in Fig. 3.

Fig. 4 – Composite and ceramic fibers produced from UV curable dispersions by the microshaping processes of Fig. 3. a) 30 vol% SiO_2/polyacrylate nanocomposite fiber extruded through a 50 µm die, b) 41.5 vol% Al_2O_3/polyacrylate nanocomposite fiber extruded through a 50 µm die, c) 41.5 vol% Al_2O_3/polyacrylate nanocomposite fibers extruded through a 160 µm die and elongated by a spooling device, d) 41.5 vol% Al_2O_3/polyacrylate nanocomposite fibers extruded through a 160 µm die and elongated by glycerol sheathing fluid, e) coextruded fiber: core - 30 vol% SiO_2/polyacrylate, shell - 41.5 vol% Al_2O_3/polyacrylate, f) Al_2O_3 fiber of example b) sintered at 1500 °C.

Nanocomposite fibers of 30 vol% SiO_2/polyacrylate and 41.5 vol% Al_2O_3/polyacrylate are shown in Figs. 4a and 4b, respectively. The fibers were produced using the method of Fig. 3d where the dispersions are extruded through a 50 μm die into a beaker filled with deionized water illuminated by an external UV lamp. The dispersion stream exiting the die is typically cured to a continuous solid fiber in ~1 second while traveling to the bottom of the beaker. The silica fibers produced at a rate of 1-2 m min^{-1} have an average diameter of ~65 μm and exhibit significant die swell effect, while the Al_2O_3/polymer fibers show diameters of 40-60 μm corresponding closely to the die diameter. This difference is likely due to the rheology of the dispersions. SiO_2 dispersions[12] show shear thickening and and Al_2O_3 dispersions[13] shear thinning behavior. Despite the relatively low viscosity of the dispersions (<1 Pas) and solubility of the monomer (2-HEA or 4-HBA) in water, the fibers have a circular cross section and smooth surfaces. With the chosen monomer composition, the resulting nanocomposite fibers are highly elastic and suitable for woven applications. Modification of fiber microstructure and properties can be achieved with other mono/multifunctional monomers/oligomers (e.g. methacrylate) and diluents.

As described in Figs. 3b and 3c, fiber elongation prior to curing can be achieved in a number of ways. Fig. 4c presents continuous 41.5 vol% Al_2O_3/polyacrylate composite fibers with diameters around 45-80 μm obtained with an extrusion die of 160 μm and a mechanical tensioning device. This represents a reduction of fiber diameter from the nominal die diameter by a factor of 2-3.5. The circular cross section of the fibers obtained indicates that the elongation occurs at the liquid state before the fibers are cured and spooled. Fiber elongation can also be achieved using hydrodynamic elongation (Fig. 3c)[15,19] Continuous fibers with diameters of 40-60 μm are obtained using water as sheathing fluid, at a Reynolds number, Re~230. Short fibers (5-10 mm) with diameters as small as 20-30 μm can be obtained at Re~ 230-300. Above Re~300 the dispersion is disintegrated due to the increased turbulence in the system. With the use of higher viscosity sheathing fluids such as glycerol, continuous fibers with diameters between 5-20 μm are obtained at similar flow rates but at Re~0.3 (Fig, 4d). We expect continuous fibers with diameters around or below 1 μm in the near future with the present technology by the combination of extrusion through smaller diameter dies and mechanical, electrohydrodynamic and/or hydrodynamic elongation.

Shown in Fig. 4e is a 250 μm co-extruded fiber consisting of a 30 vol% SiO_2/polyacrylate core surrounded by a 41.5 vol% Al_2O_3/polyacrylate shell. While the fiber cross section is somewhat deformed, the interface between the layers appears intact. It also can be seen that the Al_2O_3/polyacrylate shell can be as thin as 20 μm and still intact.

The Al_2O_3/polyacrylate nanocomposite fibers shown in Fig. 4b were sintered at 1500°C to dense Al_2O_3 fibers (>98% theoretical density). The sintered fibers show diameters in the 30-45 μm range (Fig. 4f). Similarly, 120-130 μm SiO_2/polymer nanocomposite fibers were sintered at 1250 °C to silica glass fibers with diameters in the range of 90-100 μm.

With the new shaping/structuring process it is possible to produce ceramic and ceramic/polymer fibers, tubes and tapes from a wide range of ceramic or other inorganic materials. Fabrication of functional ceramic and ceramic/polymer composite fibers is especially attractive. 37 vol% ZnO, 42 vol% mixed ZnO/Al_2O_3 (1:1), 33 vol% Hydroxyapatite/polyacrylate and 17 vol% TiO_2/polyacrylate nanocomposite fibers with diameters in the 5-100 μm range were similarly fabricated. TiO_2 and ZnO/Al_2O_3 fibers may find applications in photocatalysis or renewable energy solutions.[20] Hydroxyapatite ceramic/polymer nanocomposite fibers are candidates for applications in biology and medicine[21], potentially in combination with UV curable biodegradable polymers.[22]

The low viscosity[12,13] and fast curing of the dispersions used for the new fiber spinning process facilitate fast alignment of particles by external fields prior to curing,[23] elongation of fibers prior curing and good interfaces between layers produced by co-extrusion,[15] as also demonstrated in this work (Fig. 4d and 4e). Low viscosity dispersions, however, are prone to shape instabilities (e.g. Fig. 4e). An optimal viscosity range therefore exists for specific applications. Higher viscosity dispersions can be obtained by increasing the solids loadings, use of smaller particles or use of high viscosity monomers/oligomers. Especially interesting are high loaded pastes for direct write/robocasting of 3d structures or non-woven networks produced with high velocity air sheath (Fig. 3c). Initial feasibility studies for these applications have been made. These applications benefit from the latitude in material combination, high solid loading yet low process temperature and pressure, coextrusion and UV structuring possibilities.

Fabrication of Cellular Materials and Microspheres

Cellular materials and microspheres are used in filtration and catalysis, light structures, thermal and acoustic insulation, biomedical and renewable energy applications.[24-27] We find that high quality cellular materials are accessible when pore formers such as gases, liquids or solids are introduced into the low viscosity dispersions, producing UV curable pastes which can be shaped into fibers or multilayer ceramic articles (Fig. 5).[28] Fig. 5a shows a cellular Al_2O_3 material produced by casting and UV curing of 2 cm diameter disks with thickness of a 2-3 mm using 40 μm pore forming polystyrene spheres. The material is 72% porous and the shape of the pore former is precisely replicated in the cellular ceramic. Fig. 5b show a broken section of a 79% porous cellular Al_2O_3 material using PE pore former. Here most of the 10 μm cells are interconnected through 1-3 μm passages. 13.5% of the porosity, however, is impermeable to Hg. The average pore size found by Hg porosity measurements is about 2 μm, corresponding closely to the passage between the cells. Fig. 5b shows that the cell wall consists in many instances of only 2-3 nanoparticles. This may be aided by the crosslinked nature of the UV cured nanocomposite, as shown also for gel derived cellular materials.[25-27] The crosslinking degree, which is promoted by use of multifunctional monomers and/or curable polymers, is inherent and controlable with UV curable resins. Similar single and multilayered cellular shapes have been produced from a variety of other materials, including hydroxyapatite and TiO_2.

By proper choice of monomers and emulsification media an emulsion of dispersion microdroplets can be obtained. The UV curable droplets suspended in the UV transparent medium are cured by UV radiation under flow or static conditions. Fig. 5c shows dense Al_2O_3 microspheres about 150 μm in size, prepared by emulsification of the mildly polar colloidal dispersion in oil followed by curing under static conditions, debinding and sintering.

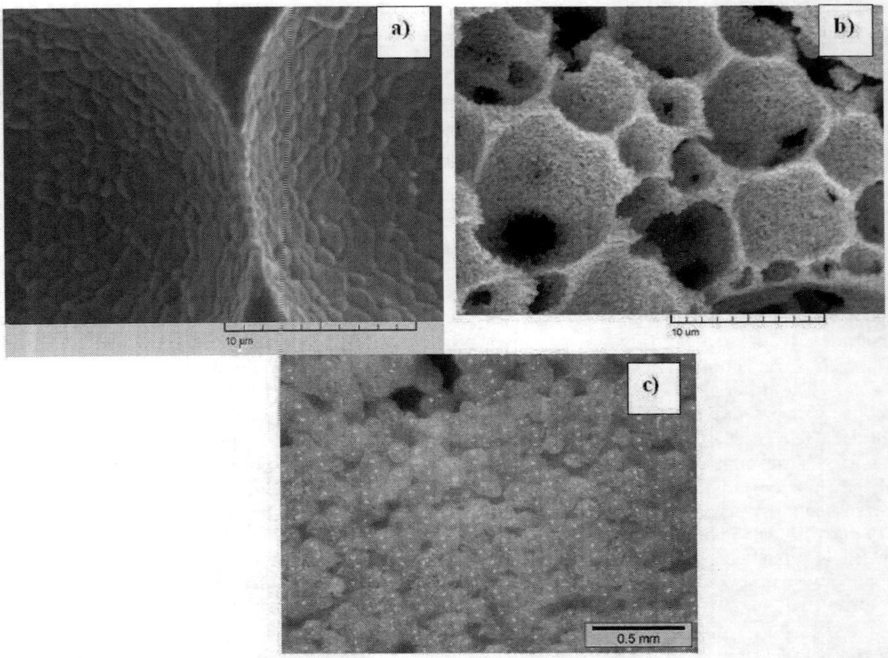

Fig. 5 – Al_2O_3 cellular articles and microspheres produced with UV curable dispersions containing pore formers or by emulsification. a) section of cellular Al_2O_3 article with 72% porosity made using PS pore former, b) section of cellular Al2O3 article with 79% porosity made with PE pore formers, c) Al_2O_3 microspheres prepared by emulsification.

CONCLUSIONS

New technologies for the fabrication of ceramic/polymer composite fibers have been developed. Continuous fibers are produced by curing of UV curable colloidal dispersions 'on the fly'. New ceramic and high loaded nanocomposite fibers with diameters in the range of 5-500 μm have been demonstrated. The method allows good control of nanocomposite composition and microstructures. The UV curable colloidal dispersions can be loaded with a high amount of pore formers and cured in the shape of fibers, microspheres or multilayered articles. Highly porous cellular ceramics with controlled cell size and shape are obtained after conventional debinding and sintering procedures.

REFERENCES

[1] C. Van Hoy, A. Barda, M. Griffith, J.W. Halloran, "Microfabrication of Ceramics by Co-extrusion," *J. Am. Ceram. Soc.*, **81**, 152-8 (1998).

[2] J. de Jong, N.E. Benes, G.H. Koops and M. Wessling, "Towards single step production of multi-layer inorganic hollow fibers," *J. Membr. Sci.,* **239**, 265-9 (2004).

[3] D. Li, Y. Xia, "Direct Fabrication of Composite and Ceramic Hollow Nanofibers by Electrospinning," *Nano Lett.,* **4**, 933–38 (2004).

[4]J.E. Smay, J.Cesarano, J.A. Lewis, "Colloidal Inks for Directed Assembly of 3-D Periodic Structures," *Langmuir,* **18**, 5429-37 (2002).
[5]J.A. Lewis, "Direct Ink Writing of 3D Functional Materials," *Adv. Funct. Mater.* **16**, 2193–204 (2006).
[6]T. Huang, M.S. Mason, G.E. Hilmas, M.C. Leu, "Freeze-form Extrusion Fabrication of Ceramic Parts," *Journal of Virtual and Physical Prototyping,* **1**, 93 – 100 (2006).
[7]E.B. Duoss, M. Twardowski, J.A. Lewis, "Sol-Gel Inks for Direct-Write Assembly of Functional Oxides," *Adv. Mater.* **19**, 3485-9 (2007).
[8]R.A. Barry, R.F. Shepherd, J.N. Hanson, R.G. Nuzzo, P. Wiltzius, J.A. Lewis, "Direct-Write Assembly of 3D Hydrogel Scaffolds for Guided Cell Growth," *Adv. Mater.,* **21**, 1–4 (2009).
[9]S. Motojima, T. Suzuki, Y. Noda, A. Hiraga, S. Yang, X. Chen, H. Iwanaga, T. Hashishin, Y. Hishihawa, "Preparation of Helical TiO2/CMC Microtubes and Pure Helical TiO2 Microtubes," *J. Mater. Sci.,* **39**, 2663-74 (2004).
[10]J.W. Halloran, M.L. Griffith, T.M. Chu, "Stereolithography Resin for Rapid Prototyping of Ceramics and Metals" *US patent 6117612* (2001).
[11]J.H. Jang, S. Wang, S.M. Pilgrim, W.A. Schulze, "Preparation and Characterization of Barium Titanate. Suspensions for Stereolithography," *J. Am. Ceram. Soc.*, **83**,1804–806 (2000).
[12]M. Wozniak, T. Graule, Y. DeHazan, D. Kata, J. Lis, "Highly Loaded UV Curable Nanosilica Dispersions for Rapid Prototyping Applications," *J. Eur. Ceram. Soc.*, **29**, 2259-65 (2009).
[13]Y. DeHazan, J. Heinecke, A. Weber, T. Graule, "High Solids Loading Ceramic Colloidal Dispersions in UV Curable Media via Comb-Polyelectrolyte Surfactants," *J. Coll. Int. Sci.,* **337**, 66-74 (2009).
[14]Y. DeHazan, T. Graule, G. Müller, *European patent applications 08017531.8, 08019324.6,* (2008); *PCT/CH2009/000322,* (2009).
[15]W. Jeong, J. Kim, S. Kim, S. Lee, G. Mensing, D.J. Beebe, ''Hydrodynamic Microfabrication via ''on the Fly'' Photopolymerization of Microscale Fibers and Tubes," *Lab Chip*, **4**, 576-80 (2004).
[16]S. Kim, H. Oh, J. Baek, H. Kim, W. Kim, S. Lee, "Hydrodynamic Fabrication of Polymeric Barcoded Strips as Components for Parallel Bio-analysis and Programmable Microactuation," *Lab Chip,* **5**, 1168–72 (2005).
[17]R.F. Shepherd, P. Panda, Z. Bao, K.H. Sandhage, T.A. Hatton, J.A. Lewis, P.S. Doyle, "Stop-Flow Lithography of Colloidal, Glass, and Silicon Microcomponents," *Adv. Mater.*, **20**, 1–6 (2008).
[18]S. Passinger, M.S.M. Saifullah, C. Reinhardt, K.R.V. Subramanian, B.N. Chichkov, M.E. Welland, "Direct 3D Patterning of TiO2 Using Femtosecond Laser Pulses," *Adv. Mater.*, **19**, 1218–21 (2007).
[19]A.S. Utada, A. Fernandez-Nieves, H.A. Stone, D.A. Weitz, "Dripping to Jetting Transitions in Coflowing Liquid Streams," *Phys. Rev. Lett.*, **99**, 094502 (2007).
[20]Y. Choi, H.G. Stenger, "Water Gas Shift Reaction Kinetics and Reactor Modeling for Fuel Cell Grade Hydrogen," *J. Power source,* **124**, 432-9 (2003).
[21]S. Michna, W. Wu, J.A. Lewis, "Concentrated Hydroxyapatite Inks for Direct-Write Assembly of 3-D Periodic Scaffolds," *Biomaterials,* **26**, 5632–9 (2005).
[22]M. Schuster , C. Turecek, F.Varga, H. Lichtenegger , J. Stampfl, R. Liska, "3D-Shaping of Biodegradable Photopolymers for Hard Tissue Replacement," *Appl. Surf. Sci.,* **254**, 1131–4 (2007).
[23]S.W. Montgomery, T.V. Aldridge, "Method and Apparatus for Producing Aligned Carbon Nanotube Thermal Interface Structure" *US patent 6921462,* (2005).
[24]M. Scheffler, P. Colombo, *Cellular ceramics, Wiley-VCH, Weinheim,* (2005).
[25]P. Colombo, *Phil. Trans. R. Soc. A*, **364**, 109-124 (2006,)
[26]U.T. Gonzenbach, A.R. Studart, D. Steinlin, E. Tervoort, L.J. Gauckler, *J. Am. Ceram. Soc.*, **90**, 3407–3414 (2007).
[27]L. Andersson, L. Bergström, *J. Eur. Ceram. Soc.*, **28**, 2815–2821 (2007).
[28]Y. DeHazan, T. Graule, J. Heinecke, *European patent application 09005854.6,* (2009).

MELT-INFILTRATION PROCESSING OF TITANIUM CARBIDE-STAINLESS STEEL CERMETS

Tyler Stewart, R. Bradley Collier, Zoheir N. Farhat, Georges J. Kipouros and Kevin P. Plucknett*
Dalhousie University, Materials Engineering Program, Department of Process Engineering and Applied Sciences, 1366 Barrington Street, Halifax, Nova Scotia, B3J 2X4, Canada.

ABSTRACT

Titanium carbide (TiC) is a common component in wear resistant ceramic-metal composites, or cermets. In the present work a variety of stainless steel binders have been employed for the production of TiC-based cermets, including the grades 304-L, 316-L and 410-L. In order to prepare the cermets, TiC preforms have been produced by aqueous slip casting. For this initial study, colloidal processing was conducted using an ammonium salt of poly(methacrylate), or PMA-NH_4, under slightly basic conditions (pH = 8.5). An anti-foaming agent, tri-n-butyl phosphate, was employed to minimize bubble formation. The cast preforms were subsequently melt-infiltrated at temperatures between 1500 and 1550°C, with steel contents varied from 5 to 30 vol. %. The post infiltration densities of the cermets were typically in excess of 98 % of theoretical. The densified cermets typically exhibited grain sizes of the order of 5 microns, with significant grain rounding and shape accommodation occurring during processing. The cermet hardness decreased with increasing stainless steel content, although there was little variation from one steel grade to another.

INTRODUCTION

Titanium carbide (TiC) ceramics exhibit a variety of properties that make them useful for elevated temperature wear applications, notably a high melting point (3065°C), low density (4.93 g/cm^3), and high Vickers hardness.[1,2] Traditionally these materials have been processed using dry pressing methods, which can lead to problems of agglomeration and reduced homogeneity, residual porosity and geometry limitations. Conversely, colloidal processing approaches can help to eliminate these problems while potentially generating an improvement in mechanical properties.[3]

Invariably TiC-based ceramics are prepared with a metallic binder, which can significantly increase the materials toughness, forming a ceramic-metal composite, or cermet. A variety of binders have been used in TiC-based cermets, including nickel, iron and cobalt.[4-6] More recently, ductile intermetallic alloys such as iron and nickel aluminides have been employed, which offer increased potential for use at elevated temperatures.[7,8] TiC-based cermets have also been prepared using stainless steel binders, although this class of material has received relatively limited attention to date.[9,10] In the present work several stainless steel grades are used to produce TiC-based cermets, notably the austenitic steels 304-L and 316-L and the martensitic steel 410-L. Melt-infiltration processing has been used to densify the cermets, utilizing TiC preforms prepared using a simple colloidal approach.

EXPERIMENTAL PROCEDURES

Raw Materials

The TiC powder used in the experiments performed was sourced from Pacific Particulate Materials Ltd. (Grade: TiC-2012), with a quoted average particle size of 1.30 μm and 0.11% retained free carbon. The as-received TiC powder was analyzed to verify the particle size distribution using acoustic attenuation spectroscopy (Zeta-APS, Matec Applied Sciences) and scanning electron microscopy (SEM; Hitachi S-4700, Hitachi Co.). The surface area of the TiC powder was measured using surface nitrogen adsorption (Horiba SA-6201). The nominal compositions of the stainless steel powders that were used in the present work, grades 304-L, 316-L and 410-L where the designation L refers to low carbon content, are provided in Table 1. The stainless steel powders were sourced from

* Corresponding Author (email: kevin.plucknett@dal.ca)

Alfa Aesar, with a nominal particle size of -100 mesh. The melting point of the stainless steel powders was determined using differential scanning calorimetry (DSC), with a heating rate of 10°C/min. to 1500°C in a flowing nitrogen atmosphere (100 ml/min.). The particle morphology was assessed using SEM, while the crystalline phase was determined using x-ray diffraction (XRD; Bruker D-8 Advance). The deflocculent used for colloidal processing was an ammonium salt of polymethacrylate (PMA-NH_4) produced in 25 wt% aqueous suspension (Darvan C, R.T. Vanderbilt Co.). Double distilled water was used throughout the colloidal characterisation and processing stages.

Table 1. Nominal compositions for the stainless steel grades used in the present work.

Type	Nominal composition (max. wt. %)									Density
	Cr	Ni	C	Mn	Si	P	S	N	Mo	(g/cm^3)
304-L	18-20	8-12	0.03	2	0.75	0.045	0.03	0.1	-	8.03
316-L	16-18	10-14	0.03	2	0.75	0.045	0.03	0.1	2-3	8.03
410-L	11-13.5	0.75	0.03	1	1	0.04	0.03	-	0.75-1.25	7.75

Colloidal Processing and Cermet Production

Aqueous colloidal characterisation of the TiC powder used in the present work has been described in detail in a prior publication,[11] and will only be briefly referred to here. Preliminary sedimentation trials were performed on 10 vol. % TiC suspensions to highlight the optimum PMA-NH_4 dispersant concentration range to be investigated in subsequent analyses. Acoustophoresis (Zeta-APS, Matec Applied Sciences) was used to assess the effects of pH and PMA-NH_4 concentration on the zeta potential of aqueous TiC suspensions (2 vol. % solids loading). Following the initial sedimentation trials, the PMA-NH_4 concentration was varied from 0 to 0.05 wt. %. Suspensions were auto titrated, using 1M NaOH and HNO_3, from pH 11 to 2. Suspension rheology was examined using a controlled stress rheometer (TA Instruments, AR-2000), using a standard concentric cylinder geometry, under steady state flow conditions using shear rates from 0.1 s^{-1} to 1500 s^{-1}.

TiC preform production was achieved by directional slip casting, using PTFE rings placed on flat plaster of Paris plates. Suspensions were prepared using the optimized PMA-NH_4 concentration, hand mixed and finally sonicated for 2 minutes. A vacuum degassing step was used to remove any entrained air in the slip, in combination with an anti-foaming agent (Tri-n-butyl phosphate, 98 % Alfa Aesar) in 0.025 wt. % concentration, prior to casting into the PTFE moulds. The cast pieces were then dried at room temperature for one hour, removed from the PTFE rings and finally oven dried for approximately 12 hours at 40°C.

Melt infiltration of the slip cast TiC preforms was performed at either 1500 or 1550°C, with steel contents varied from 5 to 30 vol. %, based on an approach previously used for preparing TiC cermets with nickel aluminide alloys.[12] Infiltration was conducted under a dynamic vacuum inside a graphite resistance furnace (MRF Research Furnaces), with heating and cooling rates of 10°C/min. and 25°C/min., respectively. The sintering temperature was held for a period of 60 minutes in all cases.

Cermet Characterization

The infiltrated composites densities were determined using the Archimedes immersion method in water. Microstructural characterization was conducted using both optical microscopy (Olympus BX-51) and SEM, on ground and polished surfaces. Grain size measurements were made using the linear intercept method, from SEM images, where the mean intercept value was multiplied by a factor of 1.5. The hardness values were obtained on the polished samples using a Vickers diamond pyramid indenter, with a 50kg applied load held for 15 seconds (V-100A, Leco).

RESULTS AND DISUSSION

TiC Powder Characterization and Colloidal Forming

The particle size analysis of the TiC powder shows a bimodal distribution (Figure 1(a)), with a narrow secondary peak at ~0.1 μm and a broad primary peak at ~1.0 μm. The average particle size was ~1.2 μm, which confirms the manufacturer's specification. The size analysis is also largely confirmed by SEM observation (Figure 1(b)). It is apparent that the TiC particles are irregularly shaped and have somewhat of a 'plate-like' habit, which is explained by the powder production route. The manufacturer ball mills the powder to provide the desired particle distribution, which would explain the presence of irregular particles and the abundance of fine fragments. The measured average surface area of the 'as-received' TiC powder was 5.3 ± 0.3 m^2/g; this moderately high value arises largely from the fine fraction of particles, and could have an impact on the dispersant concentrations required.

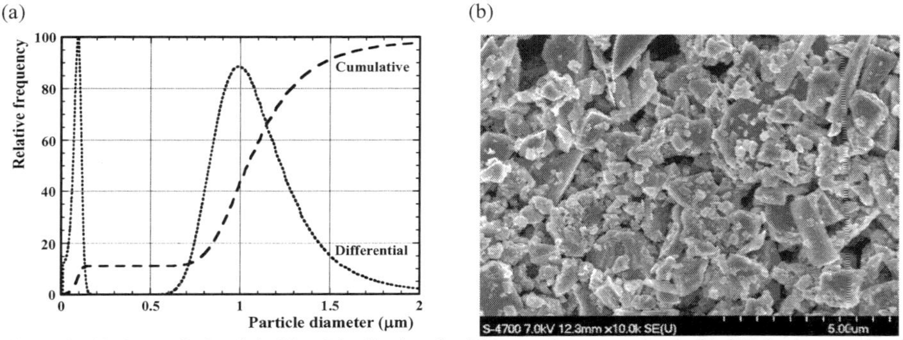

Figure 1. (a) Acoustic Particle Size Distribution for TiC powder as received. (b) SEM micrograph of the TiC powder as received.

The effects of suspension pH and PMA-NH_4 polyelectrolyte dispersant concentration on the zeta potential of the TiC powder are shown in Figure 2(a). In the absence of any dispersant, the TiC powder has an isoelectric point of pH ~2.6, which remains essentially unchanged when PMA-NH_4 is added. This is in contrast to a prior study which demonstrated the same isoelectric point without any polyelectrolyte; however with the addition of a polyelectrolyte, the suspension showed no isoelectric point at all.[13] We are currently examining alternative, cationic polyelectrolytes to assess their performance as a dispersant for TiC. It is apparent that a PMA-NH_4 concentration of 0.05 wt. % is required to provide a notable increase in zeta potential, at basic pH, above that obtained for the as-received powder in the absence of any deflocculant. Preparing slips with a pH greater than ~8.5 shows no significant benefit to stability, while operating at pH 10 (or above) is impractical as such basic conditions may have a detrimental effect on the processing and analysis equipment.

The slip rheology was assessed for a variety of PMA-NH_4 concentrations and solid loadings (all tests were performed at 20°C). The suspensions initially show shear thinning behavior, with a transition to shear thickening at higher shear rates (~600 s^{-1}), as shown in Figure 2(b). Figure 2(c) demonstrates that the minimum slip viscosity, which is also an indication of good stability, occurs in the 0.01 - 0.02 wt. % PMA-NH_4 region (measured at a shear rate of 200 s^{-1}). The viscosity then begins to increase as the dispersant concentration is increased. The subsequent reduction of the viscosity at high concentrations is likely due to the fact significant liquid is being added to the suspension, effectively reducing the solids loading and therefore the viscosity. Examining the effects of solids loading on the rheological behaviour, it is somewhat surprising that the shear thinning to shear thickening transition appears to shift to lower shear values with decreasing solids loading (Figure 2(d)), while the opposite response may be anticipated, due to dilitancy in a concentrated suspension.

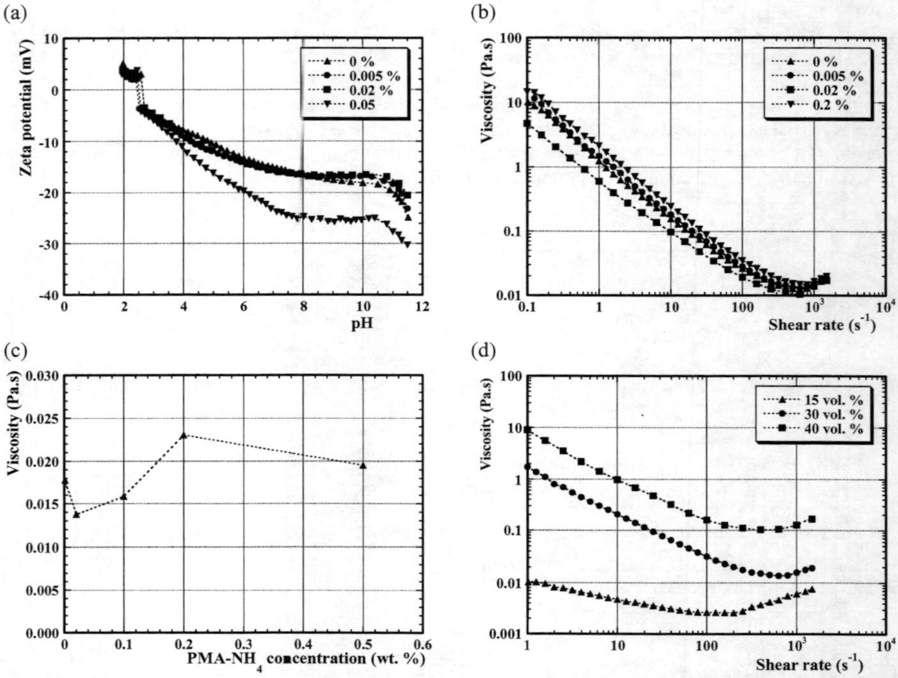

Figure 2. (a) Zeta potential titration curves for TiC in water with varying wt. % PMA-NH_4 polyelectrolyte concentrations. (b) The viscosity versus shear rate response for varying PMA-NH_4 concentrations (at pH 8.5). (c) Viscosity versus polyelectrolyte concentration at 200 s^{-1} with 30 vol. % solids loading (pH 8.5). (d) Viscosity versus shear rate for varying TiC solids loadings (shear rates from 1 - 1500 s^{-1} at pH 8.5).

A shear thinning response is initially visible for all of the suspensions that were examined, which may be explained by the particle morphology and, in the present instance, the somewhat plate-like habit of the TiC particles (see Figure 1(b)). As the shear rates increase the particles will begin to align in the shear field, thereby facilitating their slippage past one another, resulting in a reduction of the overall viscosity. Eventually a critical point is reached which the suspensions transition to shear thickening. A more detailed explanation of shear thinning and thickening in ceramic systems can be found in other work.[14] The viscosity trend for the PMA-NH_4 polyelectrolyte corroborates the previously presented data, displaying the same optimized dispersant concentration region for suspension stability. Based on the optimal slip conditions, TiC preforms were then prepared by directional slip casting into annular PTFE molds placed on plaster of Paris plates.

Melt Infiltration Behavior and Microstructure/Hardness Characterization

Densification by melt-infiltration has been successfully applied to a variety of TiC-based cermets in prior studies.[12] Preliminary DSC studies demonstrated that the onset of melting for both the 304-L and 316-L stainless steels was approximately 1410°C. Conversely, the melting point of the 410-L grade could not be accurately determined as no clear endotherm was observed (the apparent onset of an endotherm was noted at 1480°C, just below the maximum achievable DSC temperature of 1500°C) and a sintered powder structure was retained rather than a small melted ingot, as noted for 304-L and 316-L. Initial melt-infiltration tests at 1500°C were successful for both 300 series steels, with complete infiltration, but not for the martensitic 410-L, with sintered powder retained on the TiC pellet surface. The 304-L and 316-L samples also showed good dimensional stability at this temperature.

Due to the initial problems associated with infiltration when using the 410-L steel at 1500°C, all subsequent experiments were performed using an infiltration temperature of 1550°C. Figure 3 demonstrates the effect of steel volume fraction and type on the *nominal* measured density (estimated using a rule of mixtures based on the densities of TiC (i.e. 4.93 g/cm^3) and the grade of steel used (see Table 1)). It is apparent that densities in excess of 97-98 % of theoretical are achieved for samples processed with 7.5 vol. % stainless steel and higher.

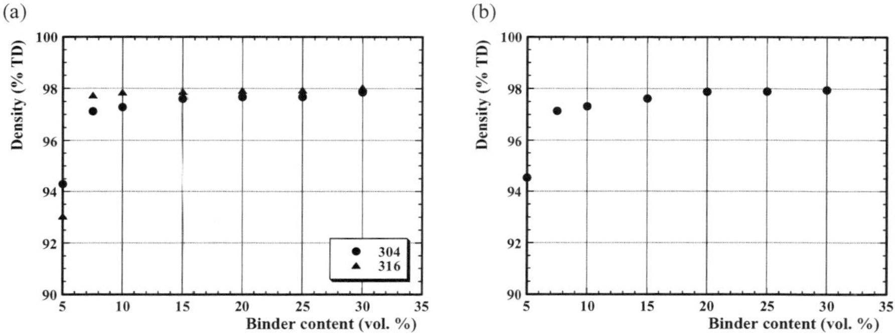

Figure. 3. The densities of TiC-stainless steel cermets processed at 1550°C as a function of steel type and binder volume fraction: (a) austenitic 304-L and 316-L and (b) martensitic 410-L grades.

While the outlined densification data indicates a small volume fraction of retained porosity, SEM examination of the microstructures of the infiltrated samples indicates that they are essentially fully dense, even with relatively low volume fractions of metallic binder (Figure 4). Several reasons may be seen to account for this apparent discrepancy. Firstly, the nominal theoretical densities are based on a simple rule-of-mixtures calculation using the individual component densities. In particular, this ignores potential modification of the stainless steel binders by dissolution of TiC. Preliminary SEM examination appears to show a higher binder volume fraction than initially specified, indicating the likelihood of binder modification (Figure 5); this can be expected based on the solubilities of both titanium and carbon into the steel lattice. More trivially, the presence of isolated bubbles in the cast preform, which are retained after melt infiltration, cannot be entirely ruled out; however it is unlikely that such coarse porosity would account for this discrepancy as TiC-Ni_3Al cermets prepared with identical slip cast TiC preforms exhibited densities well in excess of 99 % of theoretical.[11] As noted, Figure 5 highlights the microstructural evolution of the 304-L steel samples as a function of the nominal binder content. These images, taken under back-scattered electron imaging conditions, highlight the two phase nature of the materials, indicating that no phases other than TiC and steel are present (e.g. free-carbon) after infiltration. The grain size of the samples is also shown to be relatively consistent, which is reflected through measurement of the average grain size as a function of steel binder volume fraction between 10 and 30 vol. % for each of the three grades examined (Figure 6).

(a) (b)

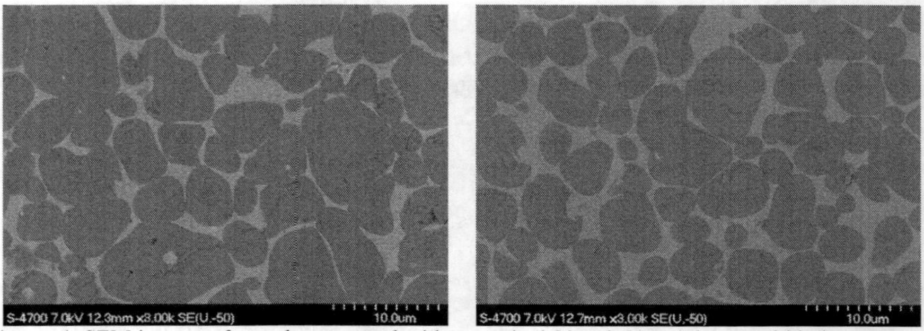

Figure. 4. SEM images of samples prepared with a nominal 20 vol. % stainless steel binder content at 1550°C: (a) 316-L and (b) 410-L grades.

(a) (b)

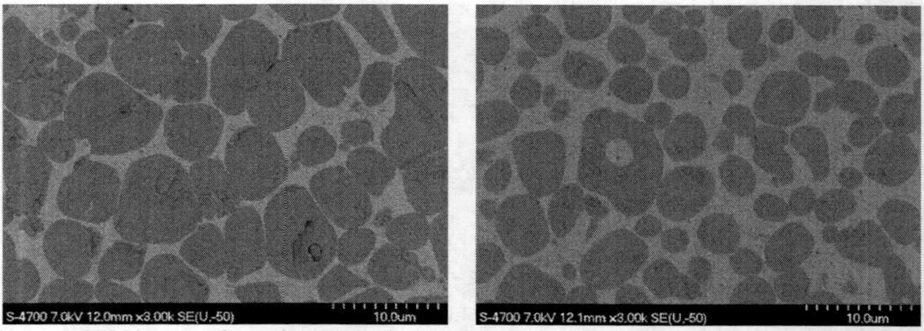

Figure. 5. SEM images of samples prepared with the 304-L stainless steel binder at 1550°C: (a) 15 vol. % and (b) 30 vol. % binder.

A preliminary assessment of the mechanical behavior of the TiC-stainless steel cermets has been conducted utilizing Vickers hardness measurements, with a 50 kg load. Figure 7 demonstrates the effects of binder volume fraction and the grade of stainless steel upon the hardness of the cermets prepared at 1550°C. It is clear that increasing the binder content has a significant effect on the hardness, with the higher binder content samples having the lowest hardness, as might be expected based on the relative elastic moduli of the two phases. While samples with 7.5 vol. % binder were not assessed in this initial study, they were essentially fully dense and can therefore be expected to exhibit high hardness. It is apparent that the selection of binder does not have a significant influence on the final hardness, which is predominantly influenced by the binder volume fraction.

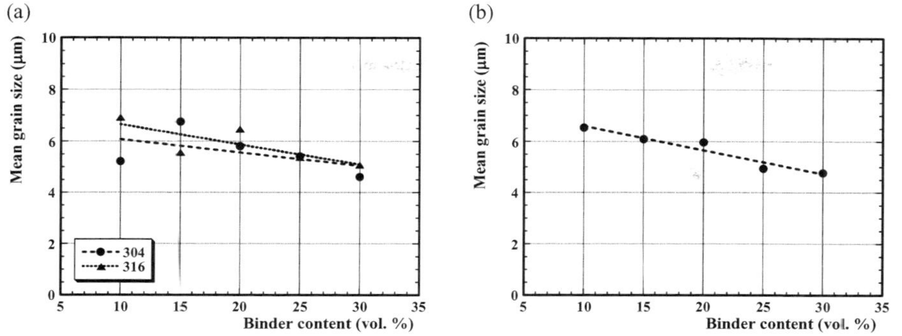

Figure. 6. The mean grain size of TiC-stainless steel cermets processed at 1550°C as a function of steel type and binder volume fraction (all samples sintered for 1 hour at the final hold temperature): (a) austenitic 304-L and 316-L and (b) martensitic 410-L grades.

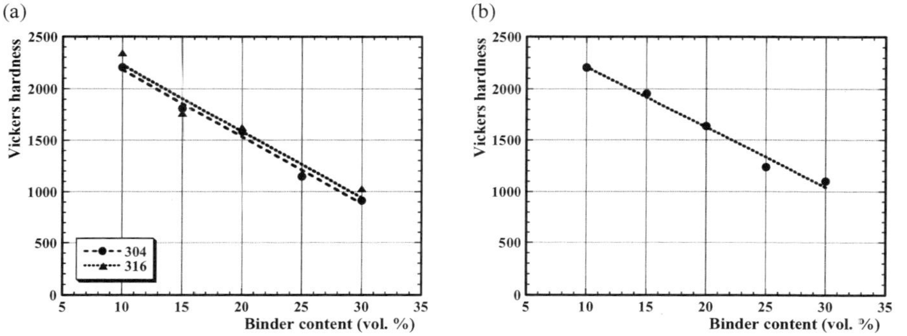

Figure. 7. The Vickers indentation hardness (HV50) of TiC-stainless steel cermets processed at 1550°C as a function of steel type and binder volume fraction: (a) austenitic 304-L and 316-L and (b) martensitic 410-L grades.

CONCLUSIONS

The present work has focused upon the development of TiC-based cermets utilizing stainless steel binders. A colloidal approach was taken for composite production, forming TiC preforms via aqueous slip casting, and then densifying the composites by melt infiltration. The rheological behavior of the TiC slips was optimized using a combination of zeta potential and shear flow measurements (along with sedimentation studies, which are not discussed in the present work). The TiC suspensions exhibited significant shear thinning throughout the majority of shear rates tested, transitioning to a shear thickening response at high shear rates, which would be unlikely to be found in most processing conditions. The slip cast preforms were successfully infiltrated with three stainless steel grades using a simple pressureless melt infiltration method; while the austenitic grades 304-L and 316-L could be readily infiltrated at 1500°C, the martensitic steel grade 410-L could only be successfully infiltrated at 1550°C. After melt infiltration, samples with steel contents between 7.5 and 30 vol. % exhibited nominal densities in excess of 97–98 % of theoretical. However, SEM examination did not reveal the evidence of any significant porosity, which was attributed to binder volume fraction modification through limited TiC dissolution into the steel. The mean grain size of the infiltrated samples was largely independent of either steel type or volume fraction, being between 5 and 6 μm for all samples prepared at 1550°C. The Vickers hardness of the resulting TiC-stainless steel composites was predominantly dependent upon the volume fraction of stainless steel used, with higher steel contents exhibiting lower hardness values. Conversely, the grade of steel utilized had minimal influence on the measured hardness for a constant binder volume fraction.

The present work has demonstrated the feasibility of preparing TiC-based cermets with a variety of stainless steels and with a wide range of steel contents. Our current emphasis is based on refining the processing, quantification of the effects of binder modification (i.e. volume fraction and composition) and assessment of mechanical, tribological and corrosion behavior.

ACKNOWLEDGEMENTS

The National Science and Engineering Research Council of Canada (NSERC) are gratefully acknowledged for funding this work through both the Discovery Grants and Undergraduate Student Research Awards programs. We are also grateful for funding from Petroleum Research Atlantic Canada (PRAC). The support of the Canada Foundation for Innovation, the Atlantic Innovation Fund, and other partners who helped fund the Facilities for Materials Characterisation, managed by the Dalhousie University Institute for Materials Research, is also gratefully acknowledged.

REFERENCES

[1]E.K. Storms, *The Refractory Carbides*; pp. 285. Academic Press, New York (1967).

[2]D.B. Miracle and H.A. Lippsitt, "Mechanical Properties of Fine-Grained Substoichiometric Titanium Carbide," *J. Am. Ceram. Soc.*, **66** [8] 592 (1983).

[3]J.A. Lewis, "Colloidal Processing of Ceramics," *J. Am. Ceram. Soc.*, **83** [10] 2341 (2000).

[4]D.J. Miller and J.A. Pask, "Liquid-Phase Sintering of TiC-Ni Cermets," *J. Am. Ceram. Soc.*, **66** [12] 841 (1983).

[5]P. Persson, A.E.W. Jarfors and S. Savage, "Self-propagating high-temperature synthesis and liquid-phase sintering of TiC/Fe composites," *J. Mater. Proc. Tech.*, **127** [2] 131 (2002).

[6]F. Arenas, C. Rondon and R. Sepulveda, "Friction and tribological behavior of (Ti,V)C-Co cermets," *J. Mater. Proc. Tech.*, **143** [S1] 822 (2003).

[7]P.F. Becher and K.P. Plucknett, "Properties of Ni_3Al-bonded Titanium Carbide Ceramics," *J. Eur. Ceram. Soc.*, **18** [4] 395 (1998).

[8]R. Subramanian and J.H. Schneibel, "FeAl-TiC cermets - melt infiltration processing and mechanical properties," *Mater. Sci. Eng. A*, **240** 633 (1997).

[9]A. Farid, S.J. Guo, F.E. Cui, F.Z. Feng and T. Lin, "TiB_2 and TiC stainless steel matrix composites," *Mater. Lett.*, **61** [1] 189 (2007).

[10]F. Akhtar and S.J. Guo, "Microstructure, mechanical and fretting wear properties of TiC-stainless steel composites," *Mater. Charact.*, **59** [1] 84 (2008).

[11]R.B. Collier and K.P. Plucknett, "Aqueous Processing of TiC Preforms for Advanced Cermet Preparation," accepted for publication in *Ceram. Trans.* (2009).

[12]K.P. Plucknett and P.F. Becher, "Processing and Microstructure Development of Titanium Carbide-Nickel Aluminide Composites," *J. Am. Ceram. Soc.*, **84** [1] 55 (2001).

[13]C. Yeh and M.H. Hon, "Dispersion and Stabilization of Aqueous TiC Suspension," *Ceram. Int.*, **21** [2] 65 (1995).

[14]L. Bergström, "Shear thinning and shear thickening of concentrated ceramic suspensions," *Colloids Surf. A*, **133** [1-2] 151-153 (1998).

OXIDATION BEHAVIOR OF ZIRCONIUM DIBORIDE-SILICON CARBIDE COMPOSITES

Ipek Akin, Filiz Cinar Sahin, Onuralp Yucel, Gultekin Goller
Istanbul Technical University, Metallurgical and Materials Eng. Dept.
Istanbul, Turkey

ABSTRACT

ZrB_2 and ZrB_2-SiC composites containing 40 and 60 mass% SiC were prepared by spark plasma sintering (SPS) at temperatures of 1810 and 1770°C for 300 s under a pressure of 40 MPa. Oxidation behavior of these samples was characterized by exposing them to 1400, 1500, and 1600°C in an ambient atmosphere for 180 min, and by measuring the weight gains of the sample. The oxide layers were characterized by field emission scanning electron microscopy with energy dispersive spectroscopy analysis. At 1400 and 1500°C, layered structures consisted of a SiO_2 rich outer layer, a thin ZrO_2-SiO_2 layer, ZrO_2-containing layer and unaffected ZrB_2-SiC. However, at 1600°C, active oxidation of SiC was observed and changed oxidation mechanism significantly. The consumption of SiC particles resulted in a formation of oxide regions mainly composed of SiO or SiO_2 and ZrO_2. Unaffected ZrB_2-SiC region was not observed at 1600°C.

INTRODUCTION

Zirconium diboride (ZrB_2) has been regarded as a promising candidate material for structural applications such as thermal protection systems, corrosion resistant materials, high temperature electrodes, and aerospace applications because of its high melting temperature (3027°C), low theoretical density (6.1 Mg/m^3), high hardness (23 GPa), high thermal (60-120 W/mK) and electrical (10^7 S/m) conductivity and excellent chemical and physical stability at high temperatures [1-2]. However, the oxidation resistance of ZrB_2 is very poor at temperatures above 1100°C due to the volatilization of B_2O_3 that results in formation of a porous, non-protective ZrO_2 layer [3]. Below 1100°C, the liquid or glassy B_2O_3 forms a continuous layer. In this temperature, diffusion controlled kinetics are dominant. Above 1100°C, the oxidation rate of ZrB_2 increases and between 1100 and 1400°C, para-linear kinetics are effective. In this temperature range, rate of mass change is a combination of weight gain due to the formation of liquid B_2O_3 and ZrO_2 and weight loss due to the volatilization of liquid B_2O_3. Above 1400°C, the rate of evaporation of liquid B_2O_3 is greater than its rate of production, resulted in a formation of porous, non-protective ZrO_2 layer [3].

Above 1100°C, the oxidation resistance of ZrB_2 can be improved by addition of SiC. Below this temperature, SiC additions do not affect the oxidation rate and composition of the protective layer. The addition of SiC allows the formation of SiO_2, which is more resistant to evaporation than B_2O_3 [3-5]. The silica rich surface is also more refractory and resistant to oxygen diffusion, making ZrB_2-SiC composites more oxidation resistant than pure ZrB_2. Also, the silica-rich surface is remained stable by the oxidized sub-layer consisting of an interpenetrating composite of ZrO_2-silica rich glass [6,7].

ZrB_2-SiC composites have been prepared by using conventional pressureless sintering, hot-pressing, and reactive hot pressing [6-7]. However, fully dense composites have been hardly obtained because of high melting temperature and strong covalent bonding of ZrB_2 and SiC [1]. The spark plasma sintering (SPS) method has provided a processing technique to rapidly densify ZrB_2-based ceramics. However, many of these studies also use sintering aids which may affect the oxidation properties. In this study, ZrB_2-SiC composites were prepared by using SPS without sintering aids.

The objectives of this study are to produce dense ZrB_2-SiC composites containing 40 and 60 mass% SiC and determine the oxide layer morphology and composition as a function of temperature.

EXPERIMENTAL STUDIES

Processing

ZrB_2 (Grade B, H.C. Starck Corp., an average particle size of 2 μm) and α-SiC (Grade UF-10, H.C. Starck Corp., an average particle size of 1 μm) powders were used as starting materials. The powders were weighed in appropriate quantities, ball milled in ethanol for 24 h and then dried. A graphite die 50 mm in inner diameter was filled with the mixture, followed by sintering using a spark plasma sintering (SPS) apparatus (SPS-7.40 MK-VII, SPS Syntex Inc.). ZrB_2 was sintered at 1810°C, ZrB_2-SiC composites containing 40 and 60 mass% SiC were sintered at 1770°C for 300 s with a heating rate of 1.7°C/s in a vacuum. A uniaxial pressure of 40 MPa and pulsed direct current (12 ms/on, 2 ms/off) were applied during the entire process. The current was controlled manually during monitoring the displacement behavior of the samples.

Oxidation Studies

ZrB_2 and ZrB_2-SiC composites were exposed to stagnant air at temperatures of 1400, 1500 and 1600°C for 180 min. The samples with size of 12 mm x 7 mm x 6 mm were heated and cooled at 10°C/min. A $MoSi_2$ resistance-heated furnace (Nabertherm C42) was used to heat the samples. Sample weights were measured before and after oxidation and dimensions were measured to ± 0.01 mm to calculate the surface area of each test specimen. The microstructures of the oxidized samples were characterized using scanning electron microscope (SEM; JEOL JSM 7000F). Chemical analysis was performed with energy dispersive spectroscopy (EDS; Oxford INCA).

RESULTS AND DISCUSSIONS

Weight Change and Mass Oxidation Rate of Composites

The normalized mass change (Table I) and the mass oxidation rate (Table II) of the composites were calculated. These terms can be defined as total mass change divided by surface area calculated from the dimensions of the rectangular prism and mass change divided by oxidation time, respectively.

Table I. Normalized mass change (mg/mm^2) of the specimens after oxidized at 1400, 1500 and 1600°C for 180 min.

Composition	1400°C	1500°C	1600°C
ZrB_2	0.32	0.45	0.34
ZrB_2-40 mass% SiC	0.11	0.15	0.09
ZrB_2-60 mass% SiC	0.01	0.05	-0.05

For all oxidation temperatures, the mass gain and mass oxidation rate for pure ZrB_2 were significantly greater than other specimens. In addition, the mass gain decreased as SiC content increased. For the composite containing 60 mass% SiC, weight loss was observed at 1600°C.

As it is well known, the oxidation of SiC is divided into two regimes. One of them is passive oxidation with the formation of a SiO_2 layer on the SiC surface according to Eq. (1). When SiC oxidizes to SiO_2 in the passive regime, there is a net mass gain.

$$SiC_{(s)} + \frac{3}{2}O_{2(g)} = CO_{(g)} + SiO_{2(s)} \quad (1)$$

The other type of oxidation is active oxidation. The active oxidation resulted in $SiO_{(g)}$ formation according to Eq. (2-4) at higher temperatures and resulted in a weight loss.

$$SiC_{(s)} + \frac{3}{2}O_{2(g)} = CO_{(g)} + SiO_{2(s)} \tag{2}$$

$$SiC_{(s)} + O_{2(g)} = CO_{(g)} + SiO_{(g)} \tag{3}$$

$$SiC_{(s)} + \frac{1}{2}O_{2(g)} = C_{(s)} + SiO_{(s)} \tag{4}$$

The weight loss for the composite containing 60 mass% SiC oxidized at 1600°C indicated that oxidation occurred in active regime. The mode of SiC oxidation strongly dependent on the nature of SiC, composition, partial and total oxygen pressures, and oxidation temperature [3,8]. Different studies have been published about transition temperature of active-passive oxidation. Villegas[8] and Goto[9] reported the transition temperature at 1400°C and 1600°C, respectively.

Table II. The mass oxidation rate results (mg/s) of the specimens after oxidized at 1400, 1500 and 1600°C for 180 min.

Composition	1400°C	1500°C	1600°C
ZrB_2	$5.96.10^{-3}$	$8.42.10^{-3}$	$6.37.10^{-3}$
ZrB_2-40 mass% SiC	$2.05.10^{-3}$	$2.89.10^{-3}$	$1.66.10^{-3}$
ZrB_2-60 mass% SiC	$1.85.10^{-4}$	$9.91.10^{-4}$	$9.26.10^{-4}$

The mass gain and rate of mass oxidation increased for all specimens with increasing oxidation temperature from 1400 to 1500°C. The increase in mass gain and mass oxidation rate may have resulted from the oxidation phenomena of ZrB_2 and SiC when exposed to air. Below 1400°C, the rate of mass change is a combination of weight gain due to the formation of glassy or liquid B_2O_3 and ZrO_2 layer and weight loss due to the volatilization of liquid B_2O_3. Above 1400°C, the rate of evaporation of liquid B_2O_3 is greater than its rate of production, leaving a non-protective porous ZrO_2 scale.

Morphology and Composition of Oxide Scales

Analysis of the oxide layers formed on pure ZrB_2 by heating at 1400, 1500 and 1600°C for 180 min showed that the reaction layers were approximately 100 μm thick. For pure ZrB_2, volatilization of B_2O_3 at oxidation temperatures resulted in a formation of a porous, non-protective ZrO_2 layer.

Figure 1 shows the cross sectional morphology and EDS mapping of the oxide scale at 1400°C for the composites containing 60 mass% SiC. Mapping of oxygen (O) and silicon (Si) concentration using EDS (Figure 1) showed that the surface layer was Si and O rich. The analysis revealed that, oxidation of specimen at 1400°C produced a SiO_2-rich surface layer (Layer 1) with a ~ 7 μm thickness. The ZrO_2-SiO_2 layer (Layer 2), ~ 17-20 μm thick, remained below the outer SiO_2 rich layer.

Above 1100°C, α-SiC oxidized by the reaction given above in Eq. (1) and resulted in formation of SiO_2. The silica containing scale on ZrB_2-SiC composites was stable at 1400°C than the B_2O_3 due to the lower volatility of liquid SiO_2 compared with liquid B_2O_3 at that temperature [3].

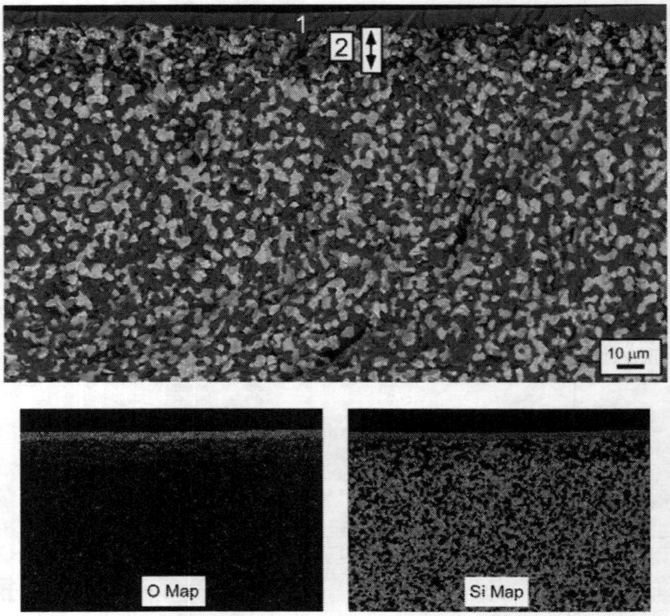

Figure 1. Cross-sectional micrographs of oxidized composite containing 60 mass% SiC at 1400°C for 180 min and elemental composition maps of O and Si.

Figure 2 demonstrates cross section SEM images and EDS mapping of ZrB_2-SiC composites containing 60 mass% SiC oxidized at 1500°C for 180 min. The structure of the specimen heated in air at 1500°C was similar to the structure of the specimen exposed to 1400°C (Figure 1), except that the oxide layers were thicker after heated at 1500°C, in agreement with the normalized mass change and mass oxidation rate results (Table I and II).

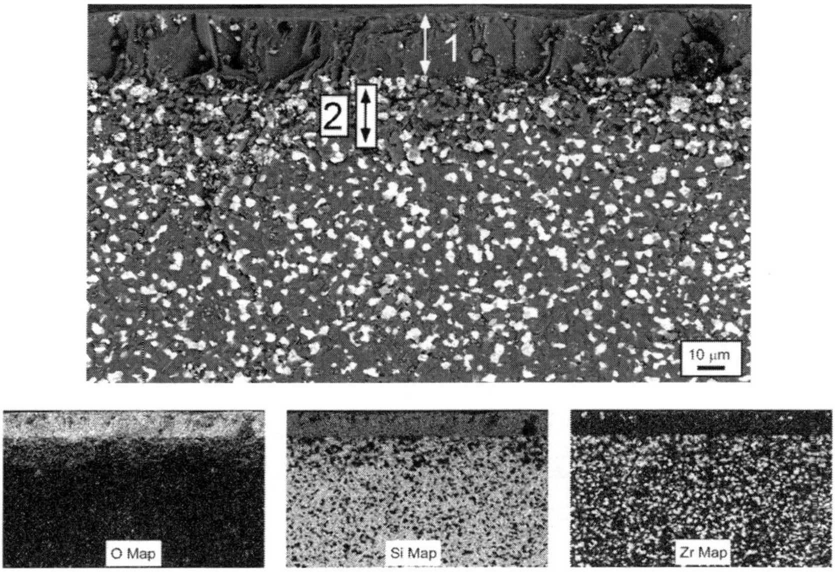

Figure 2. Cross-sectional micrographs of oxidized composite containing 60 mass% SiC at 1500°C for 180 min and elemental composition maps of O, Si and Zr.

Mapping of oxygen (O), silicon (Si) and zirconium (Zr) using EDS (Figure 2) showed that the surface layer was Si and O rich. Zirconium (Zr) was not observed in Layer 1. The SiO_2 and ZrO_2-containing layers were uniform and thicker than 1400°C. Oxidation of specimen at 1500°C produced a SiO_2-rich surface layer (Layer 1) with a ~ 25 μm thickness and ~ 20 μm thick ZrO_2-containing layers (Layer 2).

At 1400 and 1500°C, the layered structures consisted of a SiO_2-rich layer, a thin ZrO_2-SiO_2 layer, a ZrO_2-containing layer (Si particles partially depleted), and unaffected ZrB_2-SiC layer. These layered structures were similar with the reported oxidation studies for ZrB_2-SiC composites[4,5,10].

Cross section microstructure of ZrB_2-SiC composites containing 60 mass% SiC oxidized at 1600°C for 180 min and elemental composition maps of Zr, C, O and Si are shown in Figure 3.

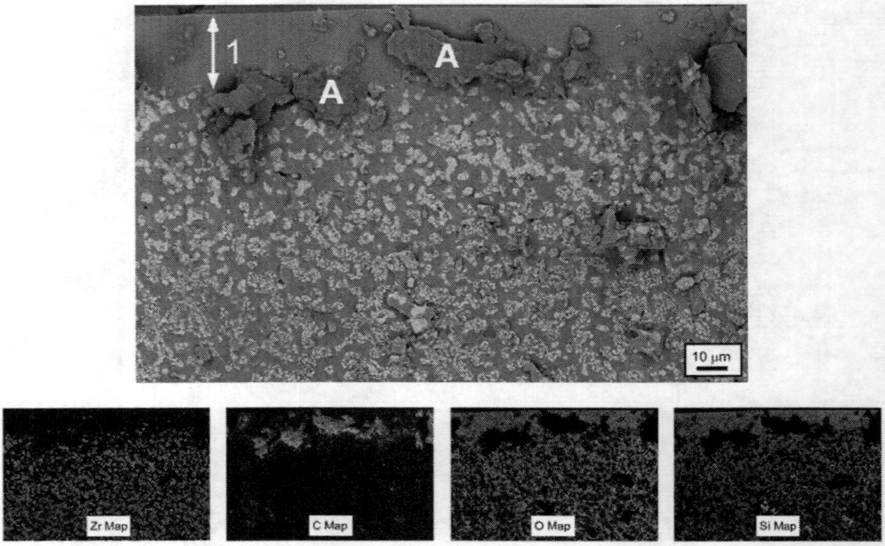

Figure 3. Cross-sectional micrographs of oxidized composite containing 60 mass% SiC at 1600°C for 180 min and elemental composition maps of Zr,C, O and Si.

It was observed that, the thickness of the SiO_2 layer (Layer 1), and the underlying ZrO_2-containing layer were not uniform over the specimen surface at 1600°C. This may be caused by wetting characteristics, local variations of the composites or surface topology [7]. In addition, gray particles deposition (shown as A in Fig.3) was also detected in the SiO_2-rich layer (Figure 3). EDS analysis confirmed that they were mainly carbon deposition. This precipitation could be mostly due to the active oxidation of SiC as shown in Eq. (4)[4]. Also, the oxide scale on SiC is no longer protective due to the reaction of liquid SiO with underlying α-SiC to form gas phases SiO and CO according to Eq. (5) [3].

$$SiC_{(s)} + 2SiO_{2(l)} = 3SiO_{(g)} + CO_{(g)} \tag{5}$$

Active oxidation of SiC was observed at 1600°C for the composite containing 60 mass% SiC and oxidation mechanism changed. At 1400 and 1500°C, the inner structure, defined as unaffected ZrB_2-SiC layer, did not contain oxygen (Figure 1 and 2). However, for the oxidation studies performed at 1600°C, unaffected layer was not observed (Figure 3). The consumption of SiC particles due to active oxidation promoted the generation of oxide regions mainly composed of SiO_2 and ZrO_2.

CONCLUSIONS

Dense ZrB_2 and ZrB_2-SiC composites containing 40 and 60 mass% SiC were produced by spark plasma sintering. Oxidation behavior of the samples was investigated by exposing them to 1400, 1500, and 1600°C in a stagnant atmosphere for 180 min.

For all oxidation temperatures, total mass gain and mass oxidation rate in air decreased with increasing SiC content except the composite containing 60 mass% SiC oxidized at 1600°C for 180 min. For this composition, weight loss was observed due to the formation of active oxidation and transition temperature was determined as 1600°C. The transition from passive to active oxidation mainly depended on composition, partial pressure of oxygen and temperature. For the composites containing 60 mass% SiC this transition temperature measured as 1600°C. The weight loss results and microstructural analysis confirmed the passive to active transition temperature.

For the oxidation temperatures of 1400 and 1500°C, composites showed similar oxidation behavior due to the formation of protective layered structures consisted of SiO_2-rich layer, a thin ZrO_2-SiO_2 layer, ZrO_2-containing layer and unaffected ZrB_2-SiC inner structure. At 1600°C, oxidation mechanism of the composites changed due to the active oxidation of SiC. The consumption of SiC particles resulted in a formation of oxide regions mainly composed of SiO_2 and ZrO_2 and unaffected ZrB_2-SiC region was not observed at 1600°C.

Acknowledgement

The authors would like to thank H. Dincer due to his contribution for SPS processes, and H. Sezer and T.T. Alpak for electron microscopy investigations.

REFERENCES

[1] I. Akin, M. Hotta, F. Sahin, O. Yucel, G. Goller, and T. Goto, Microstructure and densification of ZrB_2-SiC composites prepared by spark plasma sintering, *J. Eur. Ceram. Soc.*, **29**, 2379-2385 (2009).

[2] H. Wang, C. Wang, X. Yao, and D. Fang, Processing and mechanical properties of zirconium diboride-based ceramics prepared by spark plasma sintering, *J. Am. Ceram. Soc.*, **90**, 1992-1997 (2007).

[3] W.G. Fahrenholtz, Thermodynamic analysis of ZrB_2-SiC oxidation: formation of a SiC-depleted region, *J. Am. Ceram. Soc.*, **90**, 143-148 (2007).

[4] P. Hu, W. Guolin, and Z. Wang, Oxidation mechanism and resistance of ZrB_2-SiC composites, *Corros. Sci.*, **51**, 2724-2732 (2009).

[5] C. Carney, P. Mogilvesky, and T. Parthasarathy, Oxidation behavior of zirconium diboride silicon carbide produced by the spark plasma sintering method, *J. Am. Ceram. Soc.*, **92**, 2046-2052 (2009).

[6] Y. Yan, Z. Huang, S. Dong, and D. Jiang, Presureless sintering of high density ZrB_2-SiC ceramic composites, *J. Am. Ceram. Soc.*, **89**, 3589-3592 (2006).

[7] S. Zhu, W.G. Fahrenholtz, and G.E. Hilmas, Influence of silicon carbide particle size on the microstructure and mechanical properties of zirconium diboride-silicon carbide ceramics, *J. Eur. Ceram. Soc.*, **27**, 2077-2083 (2007).

[8] M. Villegas, T. Sierra, F. Lucas, J.F. Fernandez, and A.C. Caballero, Oxidation treatments for SiC particles and its compatibility with glass, *J. Eur. Ceram. Soc.*, **27**, 861-865 (2007).

[9] T. Goto, H. Hommai High-temperature active/passive oxidation and bubble formation of CVD SiC in O_2 and CO_2 atmospheres, *J. Eur. Ceram. Soc.*, **22**, 2749-2756 (2002).

[10] A. Rezaie, W.G. Fahrenholtz, G.E. Hilmas, Evolution of structure during the oxidation of zirconium diboride-silicon carbide in air up to 1500°C, *J. Eur. Ceram. Soc.*, **27**, 2495-2501 (2007).

Rapid Processing

NANO-CRYSTALLINE YTTRIA SAMARIA CODOPED ZIRCONIA: COMPARISON OF ELECTRICAL CONDUCTIVITY OF MICROWAVE AND CONVENTIONALLY SINTERED SAMPLES

Soumyajit Koley, Abhijit Ghosh, Ashok Kumar Sahu & Ashok Kumar Suri
Materials Group, Bhabha Atomic Research Centre, Mumbai, India.

ABSTRACT

A systematic study on powder preparation, compaction and sintering in conventional heating and microwave (MW) field has been carried out for yttria samaria codoped tetragonal zirconia polycrystal and the electrical conductivity of the sintered samples has been compared. The powder compositions, selected as $(Y_2O_3)_X$ - $(Sm_2O_3)_{3-X}$ - $(ZrO_2)_{97}$ (where, x takes the values of 0, 1, 2 and 3) were prepared by co-precipitation method. Based on simultaneous thermo-gravimetry and differential thermal analysis (TG-DTA) study the calcination temperature was selected as 800^0C. The diffractograms of the calcined powders confirmed the presence of tetragonal phase. Several techniques were used to determine particle size of the wet ground powder. Shrinkage behaviour of the sample studied by dilatometric technique indicated an increase in sintering onset temperature with samaria content. Isothermal sintering studies were carried out under conventional heating and microwave field. In both the cases high density (95-99% of theoretical density) was obtained at 1200^0C temperature for all the compositions. Sintering in MW furnace resulted in a marginally denser sample compared to its conventionally sintered counterpart. Electrical conductivity of the samples measured by two-probe AC impedance spectroscopy in the temperature range of 350^0-750^0C revealed that activation energy for ionic conduction decreases when samples are sintered in microwave field.

INTRODUCTION

Yttria stabilized zirconia (YSZ) is widely used as electrolyte materials in solid oxide fuel cells (SOFC), oxygen sensors, oxygen separation membranes and oxygen pumps owing to its high ionic conductivity, good mechanical strength and chemical stability at high temperatures[1-3]. The ionic conductivity of the YSZ electrolyte is strongly related not only to the dopant concentration but also to dopant size, the processing technology of the powders and the sintered density[4, 5].

Tetragonal zirconia compared to its cubic counterpart exhibits superior mechanical properties along with substantial ionic conductivity and considered as an potential electrolyte material for SOFC[5]. For this a proper composition along with tailor-made microstructure is required to attain maximum ionic conductivity with optimum mechanical strength[4, 6].

So far, most of the work has been concentrated on single element doping in zirconia system. However, co-doping has several advantages like better thermal and chemical stability, mechanical property and tailor-made microstructure[7]. In earlier studies, the oxides of scandium[7], ytterbium[8], titanium[9], terbium[10], cerium[11] etc were codoped with yttria in zirconia and studied.

In recent time there has been an interest in the use of microwave energy for sintering ceramic compacts. Microwave sintering has many attractive features, including rapid volumetric heating finer microstructure, better mechanical properties and low cost[12]. Higher production rate and lower energy (25-95% less) requirements make this process commercially attractive[13]. Stabilized zirconia is also a good susceptor of microwave radiation[14]. Zirconia stabilized with 3 mol% and 8 mol% yttria were densified in the microwave field[15, 16]. However, there is merely any comparative result available on the densification behaviour of the microwave and conventionally sintered material for samaria-yttria codoped zirconia system.

Electrical conduction of zirconia at elevated temperature is due to transport of oxygen ions through the lattice and grain-boundary without any electronic conduction[17]. This is particularly required for the electrolytic application in SOFC, where electronic conductivity would give rise to leakage current and reduce the efficiency of the cell. There are reports referring to the electrical conductivity of titania-yttria[18], terbia-yttria[10], scandia-yttria[7], samaria[5] and indium oxide[5] doped zirconia. In the present investigation, a systematic study of the yttria-samaria codoped zirconia has been carried out with respect to synthesis, powder processing, sintering and measurement of electrical conductivity.

EXPERIMENTAL

The powders having compositions $(Y_2O_3)_X$-$(Sm_2O_3)_{3-X}$-$(ZrO_2)_{97}$ (where, x takes the values of 0, 1, 2 and 3) were selected and named accordingly (Table I), hereinafter, these powders will be indicated by these names only.

Table I: Nomenclature given to the different compositions.

Sample Name	Zirconia (mol %)	Yttria (mol %)	Samaria (mol %)
3Y-TZP	97	3	0
1S-2Y-TZP	97	2	1
2S-1Y-TZP	97	1	2
3S-TZP	97	0	3

(1) Powder Synthesis & Characterization

All the powders have been synthesized by co-precipitation method, similar to the process adopted by Ghosh et al[19, 20] for the preparation of 3 mol% yttria stabilized zirconia. In brief, the hydroxide precursor was co-precipitated from mixed chloride solution (containing Zr^{+4}, Y^{+3} and Sm^{+3} ions in required concentration) in a basic bath (pH>9) containing ammonium hydroxide (NH_4OH). The particle agglomeration problem during precipitation was controlled by using ammonium sulphate $[(NH_4)_2SO_4]$ and polyethylene glycol (PEG) in the bath. The precipitate was thoroughly washed with deionized water to remove undesirable ions e.g. Cl^- and SO_4^{2-} subsequently rinsed with ethanol. Thermal decomposition behaviour of the vacuum dried (at 80^0C for 24hrs) precursor was studied by simultaneous thermo-gravimetry and differential thermal analysis (TG-DTA) technique (Labsys TG, Setaram, France). Based on this result the calcination temperature of the precursor was selected as 800^0C. A powder X-ray diffractometer with graphite monochromator and Cu-K_α radiations (λ = 1.54178Å) was used for x-ray diffraction study of the calcined powder. Calcined powder was ground in a planetary ball mill (Fritsch, Germany) using zirconia pots and grinding media. Particle size analysis of the as calcined and ground powder samples were carried out by laser scattering technique (Master Sizer-2000, Malvern, UK). Specific surface area of the powder was measured in Sorptometric-1990 (CE Instruments, Italy) instrument by the conventional BET method. Green pellets made by iso-statically pressing at 300MPa were used for mercury porosimetry study (Pascal 140 & 440, Thermo Electron, Italy). Shrinkage behaviour of the iso-pressed green powder compacts were studied in a dilatometer (TD-5000S, MAC Science, Japan), with a heating rate of 300^0C/hr up to 1200^0C temperature in air.

(2) Densification Study

The densification behaviour of the green pellets was studied by isothermal holding at different temperatures (1100^0-1250^0C) in both conventional and microwave furnace (soaking time was 180 and 30min for conventional and MW sintering respectively). A box type resistive furnace was used for conventional sintering operation.

A box-type microwave (MW) sintering furnace (frequency: 2.45GHz, power: 1.1KW, VB Ceramic Consultant, India) was used for densification study under MW radiation. The temperature was monitored and controlled by a radiation pyrometer. To ensure proper measurement of sample temperature the area of the sensing zone of the pyrometer was chosen bigger than the area covered by pellets kept for sintering. Porous silicon carbide slabs were used as microwave susceptor for effective heating at the lower temperature range.

(3) Electrical Conductivity Measurement

Disk shaped pellets, sintered at 1200^{0}C in conventional and microwave furnace were used as the samples in the electrical conductivity measurement experiments. The parallel faces of these pellets were polished and painted with a platinum paste and subsequently fired at 1000^{0}C in air to remove the organic binder used in the paste. A two probe AC impedance analyzer (Solartron SI 1260, UK) was used to measure the impedances at different temperatures (350^0 to 750^{0}C). At a fixed temperature, measurement was made applying a voltage bias (1V) within the frequency range of from 0.05Hz to 5MHz. The spectra obtained were fitted with the software Z-View (Version-3a, Scribner Associates, UK) to calculate the resistance and capacitance of the grain body and grain boundary.

RESULTS & DISCUSSION

The results of simultaneous TG-DTA of the four precursors of different compositions are shown in the Fig 1. A broad endotherm at ~130^{0}C could be found out in the DTA signal of all the four samples. This could be attributed to the removal of absorbed water from the surface of the precursor powder[21]. The DTA peaks for each of them show a very sharp exothermic peak in the range of 250^0-350^{0}C accompanied with significant amount of mass loss (~23%). This corresponds to the oxidation of organic materials, used in the processing stage[22-24]. At 500^{0}C and above, the mass loss was found to be insignificant irrespective of the composition. A very small exotherm could be seen at ~500^{0}C. This may be due to conversion of amorphous precursor to crystalline form. The reported values of 'crystallization temperature' of the yttria stabilized zirconia system are in the range of 440^0-470^{0}C[23, 24].

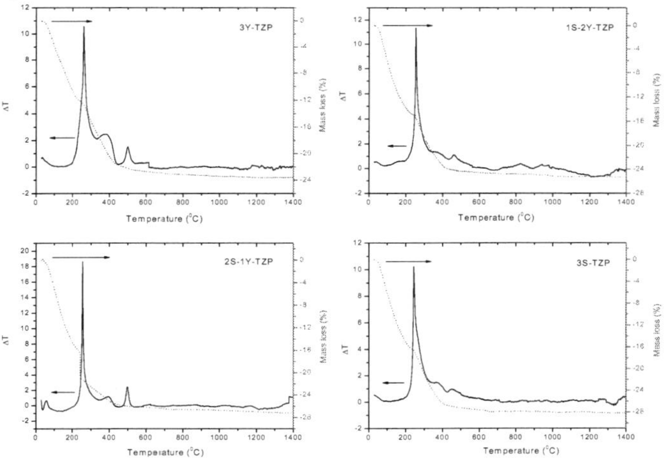

Fig 1: TG-DTA plots for four different precursors.

The X-ray diffractograms of the four calcined powders are shown in Fig 2. Each of them indicates formation of single phase. A comparison of XRD patterns obtained in the present investigation to that of JCPDS file no. 80-2187 (corresponding to 3.3 mol% yttria doped zirconia with tetragonal structure $P4_2/nmc$) revealed presence of tetragonal phase in the calcined powder[25]. The indexing of the diffractograms has been done accordingly. Earlier, stabilization of nano-crystalline tetragonal phase by 2 mol% and 3 mol% samaria doping had been shown by Cordova-Martinez et al[26] and Boulc'h et al[27] respectively. However, for samaria yttria codoped zirconia powder there was no literature indicating the phase.

In Fig 3(a) average size of the as calcined and finally ground powder were plotted as a function of the composition. The average particle size of the as calcined powder was nearly independent of the composition. However, at the end of grinding the particle size showed some dependency on the composition. The powder having composition of 3S-TZP showed the maximum particle size whereas 1S-2Y-TZP showed the minimum value i.e. the processing condition able to keep the average agglomerate size of the 3Y-TZP, 1S-2Y-TZP and 2S-1Y-TZP below 400nm. However for 3S-TZP the average particle size was found to be ~660nm.

Fig 3(b) shows the variation of specific surface area (measured by BET technique) as a function of composition. There is almost no change in this value with composition. The estimated spherical diameter (ESD) calculated from BET data were plotted simultaneously against composition in Fig 3(b). As expected it was found to be independent of the composition. The observed particle size, which was smaller than 20nm, was significantly smaller than the d_{50} values measured from laser scattering techniques. This discrepancy could be explained from the fact that laser scattering technique requires aqueous suspension/dispersion of the powder particles where formation of agglomeration/flocs is easier due to the surface tension of water. Thereby the technique inevitably measures the agglomerate size rather than the true particle size[28, 29].

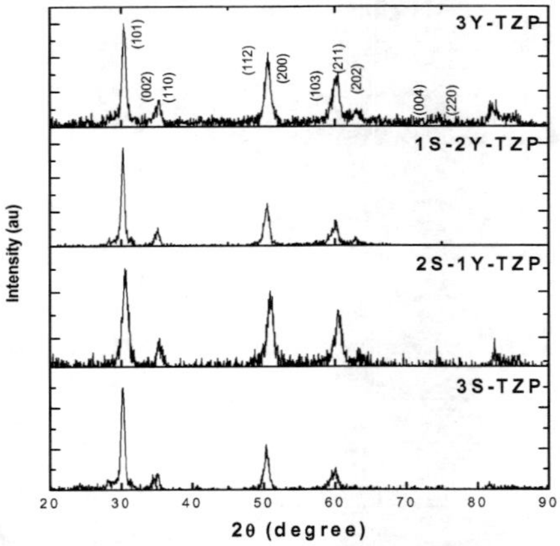

Fig 2: XRD patterns of the four different powders calcined at 800^{0}C for 3hrs.

The pore size distributions of the iso-pressed green pellets of different compositions are presented in the Fig 4. The plot of cumulative pore volume (V) as a function of pore diameter (D) (curve in "solid line") indicates how the total porosity increases with the pore size. The differential of this plot i.e. dV/d(log D) was calculated and plotted as a function of logD (the curve in the "dotted line"). This shows the pore size distribution in the green body. From Fig 5 it could be seen that a very narrow, monomodal nano sized (average pore diameter ~15nm) pore size distribution was obtained for all the powder compacts. Presence of single peak in 3Y-TZP and 1S-2Y-TZP ensures nearly agglomerate free powder. The result obtained in the present investigation is in good agreement with the reported values of Trunec & Maca and Van de Graaf et al[30, 31]. The pore size distribution of 2S-1Y-TZP and 3S-TZP show a small hump in the finer side of the distribution. This may be due to presence of ultra-fine pores within the agglomerates[32].

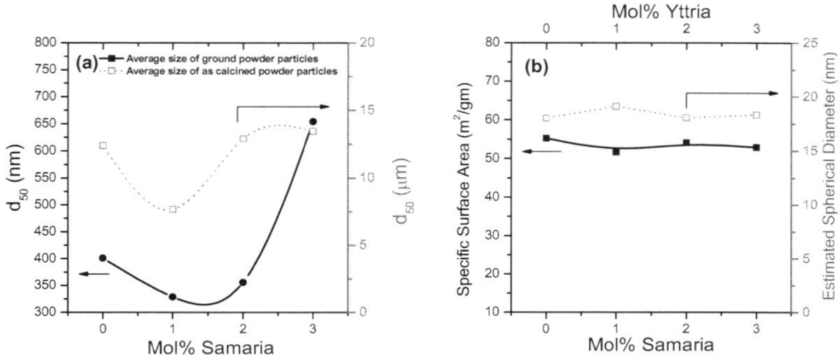

Fig 3(a): Change of average particle size (d_{50}) of the as calcined and finally ground powders with compositions. **(b)** Specific surface area measured from BET analysis and the ESD calculated from that value plotted as a function of the powder composition.

The shrinkage (Y) of the green pellets under constant rate heating and the differential shrinkage with respect to temperature (i.e. dY/dT) are plotted simultaneously as a function of temperature in the Fig 5. From these plots, it could be seen that the nature of shrinkage is smooth from the sintering onset temperature to the maximum temperature. A total of ~20% maximum shrinkage has been recorded for all the four samples.

The differential shrinkage profile of 3Y-TZP and 1S-2Y-TZP give only one peak corresponding to the maximum rate of sintering. This may be due to homogeneous, monomodal pore size distribution of the green compacts[32, 33]. But the differential shrinkage profile of 2S-1Y-TZP and 3S-TZP gives more than one peak. Earlier researchers like Luo et al [34] and L-Robert et al[35] had already showed that presence of more than one peak originates from the sintering within the agglomerates. Hence, from these differential shrinkage plots of 2S-1Y-TZP and 3S-TZP samples it may be concluded that these two powders contain agglomerates. This result is also in good agreement with the pore size distribution of these two samples, which had indicated the presence of agglomerates (Fig 4).

Fig 6 depicts the increase in density as a function of sintering temperature. Samples sintered at 1100^0C resulted in poor density irrespective of the compositions. For both microwave and conventional sintered samples, a considerable amount of increase in density was observed when the sintering temperature increased from 1100^0C to 1200^0C. The sintered density reached the maximum value at this 1200^0C in either of the sintering mode i.e. conventional and microwave. Sintering at 1250^0C produced density slightly less than the density obtained at 1200^0C. According to Lazar et al sulphate ions (used in the processing stages) could not be removed from the system by calcination below 1000^0C[28]. Later during sintering this sulphate ions create a de-densification effect. Nearly full density was obtained for 3Y-TZP (~99%) and 3S-TZP (~98%) samples sintered at 1200^0C in both sintering modes.

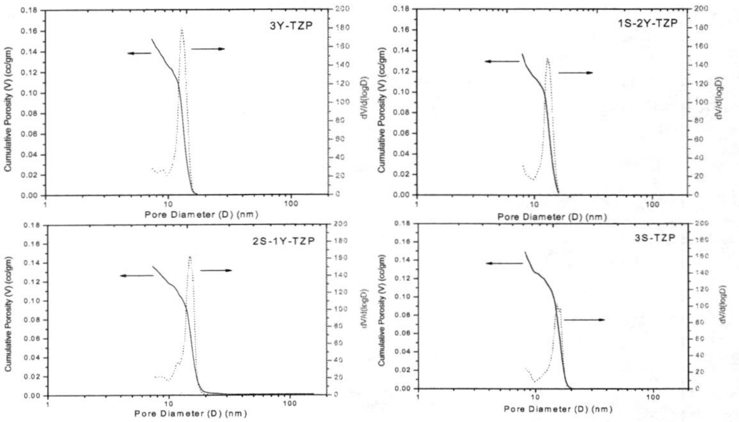

Fig 4: Pore size distribution of the iso-pressed pellets of the four different compositions.

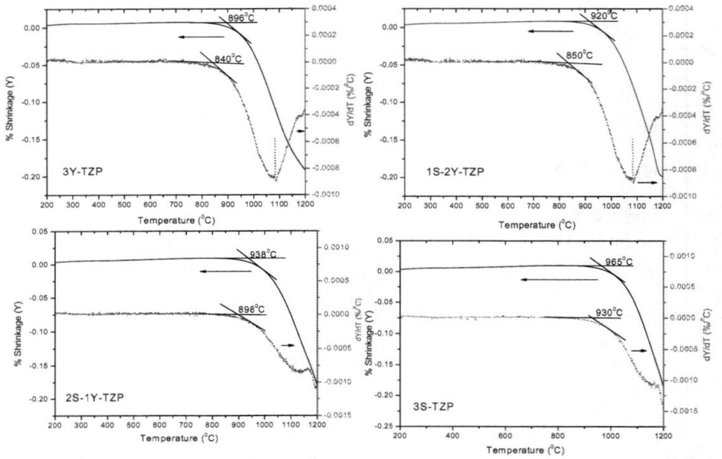

Fig 5: Shrinkage profile of the iso-pressed pellets of different compositions measured under constant rate heating experiments.

Microwave sintering being a very fast and efficient mode of energy transfer it could produce marginally denser product than the conventional sintering, while taking nearly one third of the time taken by the later one. Nearly full density was achieved by Nightingale et al[15] for 3mol% yttria stabilized zirconia by constant rate heating up to 1500^{0}C in microwave furnace. D D Upadhyaya et al also showed promising result for cubic zirconia sintered in microwave radiation[14].

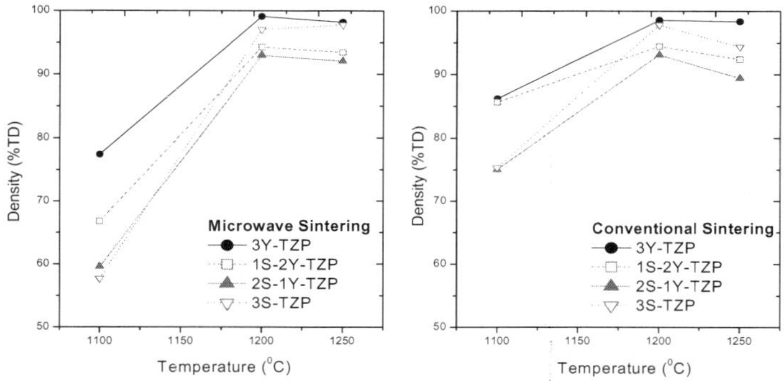

Fig 6: Change in sintered density (done in both conventional and microwave furnace as indicated in the plots) of the four different powders plotted against sintering temperature.

The complex impedance spectra for different samples at 450^0C are plotted in the Fig 7. Each spectrum consists of one very small high frequency incomplete semi-circle and another medium frequency complete semi-circular arc. The smaller semi-circle corresponds to the high frequency impedance response of the grain body/bulk material; whereas the larger complete semi circular arc is due to grain boundary contribution[36, 37]. At the end of these, a low frequency incomplete semi-circular arc was observed. This was due to electrode's impedance. This is in agreement with the result reported by Godickmeier et al[38] and Rajendran et al[39] for their 3 mol% and 2.5 mol% yttria stabilized zirconia samples respectively (sintered in conventional heating). Above 475^0C temperature, the resistance of grain/bulk was so low that it could not be detected. However, measurement of grain boundary resistance was continued up to 750^0C.

The grain body, grain boundary and total conductivity of the different samples sintered under microwave radiation were calculated from the impedance data following the approach taken by Meier[40]. The product of conductivity and absolute temperature plotted as function of inverse of absolute temperature (Arrhenius plot[37]) in Fig 8(a) & 8(b) for microwave and conventionally sintered samples respectively. In all these cases, irrespective of compositions, the plot gives perfect straight-line. As the slope of the plot indicates the activation energy of ionic conduction; it could easily be understood that there is only one mechanism of ionic transport working in the investigated temperature range.

The activation energy values for ionic conduction (calculated from Arrhenius plot) were calculated and plotted as a function of composition in Fig 9. The values of activation energy for bulk and grain boundary conduction, for 3Y-TZP were found to be 1 and 1.18 eV, which are in good agreement with the reported values[3, 6, 40]. From this plot it could also be seen that the activation energy for the grain body conduction was found to decrease from 3Y-TZP to 1S-2Y-TZP and then increase steadily with the samaria content. The trend in data was same for conventional and microwave sintered samples. Samarium ion (Sm^{+3}) is ~7% larger than the yttrium ion (Y^{+3}) and the host-dopant size mismatch is ~37% more in the case of samaria-zirconia system. Thereby it produce more lattice distortion, and to compensate that a stronger defect association is likely to takes place between Sm'_{Zr} and $V_O^{\bullet\bullet}$ [50-51], which prohibits easy transfer of oxygen ions. Thereby, the activation energy for ionic conduction increases more for the samaria rich composition. However, for the 1S-2Y-TZP sample, it may be the case where Sm'_{Zr} is preferentially segregated at the grain boundary due to larger elastic strain effect than Y'_{Zr}. Thereby the formation of weaker defect association between Y'_{Zr} and $V_O^{\bullet\bullet}$ will be more probable. Also depletion of larger Sm^{+3} in the grain body may produce less distorted lattice and provides the crucial path for oxygen ion movement and thereby helps in reducing the activation energy.

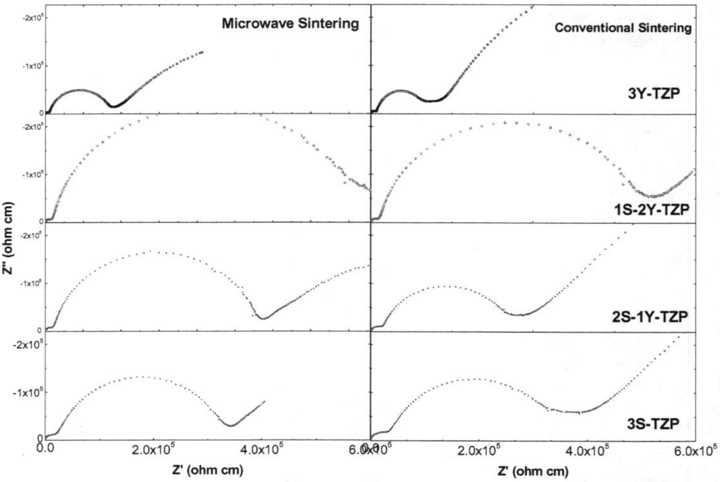

Fig 7: Geometry independent impedance spectra of the conventionally and microwave sintered samples taken at $450^{0}C$.

The activation energy of grain boundary conductivity increases steadily with samaria content. Also, the increasing trend of activation energy with dopant having larger radius matches well with the earlier reports[5, 41].

From Fig 9 it could be seen that the activation energy for grain body and grain boundary conductivity both decreases (except grain body conductivity of 3Y-TZP) when the sample is sintered in microwave field. It has already been shown by other researchers that under microwave radiation volume diffusivity of certain elements enhances[42, 43]. Janney et al have showed increased diffusivity of oxygen in sapphire while heating in microwave field[44]. In the low temperature range, the defect association in the non-dilute solution is unavoidable. In Y_2O_3 doped ZrO_2, the most commonly considered defect associate is $\left(Y'_{Zr}V_O^{\bullet\bullet}\right)^{\bullet}$. However, formation of more complex associates had also been shown by positron annihilation study[45, 46]. Enhanced diffusivity of Sm'_{Zr}, Y'_{Zr}, $O_V^{\bullet\bullet}$ or other defects in microwave field and smaller soaking time (30min) may lead to the formation of a near homogeneous distribution of dopants and defects in the grain body and grain boundary region. This eventually prohibits formation of stronger defect association and lead to cleaner microstructure which is left with sufficient oxygen vacancies. These vacancies provide an easy path for oxygen ion movement and thereby the activation energy for conduction decreases. In case of conventionally sintered samples, because of higher soaking time (180min) the defects get ample time to diffuse through the lattice and formation of stronger defect association becomes easy. This reduces the concentration of mobile oxygen ions and ultimately ionic conduction becomes more and more difficult and finally the activation energy increases.

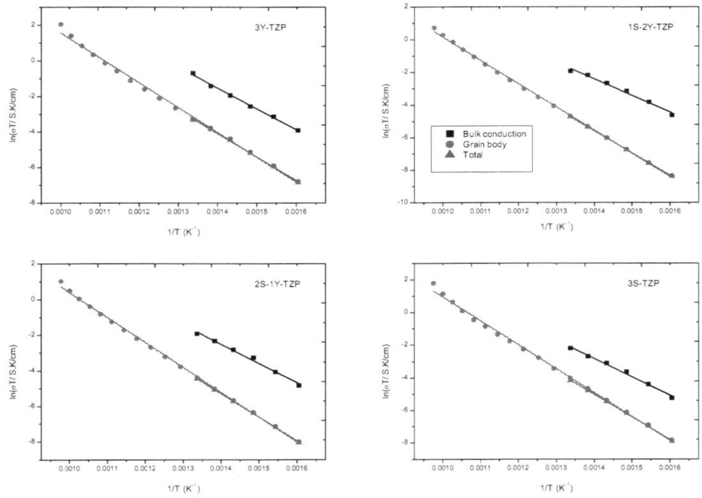

Fig 8(a): Arrhenius plot for samples of different compositions sintered under microwave radiation at 1200^0C.

CONCLUSIONS

The present investigation has led to the following conclusions:

1. Nano-crystalline yttria samaria codoped zirconia, having stabilized tetragonal phase, prepared successfully by co-precipitation technique. Nano sized powder particles (~18nm) were obtained after calcination and grinding.
2. Iso-pressed green compacts were found to possess a nearly monomodal narrow pore size distribution (7-20nm range) with nano-sized pores (average pore size ~15nm).
3. Dense samples were made by sintering at comparatively lower temperature (1200^0C) in both conventional and MW sintering. Single element doped samples reached full density (~99%) at 1200^0C, however, samples having codoped composition showed relatively less density (~95%). Sintering at higher temperature gave rise to a de-densification effect in samples other than 3Y-TZP.
4. MW sintering of the green pellets were done with lesser cycle time (~8hrs) compared to conventional sintering (~24hrs); however, samples sintered in MW field is found to produce marginally higher density than conventional sintering (for all compositions).
5. Electrical conductivity measured in the range of 350^0-475^0C was found to decrease with samaria addition. The activation energy for grain boundary and grain body ionic conduction was found to increase steadily with samaria content.
6. Samples sintered in microwave field showed marginally less activation energy for ionic conduction than the conventionally sintered samples did.

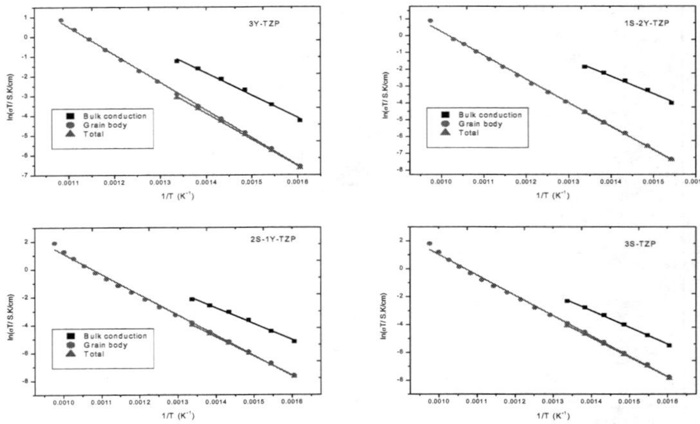

Fig 8(b): Arrhenius plot for samples of different compositions sintered in conventional furnace at 1200^{0}C

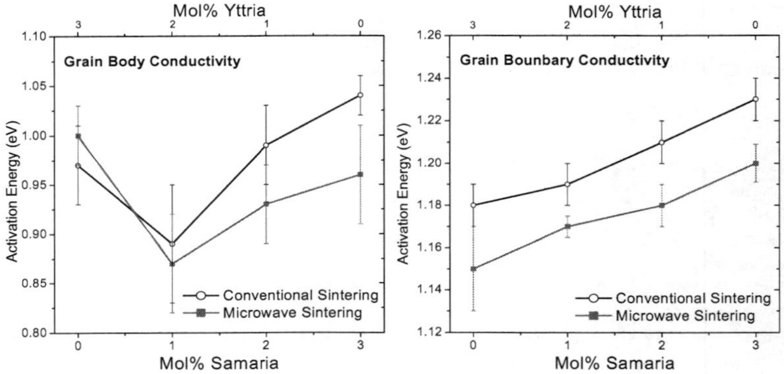

Fig 9: Comparison of activation energy obtained for the samples sintered in conventional and microwave furnace.

REFERENCES

1. N Q Minh, Ceramic Fuel Cells, Journal of the American Ceramics Society 76 (1993) 563.
2. Y Miyahara, Characterization of Sputtered Yttria Stabilised Zirconia Thin Film and its Application to a Metal -Insulator-Semiconductor Structure., Journal of Applied Physics 71 (1992) 2309.
3. R P Omorjan, V V Srdic, Electrical Conductivity of Sol-gel Derived Yttria Stabilized Zirconia, Ceramic International 27 (2001).
4. J Roller, A Hui, S Yick, X Zhang, C Deces-Petit, Y Xie, R Maric & D Ghosh, A Brief Review of the Ionic Conductivity Enhancement for Selected Oxide Electrolyte., Journal of Power Source 172 (2007) 493.

5. L Dessemond, F Boulc'h, E Djurado, Dopant Size Effect on Structural and Transport Properties of Nanometric and Single-phased TZP, Solid State Ionics 154-155 (2002).
6. A Ghosh, G K Dey & a K Suri Correlation of Electrical Conductivity with Microstructure in 3Y-TZP System: From Nano to Submicrometer Grain Size Range, Journal of the American Ceramics Society 91 (11) (2008) 3768.
7. T I Politova and J T S Irvine, Investigation of scandia-yttria-zirconia system as an electrolyte material for intermediate temperature fuel cells- influence of yytria content in system $(Y_2O_3)_x(Sc_2O_3)_{(11-x)}(ZrO2)_{89}$, Solid State Ionics 168 (2004) 153.
8. G Zhang, Y Kan, P Wang, O Van Der Biest, J Vleugels, Yb_2O_3 and Y_2O_3 Co-doped Zirconia Ceramics., Journal of the Europian Ceramic Society 26 (2006) 3607.
9. K Hebarko, W Pyda, M M Bucko and M Faryna. TZPs in the TiO_2-$Y2O_3$-ZrO_2 systems. In: S P S Badwal MJB, R H J Hannik, editor. Science and Technology of Zirconia V.
10. X Q Liu, G Z Cao, H W Brinkman, K J Vries and A J Burggraaf. Mixed conduction and oxygen permeation of ZrO_2-Tb_2O_3-Y_2O_3 solid solution. In: S P S Badwal MJB, R H J Hannik, editor. Science and Technology of Zirconia V: Technomic.
11. L Li, S G Huang, O Van Der Biest and J Vleugels, Microwave sintering of CeO2 and Y2O3 co-stabilized ZrO2 from stabiliser-coated nanopowders., Journal of the Europian Ceramic Society 27 (2007) 689.
12. W H Sutton, Microwave Processing of Ceramics, Ceramic Bulletin 68 (2) (1989) 376.
13. L Sheppard, Manufacturing Ceramics with Microwaves: The Potential for Economic Production, Am.Ceram.Soc.Bull. 67 (10) (1988) 1556.
14. A.Ghosh, D.D.Upadhyaya, K.R. Gurumurthy and Ram Prasad, Microwave Sinteringof Cubic Zirconia, Ceramic International 27 (2001) 415
15. S A Nightingale, H K Worner, and D P Dunne Microstructural Development during the Microwave Sintering of Yttria-Zirconia Ceramics, J.Am.Ceram.Soc. 80 (2) (1997) 394.
16. J Binner, B Vaihyanathan & T Corney Microwave Hybride Sintering of Nanostructured YSZ Ceramics, Advances in Science and Technology 45 (2006) 835.
17. E Ivers-Tiffee, Andre Weber, Materials and Concepts for Solid Oxide Fuel Cells (SOFCs) in Stationary and Mobile Applications, Journal of the Power Sources 127 (2004) 273.
18. H Nafe, A Kopp, W Weppner, P Kountouros and H Schubert. Ionic and Electronic Conductivity of TiO2-Y2O3 Stabilized Zirconia. In: S P S Badwal MJB, R H J Hannik, editor. Science and Technology of Zirconia V: Technomic.
19. A. Ghosh, A. K. Suri, B. T. Rao and T. R. Ramamohan, Synthesis of nanocrystalline sinteractive 3Y-TZP powder in presence of ammonium sulphate and poly ethylene glycol., Advances in Applied Ceramics In Press (2008).
20. A. Ghosh, S. Sabharwal, A. K. Suri, B. T. Rao and T. R. Rama Mohan, Synthesis and characterisation of nanocrystalline sinteractive 3Y-TZP powder, Advances in Applied Ceramics 107 (3) (2008) 170.
21. W P Hsu, B Aiken, E Matijevik, Preparation and properties of uniform mixed and coated colloidal particles: Part V Zirconium compounds, J Mat Sc 25 (1990) 1886.
22. S Pickering, D Mcgarry B Djuricic, P Glaude, P Tambuyser and K Schuster, The Properties of Zirconia Powder Produced by Homogeneous Precipitation., Ceramics International 21 (1995) 195.
23. A J G Maroto, M A Bleasa, S I Passaggio, Hydrous zirconium dioxide: interfacial properties, the formation of monodisperse spherical particles, and its crystallization at high temperature., J Mat Sc 20 (1985) 4601.
24. Abhijit Ghosh. SYNTHESIS AND SINTERING STUDIES OF YTTRIA STABILIZED ZIRCONIA NANO POWDERS. BARC, vol. PhD. Mumbai, 2006.
25. Jcpds-International Centre for Diffraction Data, JCPDS-File No. 82-1245, PCPDFWIN v. 2.00 (1998).

26. E De La Rosa-Cruz, W Cordova-Martinez, L a Diaz-Torres, P Salas, a Montoya, M Avendano, R a Rodriguez and O Barbosa-Garcia, Nanocrystalline tetragonal zirconium oxide stabilisation at low temperature by using rare earth ions: Sm^{+3} and Tb^{+3}, Optical Materials 20 (2002).
27. F Boulc'h and E Djurado, Structural changes of rare-earth-doped, nanostructured zirconia solid solution., Solid State Ionics 157 (2003) 335.
28. C A B Menezes, D R Lazar, V Ussui, a Helena, a Bressiani & J O a Paschoal, The Influence of Sulphur on the Processing of Zirconia Based Ceramics., J Eu Ceram Soc, 22 (2002) 2813.
29. J J Kingsley & K C Patil. Self Propagating Combustion Synthesis of t-ZrO_2/Al2O3 Powders. In: G L Messing, editor. Ceramic Powder Science III, Ceramic Transaction, vol. 12: The American Ceramic Soc. (Ohio), 1991. p.217.
30. M Trunec and Karel Maca, Compaction and Pressureless Sintering Zirconia Nanoparticles, Journal of the American Ceramics Society 90 (9) (2007).
31. M.A.C.G.Van De Graaf and A.J.Burggraaf., Wet chemical preparation of zirconia powders. Their microstructure and behaviour, Science and Technology of Zirconia II, Advances in Ceramics, Vol.12. (eds) N.Clausen and A.H.Heuer. The Am.Ceram.Soc. Columbus, OH, 12 (1984) 744
32. G S a M Theunissen, A J A Winnbust & a J Burggraaf Sintering Kinetics and Microstructure Development of Nano Scale Y-TZP Ceramics., Journal of the Europian Ceramic Society 11 (1993) 315.
33. J H Gao J L.Shi, Z X Lin & D S Yan, Effect of Agglomerate in ZrO2 Powder Compacts on Microstructural Development., J Mat Sc 28 (1993) 328.
34. J Luo, S Adak & R Stevens Microstructure Evolution and Grain Growth in the Sintering of 3Y-TZP Ceramics., J Mat Sc 33 (1998) 5301.
35. S W Wu & R J Brook, Kinetics and Densification of Stabilized Zirconia, Solid State Ionics 14 (1984) 123.
36. J T S Irvine, D C Sinclair & A R West, Electroceramics: Characterization by Impedance Spectroscopy, Advanced Materials 2 (3) (1990) 132.
37. J Ross Mcdonald. Impedance Spectroscopy. Chichester: Wiley, 1987.
38. B Michel, M Godickmeier, A Orliukas, P Bohac, K Sasaki, L Gauckler, H Heinrich, P Schwander, G Kostorz, H Hofmann and O Frei, Effect of intergranular glass films on the electrical conductivity of 3Y-TZP, Journal of Material Research 9 (5) (1994) 1228.
39. S Rajendran, J Drennan & S P S Badwal Effect of alumina on the grain boundary and volume resistivity of tetragonal zirconia polycrystal, J.Mater.Sci.Lett. 6 (1987) 1431.
40. X Guo & J Maier, Grain Boundary Blocking Effect in Zirconia: A Schottky Barrier Analysis, Journal of the Electrochemical Society 148 (3) (2001) E121.
41. J Kimpton, T H Randle & J Drennan Investigation of electrical conductivity as a function of dopant-ion radius in the system $Zr_{0.75}Ce_{0.08}M_{0.17}O_{1.92}$ (M=Nd, Sm, Gd, Dy, Y, Er, Yb, Sc). Solid State Ionics 149 (2002).
42. A G Whittaker, Diffusion in Microwave-Heated Ceramics, Chem. Mater. 17 (2005) 3426.
43. J D Katz and R D Blake. Microwave Enhanced Diffusion. Proceedings of the Microwave Symposium: American Ceramic Society, 1991.
44. H D Kimrey, M A Janney, W R Allen, J O Kiggans, Enhanced diffusion in sapphire during microwave heating, J Mat Sc 32 (1997) 1347.
45. C R A Catlow & B E F Fender, V Butler, The defect structure of anion deficient ZrO_2, Solid State Ionics 5 (1981) 539.
46. C. Petot, M. Filal, M. Mokchah, C. Chateau, J. L. Carpentier, Ionic conductivity of yttrium-doped zirconia and the "composite effect", Solid State Ionics 80 (1995) 27.

SPARK PLASMA SINTERING (FAST/SPS) OF NOVEL MATERIALS – TAKING THE NEXT STEP FORWARD TO INDUSTRIAL PRODUCTION

H. U. Kessel, J. Hennicke, R. Kirchner, T. Kessel*
FCT Systeme GmbH
96528 Rauenstein, Germany

ABSTRACT

State-of-the-art spark plasma sintering technology - called FAST/SPS - offers rapid consolidation of a wide range of powders from all fields of ceramic and metallic materials, enabling significantly improved or even completely novel materials. This was shown very frequently in the last decades, resulting in a huge amount of scientific papers, which report promising results at almost all times. But the main part of reported applications of FAST/SPS are still in the area of material development, although more than enough opportunities for an industrial implementation were generated. Taking the next step forward to an industrial production of novel materials by FAST/SPS is currently highly dependent on the availability of suitable equipment. After a short overview of the FAST/SPS technology this paper explains the most important industrial production requirements, which has to be fulfilled by the FAST/SPS equipment. Subsequently examples for industrial applications will be presented, concluded by a preview of future developments.

INTRODUCTION

In the last decade many scientific papers and presentations were published, dealing with the application of spark plasma sintering[1]. The rapid consolidation of powders is the most important advantage of the state-of-the-art spark plasma sintering technology FAST/SPS, offering significantly improved or even completely novel materials. A wide range of material types was investigated, metals, alloys and intermetallics as well as Borides, Carbides, Nitrides, Silicides and Oxides, plus all imaginable composites and special material systems. Most of the results are from scientific investigations, but nevertheless very promising and demonstrating interesting opportunities for a transfer to industrial production. But industrial production is really different from scientific investigation and requires suitable and individually tailored equipment, in order to realize a manufacturing process with optimum cost efficiency. The industrial success of FAST/SPS will critically depend on the availability of accordingly optimized equipment.

FAST/SPS TECHNOLOGY FOR RAPID CONSOLIDATION OF POWDERS

Spark plasma sintering as well as its state-of-the-art version, called FAST/SPS, are sintering techniques, which are deduced from conventional hot pressing[2]. Hence FAST/SPS systems likewise comprise a water-cooled vessel, a hydraulic pressing system, and a computer-aided process control by means of temperature and force measurement&control as well as a system for vacuum and atmospheric control inside the vessel respectively. The big difference compared to a conventional hot press is the absence of a heating element as well as conventional thermal insulation of the vessel. Instead a special power supply system feeds high current into the water-cooled machine rams, which that way act as electrodes simultaneously, feeding the high current directly through the pressing tool and the containing powder compact. Figure 1 presents the principle of FAST/SPS.

This special construction leads to a homogeneous volume heating of the pressing tool as well as the containing powder by means of Joule heat. Hence only small thermal gradients are generated even at high heating rates, whereas conventional sintering methods are hampered by thermal gradients, allowing only moderate heating rates and requiring long dwell time for subsequent (but mostly incomplete) homogenization. Figure 2 shows the advantage of FAST/SPS in terms of the temperature of the center T_I compared with the edge of the powder compact T_A during the sintering cycle.

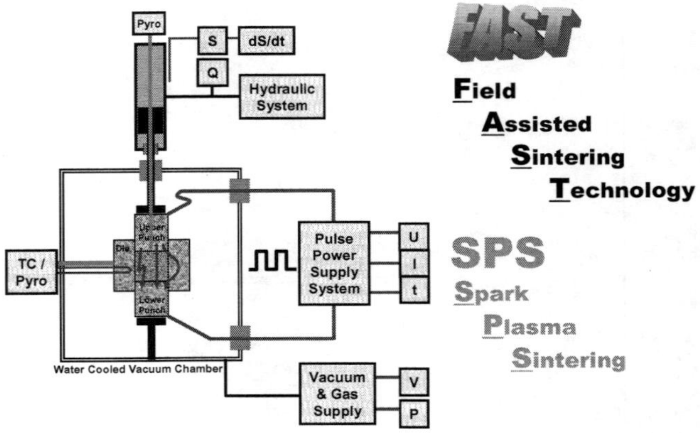

Figure 1. Schematic of FAST/SPS equipment

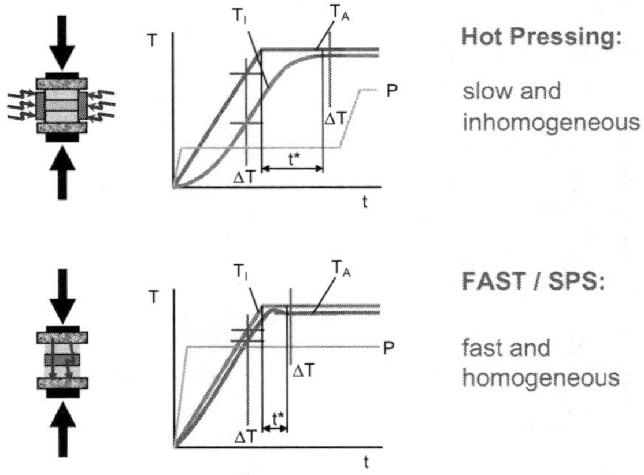

Figure 2. Comparison of FAST/SPS with conventional Hot Pressing

An additional advantage of FAST/SPS is shown in figure 3: The heating power is not only distributed over the volume of the powder compact homogeneously in a macroscopic scale, but moreover the heating power is dissipated exactly at the locations in the microscopic scale, where energy is required for the sintering process, namely at the contact points of the powder particles. This fact results in a favourable sintering behaviour with less grain growth and suppressed powder decomposition. Depending on the type of the powder, additional advantageous effects at the contact points are assumed by a couple of authors, e.g. electro-migration or microplasma-generation.

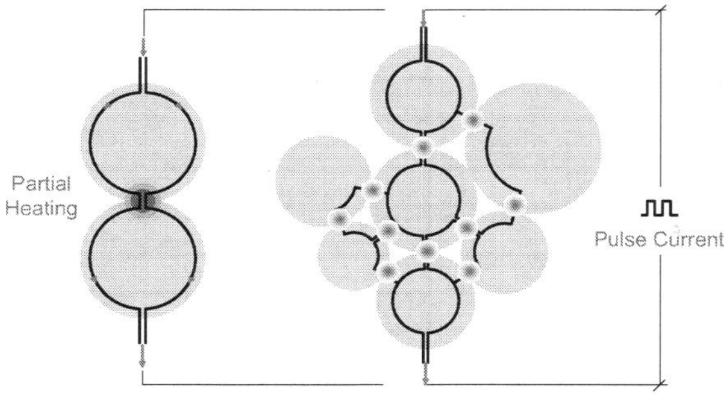

Figure 3. Energy dissipation in the microscopic scale

REQUIREMENTS FOR INDUSTRIAL APPLICATION OF FAST/SPS

As already mentioned the industrial application of the FAST/SPS sintering method for the rapid consolidation of novel materials require special features, which have to be fulfilled by the equipment and are different from the requirements of scientific work to some extent. The main issues are discussed below:

Electrical Output Power

In order to assure a cost effective production, equivalent with high throughput (amongst other things), a sufficient electrical output power must be provided by the system. It is important, that the electric losses in the system are low in order to generate high heating power at the location, where it is needed. The actual value of the required power depends on the size and material of the powder compact and the pressing tool as well as on the intended heating rates and maximum temperatures.

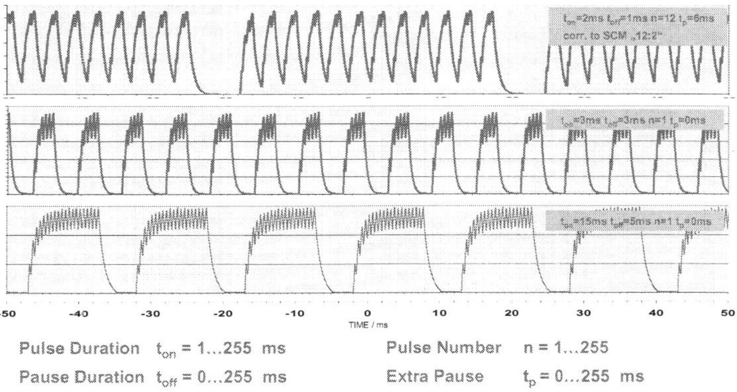

Figure 4. Flexible output of pulsed DC current

Flexible Power Supply

Depending on the type of the powder, several different sintering mechanisms are possible. Some of them can be influenced by the type of the heating current. Therefore a power supply with high flexibility are important in order to achieve optimum sintering results in terms of throughput and material quality. FAST/SPS systems are capable of generating a wide range of pulsed DC current with computer controlled, arbitrary pulse parameters to the point of pure DC current (examples see figure 4).

Precise Temperature Measurement & Control

The correct sintering temperature is the most important process parameter besides time and heating rate. Due to a special design FAST/SPS systems are measuring the temperature in the vicinity of the powder compact center, which gives a much more significant value than the measurement of the die temperature[3].

Optimized Pressing Tool Systems

As pointed out before, due to the special construction of FAST/SPS systems, the pressing tool system, consisting of the two pressing punches, the die and other auxiliary components, is the "heart" of the system, because it not only contains the powder compact but also acts as the "heater" (in interaction with the compact). Even though the temperature gradients in the system are significant lower than for conventional sintering methods, e.g. hot pressing (see figure 2), a design optimization is advantageous anyhow, especially if highest heating rates, minimized dwell time and optimum material quality are desired. A helpful tool for design optimization is the numerical simulation (finite element method "FEM") of the heating behaviour, taking into account the temperature dependent thermal and electrical properties of the applied tool materials as well as the powder compact[4]. As an example figure 5 shows the temperature distribution in a pressing tool system containing two powder compact circular disks of 200 mm diameter after heating to 1500°C within 12 min and 5 min dwell time. With the standard tool design (left) the remaining temperature difference in the compact amounts to 160 K, which can be reduced to 60 K by design optimization (right).

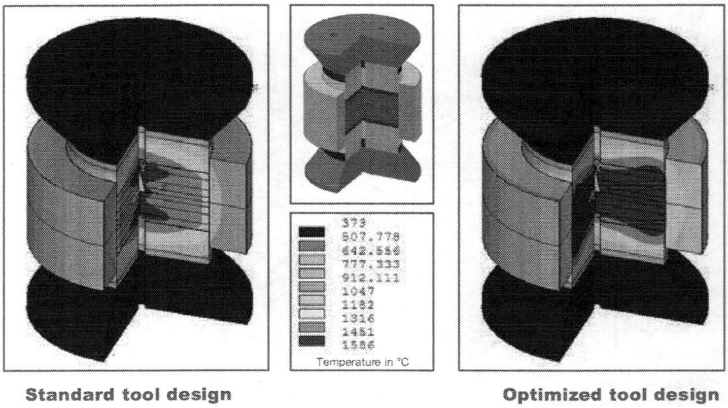

Figure 5. Temperature distributions in standard and optimized pressing tool systems

The benefit of optimized pressing tool systems is a superior material quality and homogeneity, e.g. reflected by an even distribution of high hardness values across the diameter of a 200 mm circular disk compared with the standard pressing tool situation (figure 6). Furthermore the highest heating rates made possible that way are an essential condition for the realization of nano-structured materials, which are often impossible to sinter by conventional methods due to significant longer sintering cycles.

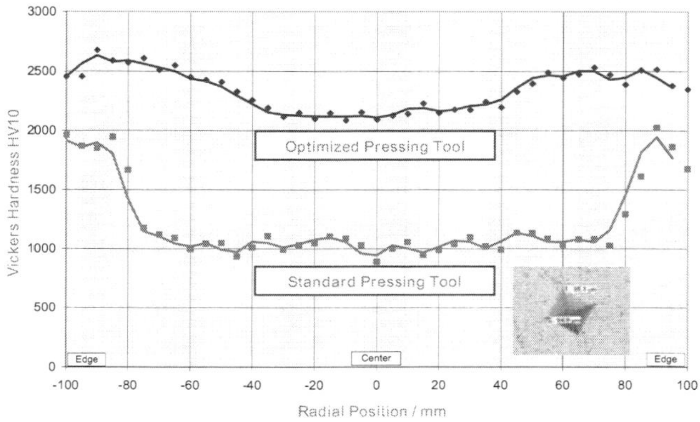

Figure 6. Hardness distributions generated by standard and optimized pressing tool systems

Hybrid Heating

The so called "Hybrid Heating" is a combination of the FAST/SPS method with one or several additional heating systems, which act usually from the outside of the pressing tool systems, as illustrated in figure 7. Thus the thermal gradients of FAST/SPS, which are directed from the interior to exterior typically can be compensated by the inversely directed gradients of the additional heating system. As shown in figure 8 the superposition of the gradients (left side) results in an extensively minimization of these gradients (right side). This allows further enhancement of the heating rates at simultaneously optimized homogeneity with all the advantages pointed out before.

A practical example showing the positive effect of hybrid heating can be found in figure 9, which compares the sintering behaviour of rectangular plates made of binderless tungsten carbide (size 150 x 175 mm). The light grey curves show the densification / densification by use of FAST/SPS, whereas the dark grey curves show the enhanced sintering behaviour by use of hybrid heating.

Fast Cooling System

The production capacity of an industrial FAST/SPS system is not only governed by the maximum possible heating rate and a minimized dwell time, but also by a fast cooling facility, which allows early discharge of the completed pressing tool. This is realized by an additional cooling chamber, separated from the actual sintering chamber by a gas/vacuum-proof, gate and equipped with special fast cooling rams (see figure 7). An automatically working handling system shifts the hot pressing tool system from the sintering chamber to the cooling chamber.

After automatic closing of the gate the sintering chamber is ready for charging the next sintering cycle during cooling of the previous pressing tool.

Automatic Operation

In order to realize a cost efficient industrial application of FAST/SPS sintering systems, the automation is an essential prerequisite. An important step is the semi-continuous operation mode mentioned above in conjunction with the fast cooling system. Due to a combination with robots and manipulators a fully automatic operation can be realized. A further example for a fully automatic production system can be found in the next paragraph.

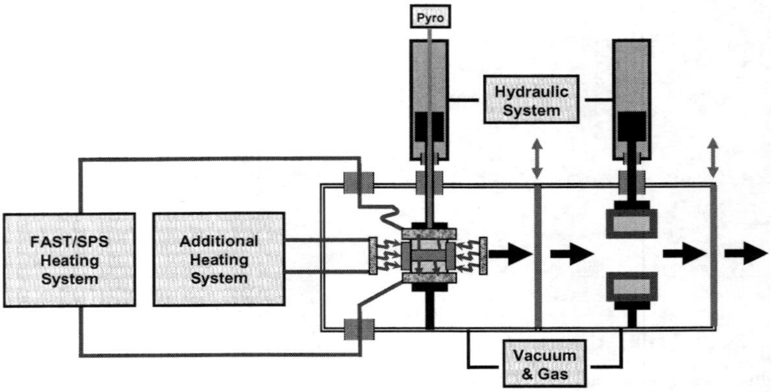

Figure 7. Schematic of an industrial high throughput sintering system with hybrid heating, separate cooling chamber and semi-continuous operation facility

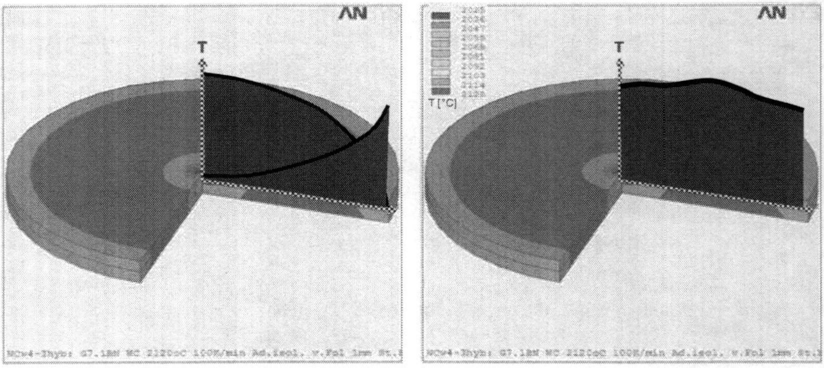

Figure 8. Compensation of residual temperature gradients by hybrid heating

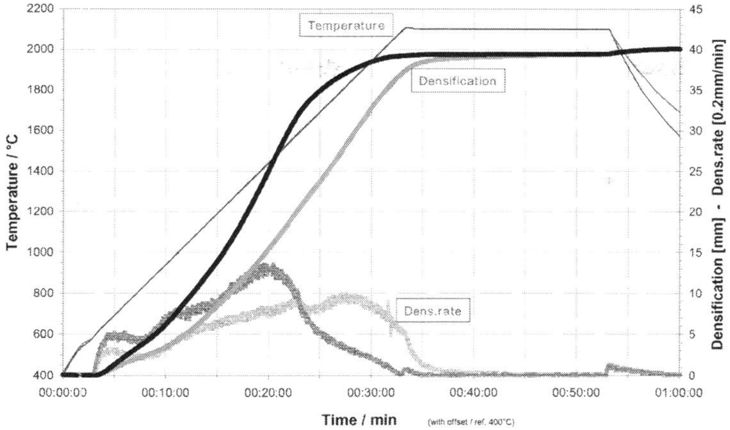

Figure 9. Comparison of sintering behaviour using FAST/SPS (light grey) and hybrid heating (dark grey)

EXAMPLES OF HIGH-THROUGHPUT FAST/SPS INDUSTRIAL APPLICATION

Hybrid - FAST/SPS Production of Large Components

One of the first industrial applications[5] of the FAST/SPS sintering technology is the manufacturing of plate-like, large-area articles (figure 10), e.g. sputtering targets for the coating of goods with a wide range of functional surface layers. The above mentioned hybrid FAST/SPS systems are ideally adapted for the high throughput manufacturing of such parts.

Large Scale Production by FAST^2

For the mass production of small parts (5 to 25 mm) a special series of FAST/SPS systems was developed called "FAST^2" (FAST square = fast FAST), which is based on state-of-the-art powder pressing technology combined with the FAST/SPS sintering method, realizing rapid and fully automatic operation including powder handling, filling of the integrated pressing tool and discharge of the readily sintered parts. The throughput of such systems can be as high as six pieces per minute, depending on the sintering characteristics of the actual material as well as the size of the parts.

Further Examples for Cost Efficient FAST/SPS

Besides all the possibilities mentioned above in order to realize an industrial production with high cost efficiency, figure 11 demonstrates one alternative method: The use of multiple pressing tools for single or manifold powder compacts can provide a number of parts by one sintering cycle, enhancing the effective throughput of the FAST/SPS system significantly.

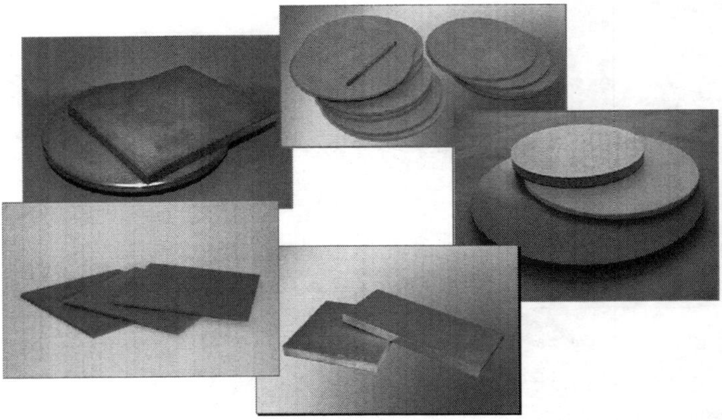

Figure 10. Plate-like, large-area articles (e.g. sputtering targets)
made of binderless tungsten carbide, noble metals, and several ceramic materials

Figure 11. Pressing tools for multiple powder compacts

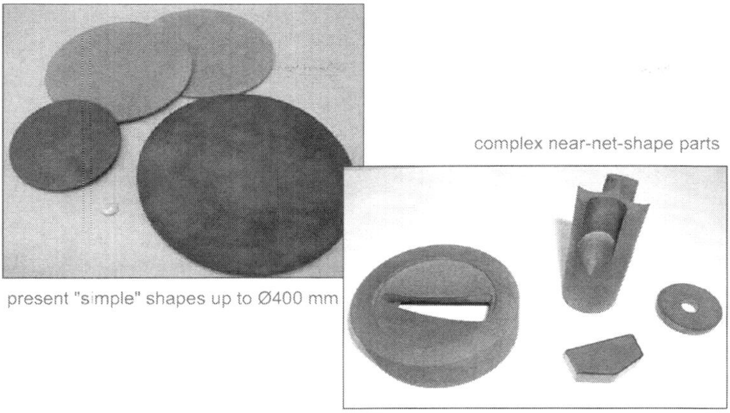

Figure 12. Currently running development from simple shapes to complex geometries

CONCLUSION

It has been shown, that the promising results of FAST/SPS reported by numerous scientific papers can be transferred to a cost efficient industrial production, if the equipment meets the respective prerequisites. These requirements have been discussed and examples for successful industrial applications have been shown. Current developments are related to the industrial production of more complex geometries (figure 12) as well as further optimization of quality and costs.

REFERENCES

[1] R. Orrù, R. Licheri, A. Locci, A. Cincotti and G. Cao, Consolidation/synthesis of materials by electric current activated/assisted sintering, *Materials Science and Engineering: R: Reports*, **63**, 127-287, (2009)
[2] J. Hennicke, H.U. Kessel, Field Assisted Sintering Technology ("FAST") for the consolidation of innovative materials, *cfi/Ber.DKG*, **81,** E14-E16, (2004)
[3] K. Vanmeensel, A. Laptev, J. Hennicke, J. Vleugels and O. Van der Biest, Modelling of the temperature distribution during field assisted sintering, *Acta Materialia*, **53**, 4379-4388, (2005)
[4] K. Vanmeensel, J. Echeberria, J.M. Sanchez, V. Martinez, L. Bourgeois, J. Hennicke, H.-U. Kessel, P. Harden, O. Van der Biest, J. Vleugels, Field Assisted Sintering of Cubic Boron Nitride Dispersed Cemented Carbide (CDCC) Composites, *EuroPM* 2006
[5] R. Kirchner, FAST (SPS) Production Systems for the Powder Metallurgy and Ceramics Industry, *Cfi/Ber. DKG*, **85**, E19, (2008)

RAPID MANUFACTURING OF CERAMIC PARTS BY SELECTIVE LASER MELTING

Wilkes, Jan; Hagedorn, Yves-Christian; Ocylok, Sörn; Meiners, Wilhelm; Wissenbach, Konrad
Fraunhofer Institute for Laser Technology
Aachen, Germany

ABSTRACT

An additive manufacturing technique for high-strength oxide ceramics is presented. Based on selective laser melting (SLM), a new approach has been developed that allows for tool-less manufacturing of complex shaped ceramic components directly from CAD data. Components are built layer by layer in a powder bed. In the new approach, a purely ceramic powder is completely melted by a laser beam. The powder does not contain any binder or glassy or metallic additives. This way, a density of the ceramic component of almost 100% is achieved directly by the SLM process, without any post-sintering, and thus without any shrinkage. The ceramic component is created by solidification from melt. In order to avoid crack formation, the ceramic is preheated to a temperature above 1600°C during the whole build-up process. Using a zirconia-alumina (ZrO_2 - Al_2O_3) material, a fine-grained microstructure can be achieved and specimens with a flexural strength of more than 500 MPa have been manufactured. Complex shaped ceramic components have been manufactured for demonstration purposes. The build-up rate is comparable to that of SLM for metallic materials. The development of the technique is not yet completed and is still ongoing. The technique has the potential to be used e.g. for manufacturing ceramic functional prototypes, all-ceramic dental restorations or complicated geometries that can not be manufactured by conventional techniques.

INTRODUCTION

Additive manufacturing techniques, also referred to as rapid manufacturing or rapid prototyping techniques, are based on adding material layerwise, computer controlled according to 3D CAD data. For polymer materials and metals, such techniques are successfully used in industry. The powder bed based laser melting techniques (Selective Laser Melting - SLM, other names used: Direct Metal Laser Sintering, Direct Laser Forming, Selective Laser Sintering) allow for direct manufacturing of fully dense metal parts. The resulting mechanical properties are similar to those achieved by conventional manufacturing techniques. Advantages of the SLM technique are for example:

- No molds or geometry-specific tools are required.
- A minimum amount of raw material is consumed. The powder that has not been melted can be reused.
- The technique allows for significantly increased design freedom. Geometric features like complex freeform shapes, undercuts, thin walls next to thick walls or internal structures are possible without additional costs.

However, for high performance ceramic materials, such as zirconia or alumina, a comparable additive manufacturing technique is not available. There are laser-based additive manufacturing approaches described in literature for less demanding materials such as silica or alumina-silica material blends. Regenfuß et al.[1,2] describe a laser sintering technique for an alumina-silica material and report a density of > 95% and a flexural strength of 120 MPa. Heinrich[3] describes a slurry based laser sintering technique, demonstrated for a porcelain and also for alumina-silica material. Approaches based on stereolithography using a resin filled with alumina powder are reported by Chartier et al.[4] and Bertsch et al.[5]. In both cases the manufactured green part needs to be post-processed, i.e. a debinding and a sintering heat treatment, associated with significant shrinkage, are required. A selective laser sintering technique based on solid state sintering of ceramic powder is described by Coulon and Aubry[6] for alumina, and by Bertrand et al.[7] for zirconia. With alumina a density of 65% and a flexural strength of 15 MPa was achieved. With zirconia a density of 56% was achieved.

The aim of the work described in the present paper is to develop an additive manufacturing technique for the high performance material system zirconia-alumina. A key application aimed at is the manufacturing of individual all-ceramic dental restorations. This application requires a flexural strength of at least 500 MPa and a high accuracy. Other applications are the manufacturing of functional prototypes, single units or small batches of a geometry in a short time, as well as the manufacturing of very complex shaped ceramic objects that can not be manufactured by conventional manufacturing techniques.

To achieve the demanding goals of high strength and high accuracy at the same time, a new approach is used. This new approach does not employ any sintering mechanisms. Instead, a purely ceramic powder is completely melted by a laser beam. With this approach, full density can directly be achieved, without any post-sintering and associated shrinkage or deformation.

SELECTIVE LASER MELTING (SLM) USING CERAMIC MATERIALS

Basic Principle

The SLM technique is a powder bed based additive manufacturing technique. Parts are built layer by layer according to CAD data. The SLM manufacturing process comprises the following steps:

1. The CAD data are pre-processed (slicing, generation of scan vector data for each layer).
2. A thin layer of ceramic powder material is deposited on the build platform.
3. The appropriate areas of the powder layer are selectively heated and melted by means of a focused laser beam.
4. The build platform is lowered by a distance corresponding to the layer thickness.
5. Steps 2 to 4 are repeated until all layers have been built up. The surrounding loose powder can then be removed and the part can be separated from the build platform.

The melting process is depicted schematically in Figure 1. Both the powder and part of the solid ceramic below the powder are melted and thus the new material is seamlessly added.
The first layer is deposited on top of a ceramic substrate. If necessary, supporting structures are built up below overhanging sections of the part (see Figure 15).

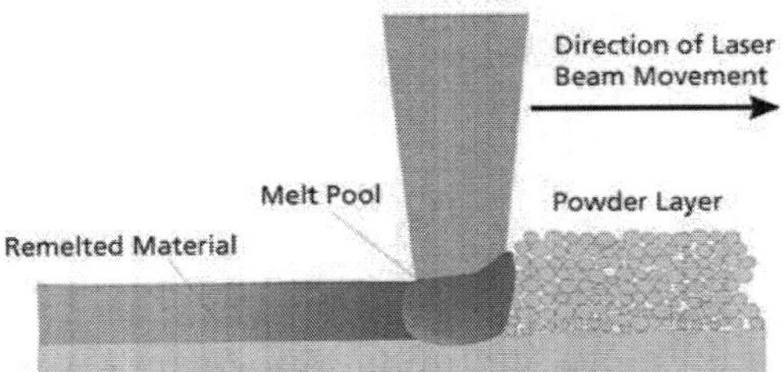

Figure 1: Schematic illustration of the melting process.

Materials

First experiments were done using pure alumina (Al_2O_3) or pure zirconia (ZrO_2) stabilized with yttria (Y_2O_3). Using these materials resulted in a coarse microstructure after solidification of the melt and it was not possible to avoid formation of cracks during the build up process, as described below in the section "Results".

In order to improve the microstructure and reduce the stresses, a mixture of alumina and zirconia has subsequently been used for the experiments. This material system has several advantages compared to the pure materials:

- The solidification temperature of the melt is lowered to 1860°C as shown in the phase diagram in Figure 2, compared to the solidification temperature of zirconia of 2710°C. This reduces temperature differences and therefore mechanical stresses in the ceramic.
- Since the two components form a eutectic system, a composite material with a fine-grained two-phase microstructure results when the melt solidifies.
- In zirconia-alumina ceramics, superplasticity has been observed at elevated temperatures (see e.g. Nieh et al.[9]). Such plasticity would help reduce stresses during the SLM build up process.
- Zirconia-alumina composite ceramics have the potential to reach significantly better mechanical properties than alumina or zirconia.

Various mixture ratios were tested, including 90 wt.% zirconia, 80 wt.% zirconia, 60 wt.% zirconia, 41.5 wt.% zirconia (eutectic mixture ratio), 20 wt.% zirconia, the remainder being alumina. The zirconia component included 6 wt.% yttria in most cases, except for the experiments without preheating (see below), where zirconia without any stabilisation was used.

Different types of zirconia and alumina powders have been used, with different particle shapes and densities. The best results were achieved with spherical fully dense powder particles with a mean diameter of 50 μm as shown in Figure 3. Both the zirconia and the alumina powder were delivered by Innalox bv, Netherlands. The powder does not contain any binder nor any other additives.

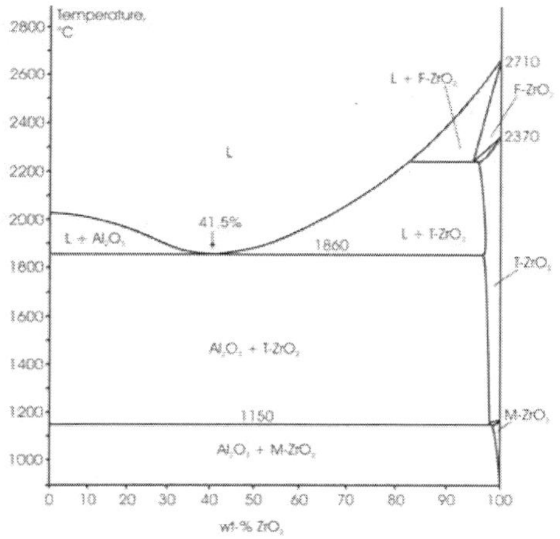

Figure 2: Phase diagram of the system Al_2O_3 - ZrO_2 (according to Lakiza and Lopato[8]). Nomenclature: L: Liquid; F-ZrO_2: Flourite-like (cubic) ZrO_2; T-ZrO_2: Tetragonal ZrO_2; M-ZrO_2: Monoclinic ZrO_2

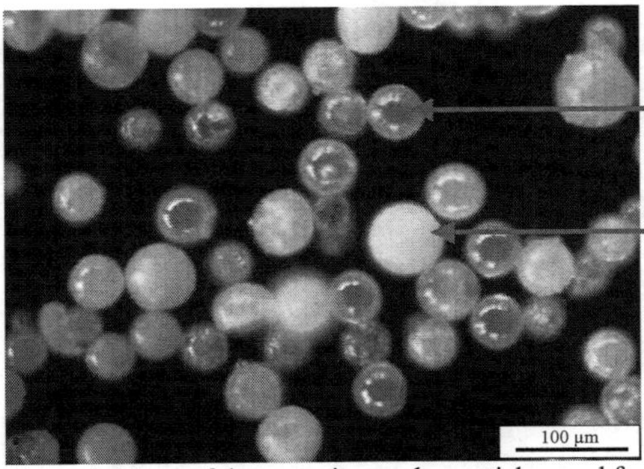

Figure 3: Photo of the ceramic powder particles used for SLM (optical microscope).

Experimental Setup

The experimental setup is shown schematically in Figure 4. Two different laser beam sources are used. An Nd:YAG-laser with a maximum output power of 150 W is used for selective melting of the ceramic powder. The beam is deflected by a computer controlled 2-axis galvanometer-scanner and focused to a spot size of approx. 200 µm. A second laser beam source, a CO_2-laser (output power approx. 1000 W), is used for preheating the ceramic. The beam is formed by homogenisation optics such that a rectangular area (size 20 mm x 30 mm) in the powder bed is irradiated with a fairly constant intensity. Thermal insulation material around the build platform ensures that the temperature distribution in the emerging ceramic part is homogeneous, even though the part is heated only from the top-surface. For measuring the preheating temperature on the surface of the ceramic, a two-colour pyrometer is employed. The build platform is positioned by a linear stage with a servomotor and an absolute optical linear encoder is used for feedback. For powder deposition, a high precision carbon fiber brush is used. The system is open to the surrounding atmosphere so that the melting process is performed in air. The experimental SLM machine is controlled by a PC, such that the complete build-up process runs automatically.

The powder bed preheated to a temperature of approx. 1700°C is shown in Figure 5. An advantage of preheating by a laser beam (compared to a furnace setup) is the very local heat input. In this case the surrounding mechanics and components stay at a low temperature and can be made out of standard materials.

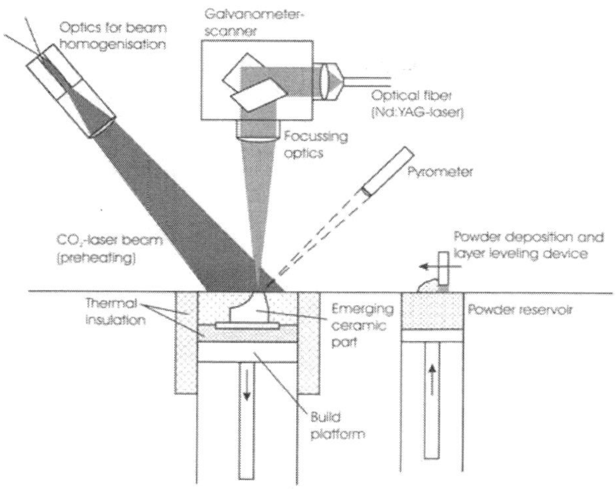

Figure 4: Schematic illustration of the experimental setup used for SLM.

Figure 5: Photo of the powder bed preheated locally to a temperature of 1700°C.

RESULTS

Avoidance of Crack Formation by Preheating

Two different alternatives have been investigated, SLM without preheating, and SLM with high-temperature preheating using the additional CO_2-laser as described above in section "Experimental Setup".

The experiments without preheating showed that it is possible to manufacture ceramic components with high density when using material compositions with 60 wt.% to 90% wt.% zirconia (unstabilized), remainder alumina. However, due to the high temperature difference between melting temperature, reached locally at the point of impact of the laser beam, and the low temperature of the surrounding solid ceramic material, extensive stresses are generated and therefore cracking occurs. Figure 6 (left) shows a cross section through a specimen manufactured by SLM without preheating.

The arrows point at cracks. The material is composed of 80 wt.% zirconia and 20 wt.% alumina. When adding e.g. yttria to stabilize the zirconia, the cracks become larger and it is not possible to build up parts higher than a few layers. A possible explanation for this phenomenon is that the volume expansion associated with the conversion into monoclinic zirconia reduces stresses and therefore improves processability by SLM (in the case without preheating). Addition of yttria prohibits the conversion into the monoclinic phase.

In order to avoid the crack formation, a successful strategy is to elevate the temperature of the ceramic material before and during the selective melting process. That means the ceramic substrate, the layers that have already been built and the powder surrounding the ceramic (the powder-bed) are constantly kept at a high temperature until the whole SLM build-up process is completed. This is called "preheating" in this paper. The closer the preheating temperature is to the melting temperature of the ceramic, the lower are the stresses in the emerging ceramic part.

Laser melting experiments at different preheating temperatures showed that a preheating temperature of at least approx. 1600°C is necessary in order to completely avoid crack formation. Figure 6 (right) shows a cross section of a crack-free specimen built by SLM at a preheating temperature of 1715°C. The following parameters were used: Laser power: 55 W; Laser beam diameter: 200 µm; Scanning speed: 200 mm/s; Layer thickness: 50 µm. The material has a composition of 41.5 wt.% zirconia (6 wt.% yttria) and 58.5 wt.% alumina. When using preheating, a (partial) stabilisation of the zirconia, e.g. by yttria, is necessary in order to manufacture crack-free parts by SLM.

Out of pure alumina or pure zirconia or yttria stabilized zirconia, it was not possible to manufacture crack-free specimens, even when using a preheating temperature of up to 1850°C.

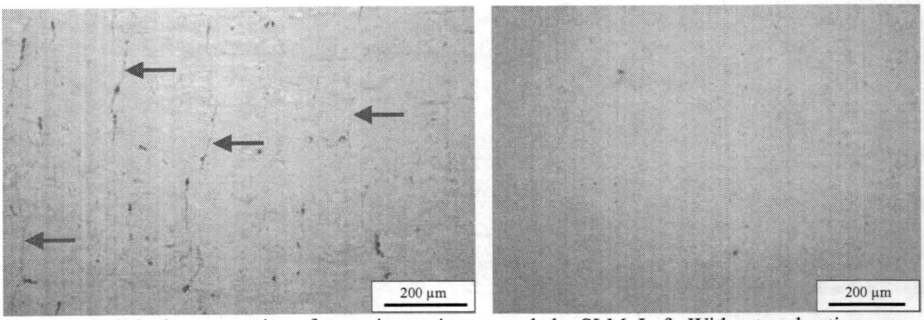

Figure 6: Polished cross section of ceramic specimens made by SLM. Left: Without preheating, arrows point at cracks. Material: 80 wt.% zirconia / 20 wt.% alumina. Right: Built up at a preheating temperature of 1715°C, no cracks in the ceramic. Material: 41.5 wt.% zirconia (6 wt.% yttria) / 58.5 wt.% alumina.

This demonstrates that crack-free ceramic parts can be manufactured by SLM when using high-temperature preheating and a material mixture of yttria stabilized zirconia and alumina.

One problem to be solved is related to the powder deposition for each layer. In the current experimental setup, cold powder (room temperature) is moved into the preheated zone and deposited on top of the 1600°C hot solid ceramic material. It is assumed that this thermoshock can again cause cracks. Especially when building up parts of larger volume (larger than the specimens shown and described below), such cracks were observed. For this reason, the experimental setup is currently being modified and a preheating mechanism for the powder is added.

Microstructure

By completely melting the ceramic powder with the laser beam, a density of almost 100% is achieved directly, without any post-processing. The specimen shown in Figure 7 contains almost no porosity. The cross section is crack-free. The following parameters were used for this specimen: Laser power: 60 W; Laser beam diameter: 200 μm; Scanning speed: 200 mm/s; Layer thickness: 50 μm. When melting pure alumina or pure yttria stabilized zirconia, a coarse microstructure with a grain size of up to 100 μm results after solidification as shown in Figure 8. With the zirconia-alumina composite material, a significantly finer microstructure can be achieved. Depending on the exact material composition and the processing parameters, different types of microstructures were observed. With a zirconia content slightly higher than the eutectic point, dendritic zirconia structures embedded in a eutectic matrix were found (Figure 9). In this case the higher zirconia content resulted from an evaporation of alumina during the laser melting. With a composition very close to the eutectic point, a microstructure as shown in Figure 10 was observed. In both photos the light phase is zirconia, while the dark phase is alumina. The structure in Figure 10 may be zirconia fibers in an alumina matrix. The zirconia precipitations have a diameter of down to approx. 200 nm. Such structures alternate with slightly coarser disordered structures and with lamellar structures.

Figure 7: Polished cross section of a ceramic specimen made by SLM. Preheating temperature: 1700°C. Material: 39 wt.% zirconia (5 wt.% yttria) / 61 wt.% alumina. The left and the right border of the specimen have a high roughness or irregular shape. See section "Components for demonstration purposes" for an explanation of this effect.

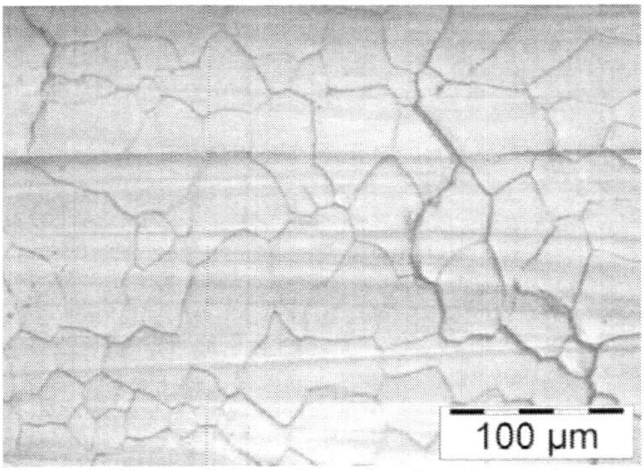

Figure 8: Photo of the surface of a zirconia specimen remelted by a laser beam (optical microscope). Preheating temperature: 1850°C.

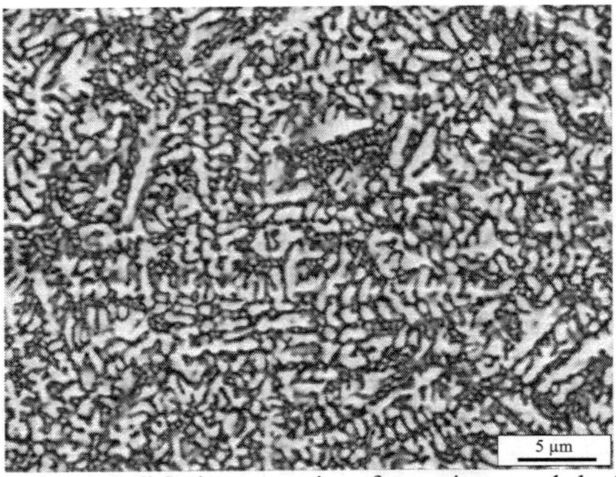

Figure 9: Polished cross section of a specimen made by SLM (Photo: SEM). Material: 41.5 wt.% zirconia (6 wt.% yttria) / 58.5 wt.% alumina. Preheating temperature: 1675°C. Light phase: Zirconia. Dark phase: Alumina.

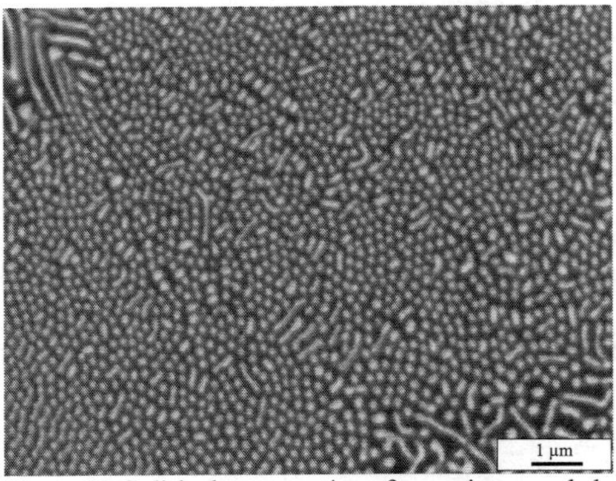

Figure 10: Polished cross section of a specimen made by SLM (Photo: SEM). Material: 35 wt.% zirconia (6 wt.% yttria) / 65 wt.% alumina. Preheating temperature: 1710°C. Light phase: Zirconia. Dark phase: Alumina.

Mechanical Properties

The flexural strength of specimens manufactured by SLM at high preheating temperatures was tested using the biaxial flexure test "ball on three balls". Disks of approx. 14 mm diameter and a thickness between 1.5 mm and 2.5 mm (30 or 50 layers) were manufactured (Figure 11). Apart from cutting the disks off the substrate using a diamond wire saw, they were not post-processed. For testing, the disks were supported on three balls and loaded by a central single ball until fracture. The specimens were oriented such that the laser melted surface (the last layer) was loaded with tensile stress. The testing was done by the company Bego Bremer Goldschlägerei Wilh. Herbst GmbH & Co. KG. Due to the time and effort required for manufacturing a specimen with the experimental system, only one

specimen was manufactured for each set of processing parameters. Therefore the measured values have to be interpreted with some caution and the effect of each parameter can not be determined with certainty. In total 14 specimens were tested. The preheating temperature was varied between 1570°C and 1800°C, the laser power between 48 W and 60 W and also the material mixture ratio and other parameters were varied. The scanning speed was kept constant at 200 mm/s, the laser beam diameter at 200 μm and the layer thickness at 50 μm.
The measured values are in a range between 173.8 MPa and 538.1 MPa. Several specimens had a flexural strength of more than 500 MPa. Values above 500 MPa have been measured for both the material composition 80 wt.% zirconia (6 wt.% yttria) / 20 wt.% alumina and the material composition 41.5 wt.% zirconia (6 wt.% yttria) / 58.5 wt.% alumina. For all detailed results see Wilkes[10].

Figure 11: Round specimen made by SLM for mechanical testing, still attached to the cuboid substrate.

Apart from the flexural strength, also the fracture toughness was measured by an indentation method. Values for the fracture toughness K_{IC} of 6.2 MPa $m^{1/2}$ for the 80 wt.% zirconia material and between 8.1 MPa $m^{1/2}$ and 9.1 MPa $m^{1/2}$ for the 41.5 wt.% zirconia material were measured.

Phase Analysis

The crystalline phases present in specimens made by SLM were determined by X-ray diffraction (XRD). The analysis was done by TNO Science and Industry, Netherlands, using a Philips X'Pert MPD X-ray diffractometer. The result of the XRD analysis of a specimen made out of 41.5 wt.% zirconia (6 wt.% yttria) and 58.5 wt.% alumina at a preheating temperature of 1700°C is shown in Figure 12. The phases detected are tetragonal zirconia and alpha-alumina. These are the desired phases that are a precondition for good mechanical properties.

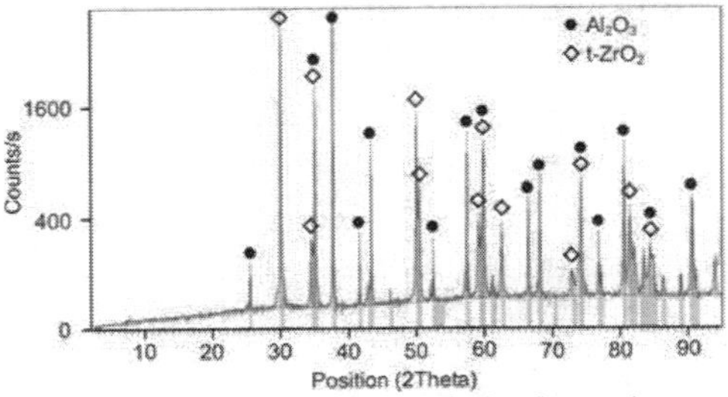

Figure 12: Result of the XRD analysis of a specimen made by SLM.

Components for Demonstration Purposes

Complex shaped ceramic parts were built to demonstrate the capabilities of the technique. The parts shown in Figure 13 and Figure 14 were built without preheating. Therefore these parts contain microcracks and have a low strength. The build up rate is approx. 1000 mm³/h to 5000 mm³/h, depending on part size and parameters used.

The part shown in Figure 15 was built at a preheating temperature of 1670°C. Even at such a high preheating temperature, it is still possible to manufacture complex shaped freeform parts. The part shown is still attached to the substrate by the supporting struts. These can be cut e.g. using a diamond-coated cutting disk. In principle, a similar build up rate can be achieved as stated above for SLM without preheating.

Figure 13: Part made by SLM without preheating. Geometry: Turbine of a turbocharger. Material: 80 wt.% zirconia / 20 wt.% alumina.

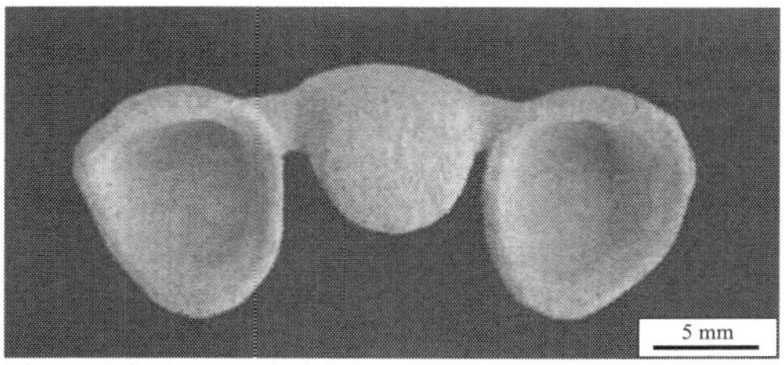

Figure 14: Part made by SLM without preheating. Geometry: Framework for an all-ceramic dental restoration. Material: 80 wt.% zirconia / 20 wt.% alumina.

Figure 15: Part made by SLM at a preheating temperature of 1670°C. Geometry: Framework for an all-ceramic dental restoration. Material: 41.5 wt.% zirconia (6 wt.% yttria) / 58.5 wt.% alumina.

When comparing the two parts in Figure 14 and Figure 15, a significantly coarser surface structure can be observed on the part made at high preheating temperature. This phenomenon was investigated using high-speed videography. The surface roughness results from melt flowing out across the boundaries of the part and is therefore strongly influenced by the size of the melt pool (see Wilkes[10] for more details). The improvement of the surface quality by reducing the melt pool size will be subject of further investigations at Fraunhofer ILT.

CONCLUSION

A new approach for the additive manufacturing of freeform parts out of zirconia-alumina composite ceramic is investigated. This new approach is based on the selective laser melting technique. A mixture of zirconia and alumina powder is completely melted by a focused laser beam. The ceramic part is created and formed by solidification from melt, without any sintering processes. When working without preheating, extensive cracking occurs in the ceramic due to high temperature gradients and stresses during the laser melting process. This crack formation can be avoided by preheating the ceramic to a temperature of at least 1600°C and keeping this temperature throughout the whole build up process. Especially when using a eutectic mixture ratio of zirconia and alumina, a fine grained microstructure (diameter of the zirconia precipitations < 1 μm) can be achieved. Manufactured parts have a density of almost 100% and specimens with a flexural strength of more than 500 MPa have been made. Remaining challenges are the formation of cracks when depositing the cold powder on top of the preheated ceramic, a problem that occurs when building parts of larger volume, and secondly the surface roughness. To solve the first problem, a preheating mechanism for the powder is currently being installed on the experimental SLM system. The surface quality can be improved by reducing the size of the melt pool. Subsequent to the currently ongoing process development, the Fraunhofer Institute for Laser Technology plans to work on a transfer of this technology into the industry. The selective laser melting technique presented in this paper can possibly also be used for other oxide ceramic materials, preferably for eutectic systems with two, three or more components, so that a fine-grained microstructure can be achieved.

ACKNOWLEDGEMENT

The research work was partially funded by the European Commission within the 6th framework programme (project "Custom-IMD", contract no. 026599/2).

REFERENCES

[1]Regenfuß, P.; Streek, A.; Ullmann, F.; Kühn, C.; Hartwig, L.; Horn, M.; Ebert, R.; Exner, H., Laser Micro Sintering of Ceramic Materials, Part 1, Interceram Vol. 56 (2007) [6], p. 420-422

[2]Regenfuß, P.; Streek, A.; Ullmann, F.; Kühn, C.; Hartwig, L.; Horn, M.; Ebert, R.; Exner, H., Laser Micro Sintering of Ceramic Materials, Part 2, Interceram Vol. 57 (2008) [1], p. 6-9

[3]Heinrich, J.G., LSD-basiertes selektives Lasersintern, cfi/Ber. DKG 86 (2009) No. 10, p. D26-D27

[4]Chartier, T.; Chaput, C.; Doreau, F.; Loiseau, M., Stereolithography of structural complex ceramic parts, Journal of Material Science 37 (2002), p. 3141-3147

[5]Bertsch, A.; Jiguet, S.; Renaud, P., Microfabrication of ceramic components by microstereolithography, Journal of Micromechanics and Microengineering 14 (2004), p. 197-203

[6]Coulon, N.; Aubry, P., Results on laser sintering system for direct manufacturing of metallic or ceramic components, Proceedings of 23rd International Congress on Applications of Lasers and Electro-Optics 2004

[7]Bertrand, Ph.; Bayle, F.; Combe, C.; Goeuriot, P.; Smurov, I., Ceramic components manufacturing by selective laser sintering, Applied Surface Science 254 (2007), p. 989 – 992

[8]Lakiza, S.M.; Lopato, L.M., Stable and Metastable Phase Relations in the System Alumina-Zirconia-Yttria, J. Am. Ceram. Soc. 80 [4] (1997), p. 893-902

[9]Nieh, T.G.; Wadsworth, J.; Sherby, O.D., Superplasticity in metals and ceramics, Cambridge University Press, Cambridge, UK, 1997

[10]Wilkes, J., Selektives Laserschmelzen zur generativen Herstellung von Bauteilen aus hochfester Oxidkeramik, doctoral thesis, RWTH Aachen, 2009, URL: http://darwin.bth.rwth-aachen.de/opus3/volltexte/2009/2994/

Joining and Machining

ACTIVE METAL BRAZING AND CHARACTERIZATION OF BRAZED JOINTS BETWEEN SILICON CARBIDE AND METALLIC SYSTEMS

Bryan P. Coddington[1], Rajiv Asthana[1], Michael C. Halbig[2] and Mrityunjay Singh[3]

[1]Engineering & Technology Department, University of Wisconsin-Stout, Menomonie, WI 54751 USA
[2]U.S. Army Research Laboratory, Vehicle Technology Directorate, NASA Glenn Research Center, Cleveland, OH 44135 USA
[3]Ohio Aerospace Institute, NASA Glenn Research Center, Cleveland, OH 44135 USA

ABSTRACT

The joining response of silicon carbide (CVD and sintered) to titanium and Kovar was investigated. The joint microstructure, composition, and microhardness were evaluated by optical microscopy (OM) and scanning electron microscopy (SEM), energy dispersive spectroscopy (EDS), and Knoop microhardness testing, respectively. Under optimal joining conditions, Ti-containing active braze alloys (Incusil-ABA, Cusil-ABA, and Ticusil) formed sound joints that revealed chemical interaction and interdiffusion, and led to good wetting and surface coverage. However, residual stresses due to expansion mismatch between SiC and metallic substrates during joint fabrication led to hairline cracks in SiC even in well-bonded joints. Use of ductile metallic interlayers to reduce the stresses was investigated and the initial results looked promising but will require further work.

INTRODUCTION

Recently, there has been a great deal of interest in the development and testing of advanced materials for applications requiring high wear resistance and heat resistance. Silicon carbide has high strength, excellent thermal and corrosion resistance, and high resistance to abrasive wear. Silicon carbide ceramics are used in engine valves, nozzles, and seals, in furnace linings and exhaust manifolds of industrial furnaces, and are being considered for nuclear reactors. Because of its brittleness, SiC has poor machining characteristics, and requires robust techniques to create complex parts by joining simpler units. Brazing and diffusion bonding are the preferred methods to join SiC.

The self-joining behavior of silicon carbide ceramics has been reported in earlier studies [1-7]; however, research studies on joining of SiC to high-temperature alloys are scant. This paper focuses on the joining response of bulk silicon carbide ceramics to a controlled expansion alloy, Kovar[1], and titanium. Kovar (density: 8,360 $kg.m^{-3}$) and titanium (density: 4,510 $kg.m^{-3}$) have a moderate level of mismatch between their coefficients of thermal expansion (CTE) and that of SiC. This can reduce residual stresses introduced from joining. The CTE of SiC, Kovar and Ti are $4.1x10^{-6}$ K^{-1}, $5.1x10^{-6}$ K^{-1} and $8.6x10^{-6}$ K^{-1}, respectively.

The objective of the study was to demonstrate the joining of silicon carbide to titanium and Kovar, and investigate the integrity, microstructure, chemical interaction, and microhardness of the joint with the aid of optical microscopy (OM), scanning electron microscopy (SEM), energy dispersive spectroscopy (EDS), and Knoop microhardness testing.

EXPERIMENTAL PROCEDURE

Three active braze alloys (ABAs), Incusil-ABA, Cusil-ABA, and Ticusil, each containing different percentages of Ti as an active metal, were selected for brazing runs. The chemical composition, liquidus temperature, and selected physical and mechanical properties of these braze

---

[1] *Kovar is a nickel-cobalt ferrous alloy with thermal expansion characteristics similar to borosilicate glass. Kovar is a trademark of Carpenter Technology Corporation. The nominal composition (in wt %) of Kovar is53Fe-29Ni-17Co(<1.0% C, Si, Mn).*

alloys are listed in Table I. These brazes were obtained from Morgan Advanced Ceramics, Hayward, CA, in either foil or powder form. Two types of silicon carbide substrates were used for brazing: chemical vapor deposited (CVD) silicon carbide, and sintered silicon carbide (called Hexoloy SiC). Unlike CVD SiC, Hexoloy SiC (a product of St. Gobain) is a sintered silicon carbide (α~phase) material. The material is designed to have a homogeneous composition and is produced via pressure-less sintering of fine (submicron) silicon carbide powder.

Table I. Selected Properties of Active Braze Alloys and Substrate Materials used for Brazing Studies.

Braze	**Braze Composition (wt%)**	**T_L, C**	**E, GPa**	**YS, MPa**	**UTS, MPa**	**CTE×10^6, K^{-1}**	**% El**	**K, W/mK**
Incusil-ABA®	59Ag-27.3Cu-2.5In-1.25Ti	715	76	338	455	18.2	21	70
Cusil-ABA®	63Ag-35.3Cu-1.75Ti	815	83	271	346	18.5	42	180
Ticusil®	68.8Ag-6.7Cu-4.5Ti	900	85	292	339	18.5	28	219
Substrate	**Substrate Composition**							
Hexoloy SiC	100% α-phase	-	466	-	-	4.1*	-	-
CVD SiC^	100% β-phase	-	461	-	-	4.0	-	-
Kovar	53Fe-29Ni-17Co (<1.0% C, Si, Mn)		137	344	516	5.1	30	17.3
Ti	Commercial purity		105	480	550	8.6	15	17.2

E: Young's modulus, YS: yield strength, TS: tensile strength, CTE: coefficient of thermal expansion, %El: percent elongation, K: thermal conductivity. *4.1 (sintered), ^Rohm and Haas, ® Morgan Advanced Ceramics, Hayward, CA.

Silicon carbide and metal substrates were sliced into 2.54 cm x 1.25 cm x 0.25 cm pieces using either a diamond saw (for SiC) or a ceramic blade (for Ti and Kovar). The braze foils (~50 μm thick) were cut into 2.54 cm x 1.25 cm pieces. All materials were ultrasonically cleaned in acetone for 15 min. prior to joining. The braze foils were sandwiched between the metal and the SiC, and a normal load of 0.30-0.40 N was applied to the assembly. Braze foils are easier to use than braze powders, especially for small gaps in which powder paste application could be difficult. Additionally, the residual organic solvents in powder pastes could cause soot formation and furnace fouling. However, as braze powders are used in industrial work, a few braze runs were made using braze powders in place of foil in order to examine the differences, if any, when using foils and powders. For this purpose, braze powders were mixed with glycerin to make a thick paste with dough-like consistency, and the paste was applied using a spatula to the surfaces to be joined. The assembly was heated in an atmosphere-controlled furnace to the brazing temperature (typically 15-20 °C above the braze liquidus) under vacuum (10^{-6} - 10^{-5} torr), held for 5 min. at the brazing temperature, and slowly cooled to room temperature. A total of 35 separate joints were created.

The joined samples were visually examined, then mounted in epoxy, ground and polished on a Buehler automatic polishing machine using the standard procedure, and examined using optical microscopy (Olympus DP 71 system) and scanning electron microscopy (SEM) (JEOL, JSM-840A) coupled with energy dispersive x-ray spectroscopy (EDS). The elemental composition across joints was assessed with the EDS and presented as relative atomic percentage among the alloying elements at point markers on SEM images. The polished joints were subjected to microhardness testing with a Knoop micro-indenter on a Struers Duramin-A300 machine under a load of 200 g and loading time of 10 s to develop hardness profiles across joints. Multiple hardness scans were performed across joints to check the reproducibility of the measurements.

RESULTS AND DISCUSSION

All SiC substrates self-joined with the three brazes revealed excellent, crack-free joints. Similar baseline data on self-joined Ti and Kovar substrates confirmed the bonding capabilities of all braze alloys. Figure 1 shows the joint microstructure in self-joined CVD SiC (Fig. 1a & b) and self-joined Kovar (Fig. 1c) made using Ticusil braze. The braze interlayers in self-joined SiC were well-defined and very consistent. The self-joined Kovar revealed some voids, possibly due to solidification shrinkage. In self-joined Ti, the braze layer appeared to reconstitute itself with the Ti substrates and obliterated the braze layer boundary which was no longer visible, thus yielding a homogenous material (some shrinkage voids formed during braze solidification and decorated the boundary). Overall, these self-joining trials confirmed that the active metal Ti induces a surface modifying reaction and promotes wettability (contact angle < 90°). These baseline tests were used to confirm the wettability enhancing role of titanium in ABAs in the absence of residual stresses resulting from a mismatch between the coefficients of thermal expansion (CTE) of joined substrates.

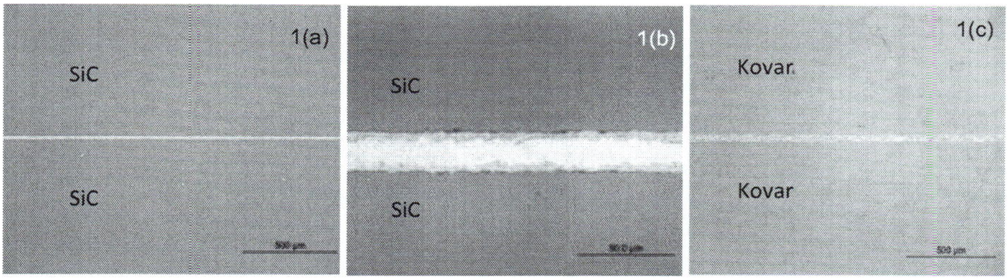

Figure 1. (a) & (b) A SiC/Ticusil/SiC joint, and (c) a Kovar/Ticusil/Kovar joint (micron bars are 500, 50, and 500 micrometers, respectively).

In CVD SiC/Cusil ABA (1 foil)/Ti joints, SiC did not bond with Ti because of incomplete wetting of SiC by the braze. However, there was good wetting of Ti by the braze. Using two braze foils in place of one did not yield improved surface coverage; additionally, three braze layers were also used. With 2- and 3- braze foils, the braze region was consistent but with some voids present, and a reaction layer (possibly titanium silicide) had formed. However, the SiC substrate in both samples exhibited significant cracking, both parallel and perpendicular to the braze region. For the self-joined Kovar, significant areas near the joint showed evidence of melting of the Kovar substrate. Some of the braze constituents may have diffused and dissolved into Kovar thus lowering its liquidus temperature. Large voids visible throughout the braze layer also suggested possible melting and solidification.

With Cusil-ABA paste in place of foils (Fig. 2), however, no cracking was visible. The braze region was very consistent (Fig. 2a), and there was no separation between braze and SiC. Optical microscopy and SEM showed a distinct dark layer (~3-5 μm thick) at the braze/SiC interface (Figs. 2b & c); presumably a titanium silicide reaction layer based on the elemental concentrations assessed via EDS (Figs. 2d & e). The Knoop microhardness profiles (Fig. 2f) displayed the expected behavior with a sharp discontinuity at the SiC/braze interface. Although the joint exhibited good wetting, preliminary mechanical testing showed that the joint was not very strongly bonded. Research on diffusion bonding of SiC laminates using titanium interlayers has shown that titanium silicides exhibit strong thermal expansion anisotropy [8]; this behavior could lead to uneven shrinkage during cooling within the joint leading to residual stresses that would weaken the joint in spite of good chemical wetting and bonding. Reaction layers in self-joined SiC using Ag-Cu-Ti filler have been shown to be composed of Ti_5Si_3 and TiC[1].

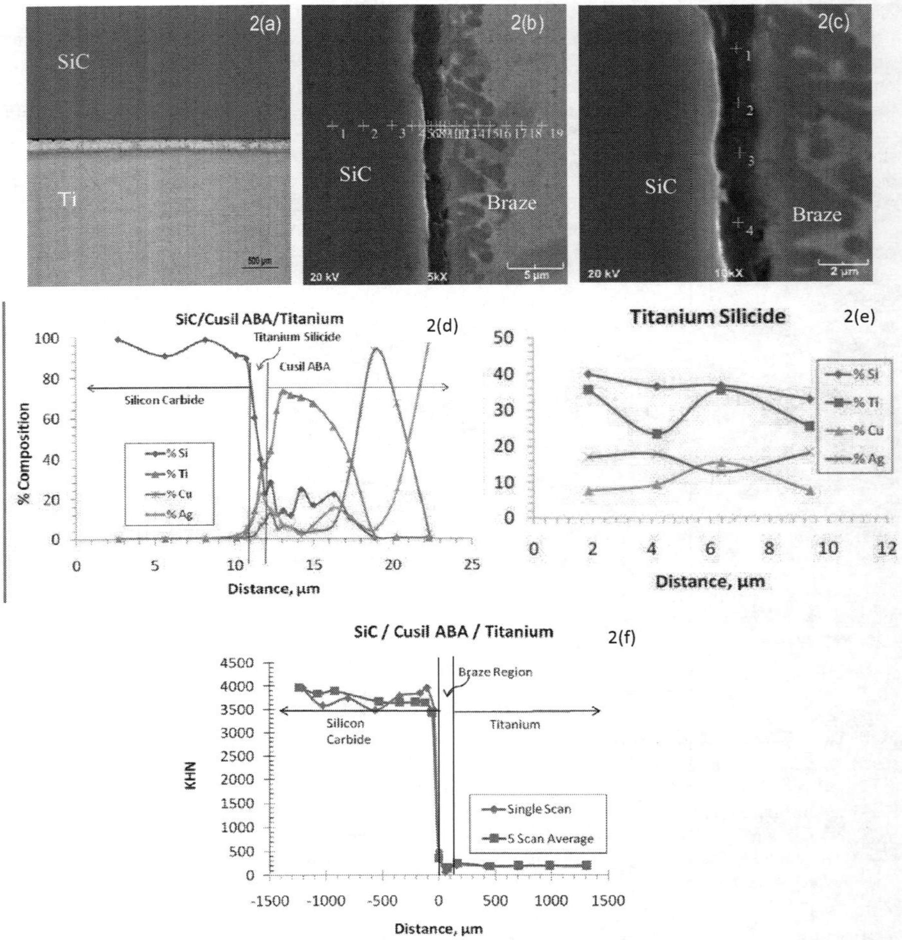

Figure 2. (a) Optical photomicrograph and (b) & (c) secondary electron SEM images of a SiC/Cusil-ABA (paste)/Ti joint, (d) & (e) show the elemental concentrations at point markers in (b) and (c), respectively, and (f) Knoop microhardness distribution across the joint.

In Hexoloy SiC/Cusil ABA (1 Foil)/Kovar joints, significant cracking in SiC substrate perpendicular and parallel to the braze region occurred. Additionally, micro-cracking occurred within SiC/braze reaction layer and Kovar/braze reaction layer. The reaction layer plus the braze region was ~180 µm thick. The Hexoloy SiC/Cusil-ABA/Kovar joints with double braze foils also displayed cracks within SiC both transverse and parallel to the joint (Fig. 3a). The braze region appeared to be microstructurally consistent (Fig. 3b) and displayed a reaction layer at the SiC/braze interface. Nickel and silicon enrichments were detected in this reaction layer using EDS; this could suggest possible formation of nickel silicide. The SiC/Cusil-ABA/Kovar joint made using braze paste displayed significant cracking in SiC both perpendicular and parallel to the braze region. The braze region exhibited considerable variation in its microstructure together with significant interaction between the braze and both SiC and Kovar. These joints were fabricated using SiC and Kovar substrates of the same thickness (3mm) to reduce warping; however, residual stresses in the joint were apparently high as was evidenced by the significant amount of cracking in the SiC substrates. In spite of large residual stresses, Cusil-ABA formed an adherent joint with SiC and Kovar and revealed a reaction layer (possibly nickel silicide) further supporting the conjecture that nickel silicide possibly forms a stronger bond with SiC than does titanium silicide.

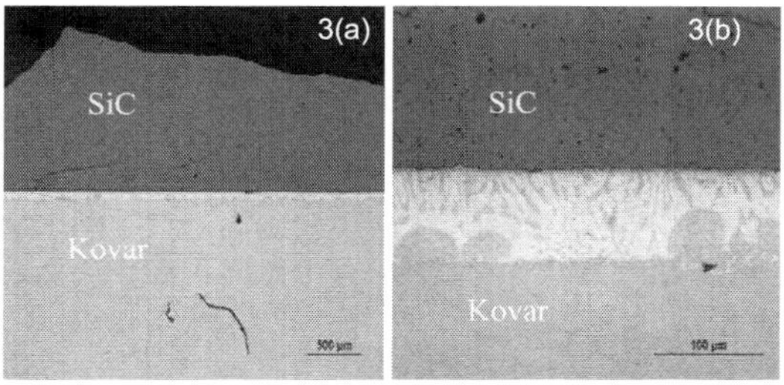

Figure 3. (a) Overall view of the joint, and (b) joint interfaces in a Hexoloy/Cusil-ABA (2 foils)/Kovar joint (the micron bars are 500 and 100 micrometers, respectively).

In SiC/Incusil-ABA/Kovar joints with either single- or double braze foils, cracking parallel and perpendicular to the braze region occurred in the SiC substrate (Fig. 4a). The braze region was consistent and showed chemical interaction with the substrate together with the formation a dark layer at the joint interface (Fig. 4b). The microhardness distribution for this joint is shown in Fig. 4c. In SiC/Incusil-ABA (1 foil)/Titanium joints, major cracking parallel and perpendicular to the braze region occurred throughout SiC and a large gap appeared between SiC and the braze. A diffused reaction layer formed at the Ti/Incusil-ABA interface. Use of 2 foils eliminated cracks parallel to the joint; however, cracks transverse to the joint persisted in the SiC substrate. There was a ~20 µm gap between SiC and braze region but the bonding between the braze layer and Ti was excellent. There shall be less effective stress accommodation in Incusil-ABA than Cusil-ABA owing to the former's higher yield strength (338MPa) versus the yield strength (271 MPa) of Cusil-ABA; the lower yield strength of Cusil-ABA will facilitate stress accommodation via plastic flow. Initial results suggest that, although the SiC/Incusil ABA/Kovar bonds are strong, residual stresses are higher and produce more significant cracking than when using Cusil ABA as a braze material.

These preliminary research outcomes suggest that as far as wettability enhancement and surface coverage are concerned, the selected ABAs (Cusil-ABA, Ticusil and Incusil-ABA) are adequate to join SiC to Ti and Kovar. However, extensive substrate cracking without joint failure observed in some joints suggests that residual stresses during joint fabrication possibly override and inhibit the beneficial effects of chemical reaction-induced wettability and bonding. Thus, engineering considerations related to residual stresses and strains could take precedence over chemical factors that facilitate joining.

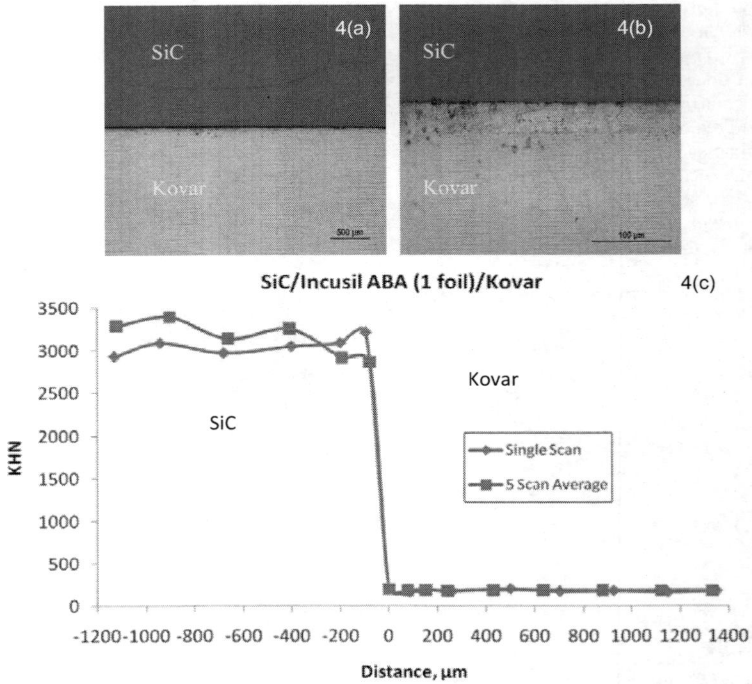

Figure 4. A SiC/Incusil-ABA/Kovar joint: (a) Overall view, (b) braze region, and (c) microhardness distribution with the distances relative to the joint location (the micron bar in 4a is 500 micrometers and in 4b it is 100 micrometers).

Large residual stresses from brazing could cause the ceramic to fracture without failure of the brazed joint. Residual stresses are a major concern in ceramic-to-metal joints because of the inherent brittleness of ceramics and fracture may occur if these stresses cannot be managed. One strategy to manage residual stresses is by judiciously arranging compliant interlayers within the joint prior to brazing [9]. Residual stresses in joints are more effectively accommodated by multiple interlayers than single layers. However, multiple interlayers also increase the number of interfaces and the probability of the formation of defects besides increasing the joint thickness.

In order to investigate how residual stresses influence and are influenced by multilayer joints, the combined effects of metal (M), ceramic (C), and interlayer material (I) must be considered. Analytical and numerical models of residual stress distribution have been developed [10,11]. A numerical model for strain energy in ceramics in brazed joints allows analytical approximations to the numerical model. For well-bonded ceramic-metal joints, the elastic strain energy, U_{eC}, in the ceramic substrate for a flat-back (e.g., disc-shaped) joint configuration can be estimated from [10,11]

$$U_{e,c} = (\sigma_{YI}^2 r^3/E_C)\{0.03\Pi_I + 0.11\varphi + 0.49\}$$

$$\text{where } \Phi = 1 - \left(\frac{\alpha_M - \alpha_I}{\alpha_C - \alpha_I}\right)^m, \text{ and } \Pi_I = \frac{(\alpha_M - \alpha_C)\Delta T E_I}{\sigma_{YI}}.$$

Here σ_{YI} is the yield strength of the interlayer, r is the distance from the center of the joint, E_C and E_I are the elastic moduli of the ceramic and the interlayer, respectively, and α is the CTE of the subscripted phases (i.e., C, M and I). The exponent m=1 for $\alpha_I > (\alpha_M + \alpha_C)/2$, and m=–1 for $\alpha_I < (\alpha_M + \alpha_C)/2$.

Using the preceding model equations and the property data in Table I (and handbook data for Ni and Cu interlayers), strain energies were calculated for joints made using the three ABAs along with ductile interlayers. The results summarized in Table II show that the strain energy varies between 1.44-7.16 mJ. Similar calculations for joints made without ductile interlayers yield higher strain energy values; for example, for SiC/Kovar joints made using Cusil-ABA, the calculated strain energy, U_{eC}, is 17.9 mJ, which is appreciably greater than the strain energy obtained with the use of ductile Cu and Ni interlayers (Table II). Thus, judicious arrangement of stress-absorbing compliant layers of ductile metals such as Cu and Ni within the joint could reduce the strain energy and propensity for joint failure. Research on the use of single and multiple Ni and Cu layers to join SiC was also undertaken. Several joints with multiple interlayers of Ni, Cu and other metals were created and characterized to test the effectiveness of this strategy to manage residual stresses.

Table II. Calculated Strain Energy in SiC in SiC/Ti and SiC/Kovar Joints made using interlayers of ABA and Ni and Cu

Joint with Interlayer	**Strain Energy in Silicon Carbide (mJ)**		
	Incusil-ABA	**Cusil-ABA**	**Ticusil**
SiC/Cu/Kovar	1.59	1.66	1.73
SiC/Cu/Ti	1.44	1.49	1.54
SiC/Ni/Kovar	6.74	6.97	7.16
SiC/Ni/Ti	6.17	6.34	6.49

In SiC/Nickel/Kovar joints brazed with Cusil-ABA cracking parallel to the braze interface continues to appear although perpendicular cracking has been reduced (Fig 5a). Significant residual stresses are still apparent in the joint utilizing 1 nickel interlayer as is evidenced by the warping and slight separation of the Kovar and nickel substrates at the location where a brittle reaction layer appears to have terminated (Fig. 5b).

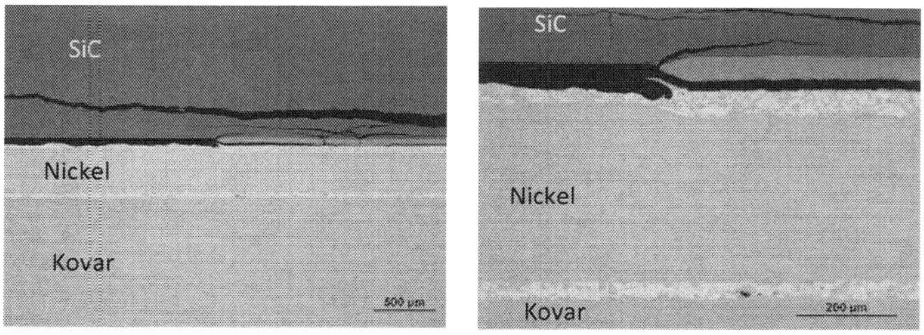

Figure 5. A SiC/Cusil-ABA/Nickel/Cusil-ABA/Kovar joint: (a) Overall view and (b) braze region.

Adding a second nickel interlayer drastically increased the amount of residual stresses that were absorbed by the interlayer approach. Perpendicular cracking in the SiC substrate was virtually eliminated while parallel cracking within the SiC substrate was limited to locations adjacent to cracking in the brittle phase formed at the braze/SiC interface (Figs. 6a, b). The use of nickel interlayers seems to have helped absorb residual stresses induced from brazing. Experimental results concerning interlayers, appear to support strain energy calculations performed on these joints. Cracks are now forming at the SiC/braze interface most likely due to a significant nickel silicide layer formation. SEM performed on the brittle phase shows that it is made up of high amounts of nickel and silicon, reinforcing the previous suggestion that nickel and silicon are reacting to form a brittle nickel silicide (Figs. 6c, d). The increased amount of nickel available for diffusion and thus silicide formation has increased the thickness of the formed layer. Knoop microhardness data from Fig 6e shows that the nickel silicide is indeed very hard when compared to the metallic phases in the system. Fig 6e also shows a slight reduction in the hardness values across the SiC substrate suggesting that the nickel interlayers have indeed helped lessen residual stresses within the ceramic. In previous samples the amount of nickel available for silicide formation was much lower and appeared to promote wetting and bonding at the SiC/braze interface; however, the increased amount of nickel silicide appears to be detrimental to the brazed joint due to its brittle nature.

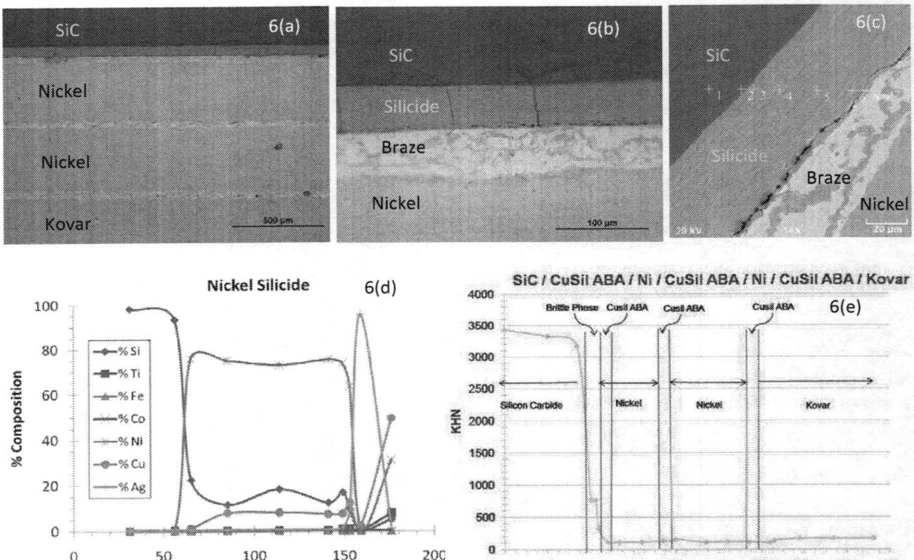

Figure 6. A SiC/Cusil-ABA/Nickel/Cusil-ABA/Nickel/Cusil-ABA/Kovar joint: (a) Overall view, (b) SiC/Nickel braze region, (c) secondary electron SEM images of a brittle phase at the SiC/Nickel braze interface, (d) shows the elemental concentrations at point markers in (c), and (e) Knoop microhardness distribution across the joint. The micron bars in 6a, 6b, and 6c are 500, 100, and 20 micrometers respectively).

Similar results were obtained in SiC/Cusil-ABA/Nickel/Cusil-ABA/Titanium systems. A nickel silicide formed at the SiC/braze interface causing parallel cracking within the SiC substrate (Fig 7a). Residual stresses were reduced which eliminated the formation of perpendicular cracks at the outside face of the SiC substrate. A new reaction layer was observed at the Nickel/Cusil-ABA/Titanium interface (Fig. 7b). The reaction layer is so extensive that braze boundaries are not distinguishable. EDS analysis of the reaction layer shows that high levels of titanium, nickel and silver are present (Fig. 7c). The reaction layer has not inhibited bonding as is seen by the excellent wetting and bonding of the titanium substrate.

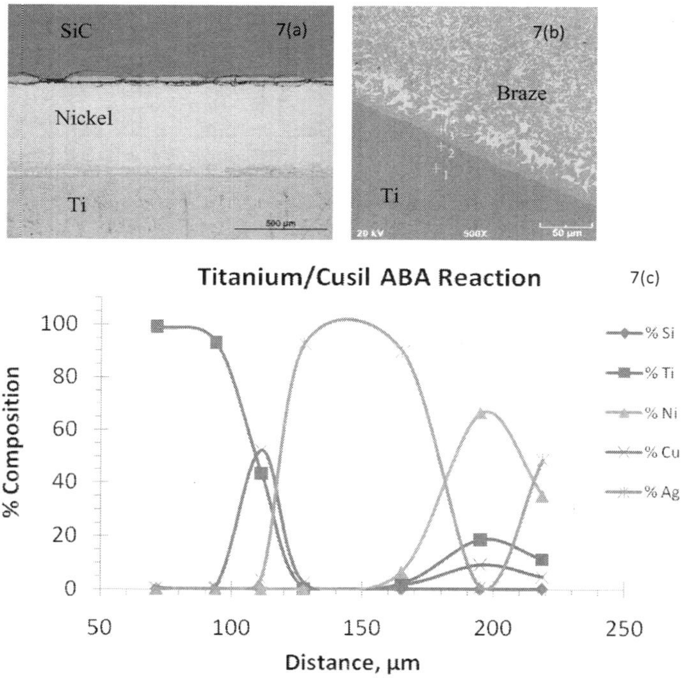

Figure 7. A SiC/Cusil-ABA/Nickel/Cusil-ABA/Titanium joint: (a) Overall view, (b) secondary electron SEM images of a reaction layer at the Nickel/braze/Titanium interface, (c) shows the elemental concentrations at point markers in (b).

Results from samples joined with copper interlayers are consistent with previous SiC/Kovar systems. Cracking perpendicular and parallel to the braze interface is readily apparent (Fig. 8a). There appears to be a nickel silicide layer forming at the SiC/braze interface (Fig. 8b). The bonding capabilities of the joint are, again, attributed to the formation of this nickel silicide layer. EDS analysis on previous samples has shown that diffusion of Kovar constituents, mainly nickel and iron, through solid metal interlayers has not occurred. In this sample the copper interlayer was very thin, 5µm, and appears to have melted and reconstituted within the braze region despite having a liquidus temperature much higher than the braze temperature. The melting of the copper interlayer would have allowed nickel from the Kovar substrate to diffuse through the braze copper layer and react with the SiC substrate to form a bond promoting nickel silicide. In identical systems using a titanium substrate instead of Kovar, eliminating the source of nickel, the samples failed to bond. The use of thin copper foils in an effort to absorb residual stresses was unsuccessful most likely due to the fact that the foils melted and reconstituted within the braze. The use of copper interlayers to absorb residual stresses will require a copper interlayer of sufficient thickness to prevent melting at the braze temperature.

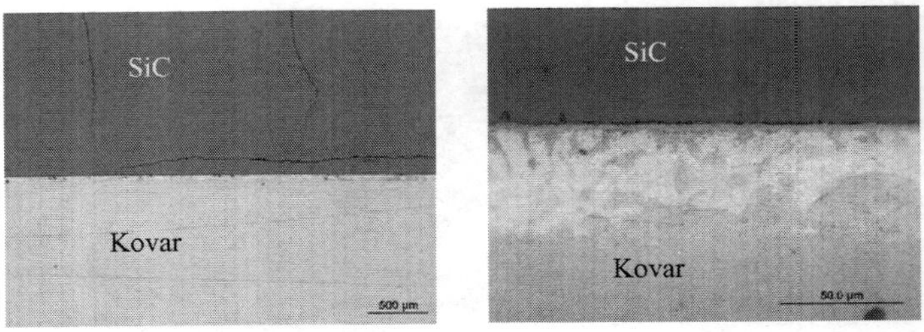

Figure 8. A SiC/Copper/Cusil-ABA/Copper/Kovar joint: (a) Overall view and (b) braze region.

CONCLUSION

All three brazes, Incusil-ABA, Cusil-ABA and Ticusil, utilizing the active metal titanium were capable of producing well formed joints with excellent wetting and bonding of the ceramic. The overriding factor was absorption of thermal induced stresses as cracking within the ceramic substrate became the biggest challenge to overcome. Cusil-ABA showed the greatest potential in forming strong bonds while absorbing residual stresses within the joints. Titanium/Cusil-ABA/SiC joints appeared to be very good but initial mechanical testing suggests weak bonds, possibly due to the absence of bond promoting nickel and iron silicides. Brazing SiC to Kovar produced strong bonds due to the availability of nickel and iron for silicide formation. Brazing SiC to Kovar introduced significant residual stresses and caused extensive warping and cracking in the SiC substrate due to the CTE mismatch. The use of nickel interlayers in a SiC system produced extensive diffusion of the nickel forming a brittle nickel silicide phase. The high amount of nickel silicide that formed was very brittle and was deemed to be detrimental to successful bond formation. Despite this, initial research with interlayers shows promising results concerning the absorption of residual stresses. Future efforts will further evaluate this bonding method to determine if it is fully capable of meeting the needs of the proposed application – uniform, leak-free bonds with stability and strength retention at temperatures up to 800°F.

ACKNOWLEDGMENT: B.P. Coddington would like to express appreciation to the Lewis' Educational and Research Collaborative Internship Program (LERCIP) for a Summer Internship Award.

REFERENCES

1. Y. Liu, Z.R. Huang, X.J. Liu, Joining of sintered SiC using AgCuTi active brazing alloy, Ceramics International, 25(8), 2009, 3479-3484.
2. P.Prakash, T.Mohandas, P.Raju, Microstructural characterization of SiC ceramic and SiC–metal active metal brazed joints, Scripta Materialia, 52(11), 2005, 1169-1173.

3. J. V. Marzik, T. Oyama, W. J. MoberlyChan, and W. J. Croft, Characterization of a ceramic-metal-ceramic bond: chemical vapor deposited (CVD) silicon carbide joined by a silver-based active brazing alloy (ABA), Silicon Carbide 2002-Materials, Processing and Devices, MRS Proceedings, vol. 742, S. E. Saddow, D. J. Larkin, N. S. Saks, A. Schoener (eds.), Proceedings of MRS Symposium K, 2002 Fall meeting, MRS.
4. M. Singh and M.C. Halbig, Bonding and integration of silicon carbide based materials for multifunctional applications, Innovation in Ceramics Science and Engineering, Volume 352, 2007, 201-206.
5. M.C. Halbig and J. Singh, Development and characterization of the bonding and integration technologies needed for fabricating silicon carbide based injector components, Advanced Processing and Manufacturing Technologies for Structural and Multi-functional Materials II, Volume 29, Issue 9, 2009, 1-14.
6. M.C. Halbig, M. Singh, T.P. Shpargel, and J.D. Kiser, Diffusion bonding of silicon carbide ceramics using titanium interlayers , Ceramic Engineering and Science Proceedings, Vol. 27, Issue 2, 2007, 133-144.
7. J. A. Fernie, R. A. L. Drew, and K. M. Knowles, Joining of engineering ceramics, International Materials Reviews, Vol. 54, No. 5, 2009, 283-331.
8. M.C. Halbig, M. Singh, Development and characterization of bonding and integration technologies for fabricating silicon carbide based injector components. In T. Ohji & M. Singh (Eds.), Wiley, 2009, 1-14.
9. M.R. Locatelli, A.P. Tomsia, K. Nakashima, B.J. Dalgleish, an A.M. Glaser, 'New strategies for joining ceramics for high-temperature applications', in Key Engineering Materials, vols 111-112, 1995, pp 157-190.
10. J.-W. Park, P.F. Mendez and T.W. Eager, Strain energy distribution in ceramic-to-metal joints, Acta Mater., 50(5), 2002, pp. 883-899.
11. J.-W. Park, P.F. Mendez and T.W. Eager, Strain energy release in ceramic-to-metal joints by ductile metal interlayers, Scripta Mater., 53(7), 2005, pp. 857-861.

JOINING OF SILICON NITRIDE WITH GLASS OR POWDER UNDER MECHANICAL PRESSURE

Naoki KONDO, Hideki HYUGA, Takaaki NAGAOKA, Hideki KITA
National Institute of Advanced Industrial Science and Technology (AIST)
Shimo-shidami 2266-98, Moriyama-ku, Nagoya 463-8560, Japan

ABSTRACT

Bulk silicon nitride was joined using glass or silicon nitride powder as insert materials. The glass composition was similar to that of the grain boundary glassy phase of the bulk silicon nitride, and the silicon nitride powder composition was the same as that of the bulk. The insert material was placed between the bulk silicon nitride pieces, and they were soaked at elevated temperatures near the sintering temperature of silicon nitride for joining. Mechanical pressure was applied for reliable joining. Successful joining could be achieved for both insert materials.

INTRODUCTION

Joining technology is critically important for fabricating large-scale or complex-shaped ceramic components. This is also a key approach for combining and integrating multiple parts with different phases or made from different materials. A large number of joining techniques have been developed for silicon nitride (SN).[1]

One conventional route to join silicon nitride is the use of glass as an insert material. Xie et al. selected suitable glass phases, which were similar in composition to that of the grain boundary glassy phase of silicon nitride.[2] Using this technique, they achieved a superior bending strength of 668 MPa in joined and post-hot isostatically pressed (post-HIPed) silicon nitride.[3] Here, the purpose of the HIP procedure was to reduce residual pores or flaws at the joined interface. Gopal et al. achieved a high strength of 937 MPa in silicon nitride joined using glass.[4] This was conducted under a low pressure of 0.1 MPa. Solid phase inserts also have been evaluated for joining. Xie et al. used superplastic SiAlON as an insert material and achieved a bending strength of 682 MPa in joined and annealed silicon nitride.[5]

The purpose of the present study was to join silicon nitride with either glass or silicon nitride powder as an insert material, based on these previous reports. The glass used had the same composition as that reported by Xie.[3] Silicon nitride was used as a raw powder with the same composition as the bulk silicon nitride. Joining was conducted at elevated temperatures and mechanical pressure was applied to enhance plastic deformation. The microstructure of the insert materials and the joined interface was investigated.

EXPERIMENTAL

Silicon nitride (SN-E10, Ube), yttria (RU, Shin-Etsu), alumina (AL160SG4, Showa Denko) silica (0.8 μm, Kojundo Chemical) were used as raw powders. Bulk silicon nitride (Si_3N_4 - 5wt.%Y_2O_3 - 3wt.%Al_2O_3, E10-5Y3A) was sintered at 1800°C for 8 h in a 0.5 MPa N_2 atmosphere. Surfaces for joining were first ground using a #200 whetstone.

The glass to be used as an insert material had a composition of Si_3N_4 - 43.4wt.%Y_2O_3 - 11.8wt.%Al_2O_3 - 14.7wt.%SiO_2, which is the same as that reported by Xie.[3] This composition forms a glassy phase similar to the grain boundary glassy phase of silicon nitride. The silicon nitride powder to be used as an insert material had the same composition as the bulk silicon nitride. These insert materials were placed between the bulk silicon nitrides pieces to be joined.

The joining (soaking) temperatures were 1600 and 1700°C and the process was carried out in an atmosphere of 0.1 MPa N_2. For the bulk silicon nitride, 1600°C was the minimum temperature required for densification, and 1700°C was the maximum temperature to avoid decomposition under 0.1 MPa N_2. Mechanical pressures of 5.4 or 25.9 MPa were used during the joining process, the former being the lowest pressure available in the furnace, and the latter being close to the usual hot press condition. The mechanical pressure was applied using two different methods, either throughout the entire joining process (heating up and soaking), or only after heating up. The soaking time was 30 min.

The sample notation is shown in Table. 1. The sample labeled X16HT, for example, has the following characteristics: composition = Xie's glass, temperature = 1600°C, mechanical pressure = high, pressure applied throughout joining.

Table 1 Sample notation

Composition of insert material	X : Xie's glass E : E10-5Y3A powder
Joining temperature	16 : 1600°C 17 : 1700°C
Mechanical pressure	H : High, 25.9 MPa L : Low, 5.4 MPa
Pressure application	T : Throughout joining A : After heating

The microstructure and composition of the joined samples were investigated by scanning electron microscopy (SEM) and energy dispersive X-ray spectroscopy (EDX). The plastic deformation characteristics of the joined samples were also measured.

RESULTS AND DISCUSSION

The microstructure of the joined interface of samples joined using a glass insert at temperatures of 1600 and 1700°C is shown in Fig. 1. The interface appears as a horizontal band near the center of the images. In this case, the application of mechanical pressure throughout the entire joining process caused softened glass to be extruded from the interface before formation of a reacted region between the glass and bulk silicon nitrides, resulting in the presence of residual flaws.

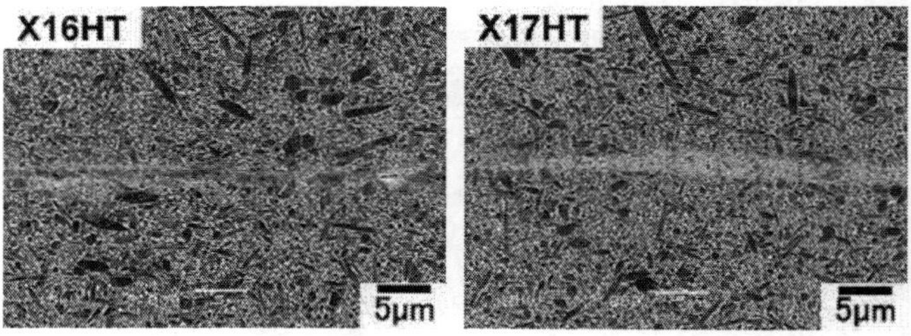

Fig. 1 Joined interface using glass insert material.
Upper and lower regions are bulk silicon nitride.
Horizontal band near the center is the joined interface.
Temperature was 1600 or 1700°C.
Mechanical pressure of 25.9 MPa was applied throughout joining.

Next, the press timing was delayed so that the mechanical pressure was applied only after the sample had been heated to the joining temperature. In this way, the formation of a reacted region between the softened glass and bulk silicon nitride was expected before pressure was applied. The resulting microstructure is shown in Fig. 2. It can be seen that a reacted region was successfully formed and the bulk silicon nitrides were joined. When pressure was applied, excess glass became extruded from the interface. The thickness of the reacted region was about 5 and 10 μm for applied pressures of 25.9 and 5.4 MPa, respectively. Note that the small black pores in the reacted region are traces of silicon nitride particles, which were etched away by plasma etching. Small rod-like silicon nitride grains were formed due to grain growth at 1700 °C. For a pressure of 5.4 MPa, particles with compositions different to the bulk silicon nitride were found. From EDX analysis, no accumulation of Y or Al was found in these particles. By referring to the phase diagrams, the particles were considered to be Si_2N_2O. These results indicate the importance of delaying the application of pressure when glass is used as an insert material.

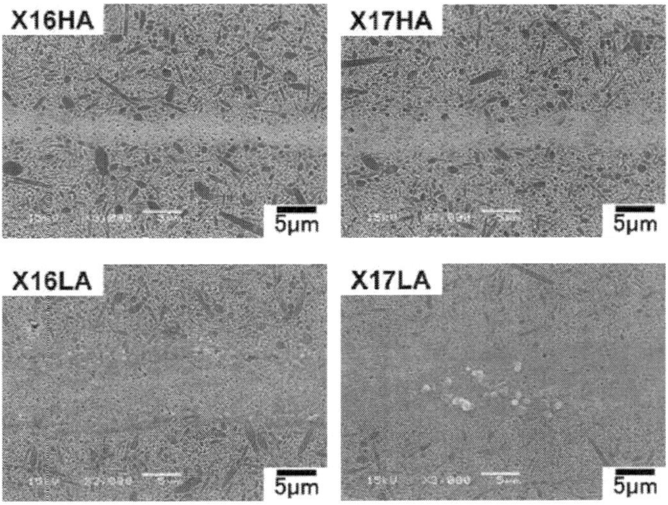

Fig. 2 Joined interface using glass insert material.
The reacted region appears as horizontal band near the center of the images.
Temperature was 1600 or 1700°C.
Mechanical pressure of 25.9 or 5.4MPa was applied.
Press timing was delayed and mechanical pressure was applied after heating.

Unlike the case for glass, when silicon nitride powder was used as the insert material, it did not soften and become extruded. Therefore, mechanical pressure was applied throughout the entire joining process. The results are shown in Fig. 3. The insert silicon nitride powder was successfully densified. No separation between the insert and bulk silicon nitride was found, nor were there any flaws within the insert material. The small black pores which appear in the insert are traces of silicon nitride particles. Small rod-like silicon nitride grains were formed due to grain growth at 1700 °C. No difference was found between the 25.9 and 5.4 MPa specimens.

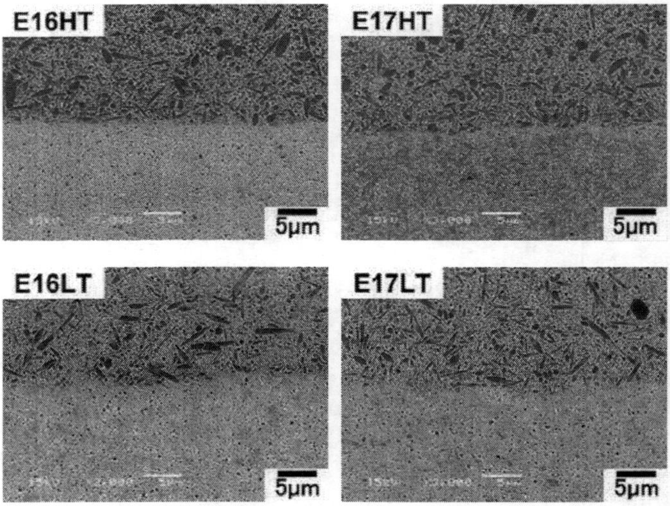

Fig. 3 Joined interface using silicon nitride powder as an insert material.
Upper part is bulk and lower part is densified insert silicon nitride.
Temperature was 1600 or 1700°C.
Mechanical pressure of 25.9 or 5.4 MPa was applied throughout joining.

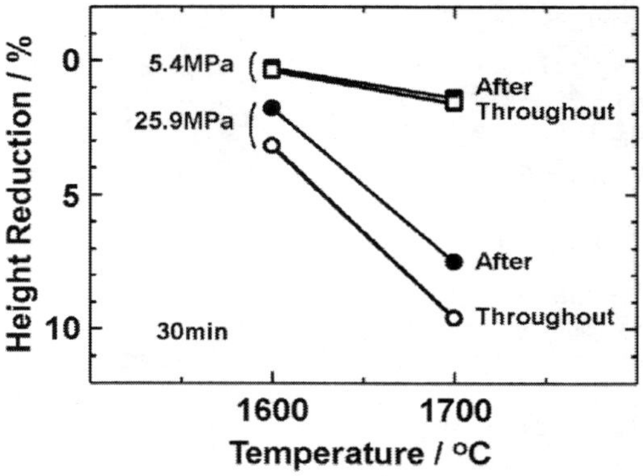

Fig. 4 Height reduction of bulk silicon nitride after joining.
"After" and "Throughout" were measured from samples using glass and silicon nitride powder as insert materials, respectively.

Xie et. al. attempted to perform joining using superplastic SiAlON as an insert material.[5] Since SiAlON exhibits better deformability than bulk silicon nitride, deformation occurred mainly in the interlayer region, which is advantageous for joining.

In this work, the silicon nitride powder exhibited better "apparent" deformability than a dense insert, involving not only the usual plastic flow but also material flow due to densification. This behavior is similar to that observed during sinter-forging of silicon nitride.[6] Additionally, since the insert powder had the same composition as the bulk silicon nitride, a homogeneous composition of the joined product was achieved. Thus, silicon nitride powder offers advantages over dense insert.

The height reduction of the bulk silicon nitride after joining is shown in Fig. 4. A larger reduction was found for the higher temperature and pressure, and when the pressure was applied throughout the entire joining process. The reduction was only 0.5% at 1600°C and 5.4 MPa. However, it increased to almost 10% when joining was carried out at 1700°C and a pressure of 25.9 MPa was applied throughout the joining process. Although there are few previous reports on deformation of the bulk material, such deformation must be considered when mechanical pressure is used during joining.

SUMMARY

1) Good joining of silicon nitride was achieved with glass or silicon nitride powder as insert materials.
2) When glass was used, the procedure of delaying the press timing to allow a reacted region to first form, and then pressing to extrude the excess glass, was found to be effective in producing a good joint.
3) The use of silicon nitride powder as an insert material offers the advantages of a large "apparent" deformability and homogeneous joining (uniform composition).
4) Deformation of the bulk must be considered when mechanical pressure is used for joining.

REFERENCES

[1] Loehman RE, "Recent Progress in Ceramic Joining," *Key Eng. Mater.*, **161-163**, 657-661, (1999).

[2] Xie RJ, Huang LP, Fu XR, and Chen Y, "Effects of Adhesive Composition on Bond Strength of Joined Silicon Nitride," *J. Eur. Ceram. Soc.*, **18**, 901-905, (1998).

[3] Xie RJ, Mitomo M, Huang LP, Fu XR, "Glass High Temp Joining of silicon nitride ceramics for high-temperature applications," *J. Mater. Res.*, **15**, 136-141, (2000).

[4] Gopal M, Sixta M, De Jonghe L, Thomas G, "Seamless Joining of Silicon Nitride Ceramics," *J. Am. Ceram. Soc.*, **84**, 708–12, (2001).

[5] Xie RJ, Mitomo M, Zhan GD, Huang LP, Fu XR, "Diffusion Bonding of Silicon Nitride Using a Superplastic b-SiAlON Interlayer," *J. Am. Ceram. Soc.*, **84**, 471–73, (2001).

[6] Kondo N, Suzuki Y, Ohji T, "Superplastic Sinter-Forging of Silicon Nitride with Anisotropic Microstructure Formation," *J. Am. Ceram. Soc.*, **82**, 1067–69, (1999).

ACKNOWLEDGEMENT

This research was supported by METI and NEDO, Japan, as part of the Project "Development of innovative ceramics manufacturing technologies for energy saving".

FABRICATION OF THERMODYNAMIC CRYSTALS BY STRUCTURAL JOINING

Soshu Kirihara
Smart Processing Research Center
Joining and Welding Research Institute
Osaka University
11-1 Mihogaoka Ibaraki, Osaka 567-0047, Japan

Yasunori Uehara and Youhei Takinami
Division of Sustainable Energy and Environmental Engineering
Graduate School of Engineering
Osaka University
2-1 Yamadaoka Suita, Osaka 565-0871, Japan

ABSTRACT

Thermodynamic crystals with periodically percolated morphologies of ceramics and metals were newly developed to control heat and stress flows intentionally. The artificial crystals were processed by using stereolithography of a spatially structural joining. In this investigation, the thermodynamic crystals with body centered cubic structures were formed in sub-millimeter orders. Pure copper spheres were closely arranged in alumina matrices to realize the periodic structures. Moreover, graded crystal structures were created successfully to realize continuous variations in volume fractions of the ceramic and metal phases through computer aided design, manufacturing and evaluation processes. The fabricated distributions of the ceramic and metal phases were observed by using a digital optical microscope. Thermal conductivities and mechanical properties for the thermodynamic crystals were and simulated by using a finite difference time domain method.

INTRODUCTION

Thermodynamic crystals with periodic arrangements of ceramic and metal phases were fabricated artificially to control heat and stress flows intentionally through computer aided design, manufacturing and evaluation. Graded structures in volume fractions of ceramic and metal phases were created to realize directional controls of heat and stress energies distributions. The alumina structures including air spheres with body centered cubic patterns were processed successfully by using stereolithography of a structural joining[1,2]. In our previous investigation, photonic crystals with periodic arrangements of ceramic lattices were fabricated in millimeter and micrometer orders to control the electromagnetic waves in wide range frequencies and wavelengths by using the above mentioned stereolithographic processes[3-5]. To create the thermodynamic crystals, resin slurries with copper particles were infiltrated into the ceramics objects formed by the stereolithography. After heat treatments, the thermodynamic crystals with the metals spheres arrangements in the ceramics matrices could be obtained. Moreover, the functionally graded materials with continuously variations in volume fractions of the ceramics and metals were fabricated. The intermediate regions with the percolated morphologies were formed between the copper and alumina layers. The artificial microstructures composed of the alumina and pure copper phases were observed by using a digital optical microscope and a scanning electron microscope. The part accuracies of the periodic structures are measured. Thermal and mechanical properties were simulated by a finite difference time domain method. The directional distributions of the heat and stress flows in the periodically percolated materials morphologies will be discussed.

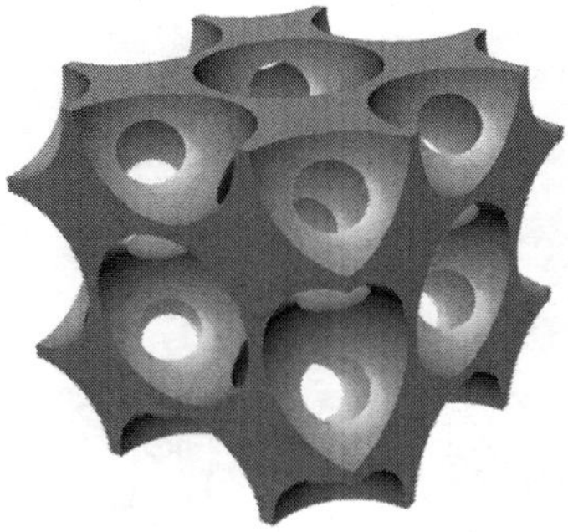

Figure 1 A computer graphic model of a thermodynamic crystal with an inverse body centered cubic lattice. Air spheres are closely arranged in a bulk material matrix to create a spatial periodic structure.

EXPERIMENTAL PROCEDURE

The three dimensional body center cubic structures were created by using a computer aided design application (Toyota Keram Ltd., Think Design Ver. 8.0) as shown in Fig. 1. The inverse crystal model of 10×10×10 mm in size included the air spheres of 4.0 mm in diameter. The lattice spacing for <100> direction was 4.5 mm. The volume ratio of the matrix was 26.5 %. The ceramics models were fabricated automatically by using the stereolithographic machine (D-MEC Co. Ltd., SI-C1000)[6]. The designed models were converted into the rapid prototyping format of the stereolithography files and sliced into thin sections. The alumina particles of 1.8 μm in average diameter were dispersed into the photo sensitive acrylic resin at 70 % in volume contents. The resin paste with the alumina particles was spread on a grass substrate with 50 μm in layer thickness by using a mechanically moved knife edge.

An ultraviolet laser beam of 355 nm in wavelength and 100 μm in beam spot was scanned on the resin surface to create two dimensional images with 10 μm in part accuracy. Through the layer stacking processes, sub-millimeter order three dimensional composite models were formed successfully. After dewaxing process of alumina dispersed resin precursor at 600 °C for 2 hs with heating rate of 1 °C/min, the samples were sintered at 1500 °C for 2hs with heating rate of 8 °C/min. Subsequently, the polyester resin past mixed with the pure copper particles of 75μm in diameter at 55 % in volume contents were percolated into the inverse ceramic structures. After the dewaxing for the surly infiltrated samples at 800 °C for 2 hs with heating rate of 1 °C/min, the pure copper particles were sintered at 1000 °C for 2hs with heating rate of 8 °C/min in an argon atmosphere. The densities of the sintered samples were measured by Archimedes method. The part accuracy of the metal spheres arrangements in the ceramic matrices were measured by using the digital optical microscope (Keyence Co., VHX-200) system. And the microstructures of ceramic and metal phases were closely observed by using the scanning electron microscope (JEOL Ltd., JSM-6060). The heat and stress flows were simulated in the thermodynamic crystals by a finite difference time domain application (Cybernet Systems Co. Ltd., Ansys Ver. 12).

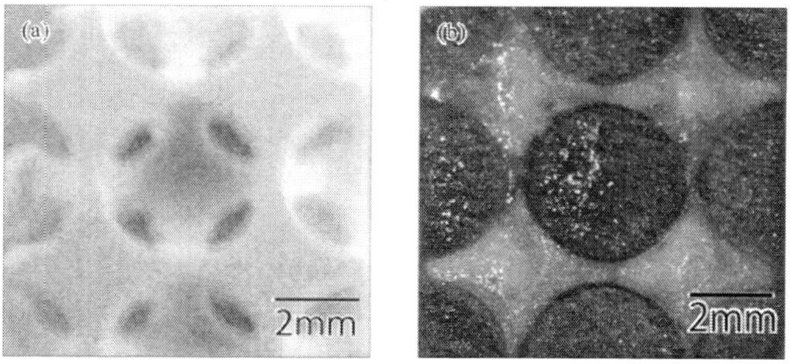

Figure 2 A sintered alumina lattice with the inverse body centered cubic structure fabricated by using stereolithography of a structural joining process (a) and a formed thermodynamic crystal with periodic arrangements of pure copper spheres in an alumina matrix though powder sintering treatments (b).

RESULTS AND DISCUSSION

The three dimensional lattice structures composed of the alumina dispersed acrylic resin were processed exactly by using the stereolithography. The spatial resolution was approximately 0.5 % in size. Figure 2-(a) shows (100) planes of the sintered body centered cubic structure composed of the sub-millimeter order alumina lattices including the air spheres. The deformation and cracking were not observed. The linear shrinkage on the horizontal axis was 6.5 % and that on the vertical axis was 7.8 %. It is possible to obtain the uniform shrinkage by designing an appropriate elongated structure in the vertical direction for compensation to the gravity effect. The relative density reached 98.5 %. Dense alumina microstructure was formed, and the average grain size was approximately 2 μm. Figure 2-(b) shows the (100) plane in the sintered thermodynamic crystal composed of the pure copper spheres and alumina matrix. Diffraction peaks indicating the formation of copper oxide phases did not appeared in X-ray diffraction spectra. The cracks and pores as the sintering defects were not observed in the copper phase microstructures. The smooth interfaces between the copper spheres and alumina matrix were obtained successfully. These material phases were joined continuously without intermediate layers.

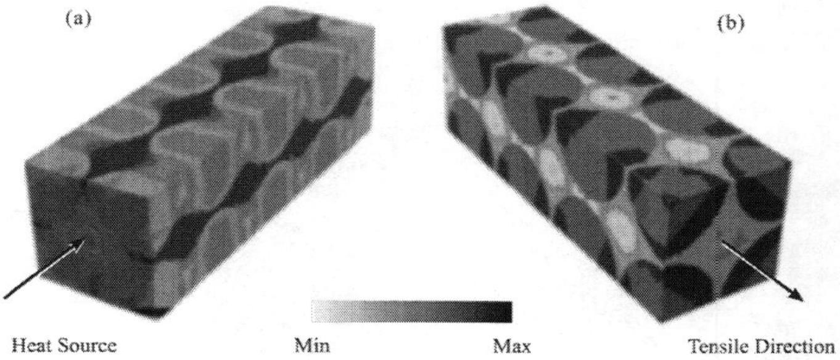

Figure 3 The computer graphic models of simulated heat and stress distributions in the fabricated thermodynamic crystals with the body centered cubic structure. The graphics (a) and (b) show the heat and stress flows through the connected metal spheres and the continuous ceramic matrix, respectively.

The three dimensional heat and stress distributions in the fabricated thermodynamic crystals were simulated as shown in Fig. 3-(a) and (b), respectively. The heat flow transmitted from the source plane to the opposite side through the connected copper spheres of the percolated metal phases with high thermal conductivities. Compare with this, the tensile stress were distributed periodically into the alumina lattices of the distributed ceramics phases with high Young's modulus. Through introductions of point or plane defects in to the perfect periodic structures, heat and stress flows can be localized into the specific regions to control the thermal and mechanical properties of the artificial crystals. Figure 4 shows the simulated result of the temperature distribution in the graded body centered cubic structure composed of the copper spheres with the varied diameters form 1.75 mm to 1.9 mm in the alumina matrix. The heat flow transmitted from the source plane to the opposite side can be curved for the metal spheres arrangements with the larger diameter along the structural gradient. The thermodynamic crystals with the graded structures are considered to be applied to the effective heating and cooling devices in the mechanical tools, internal combustion engines and thermoelectric conversion elements.

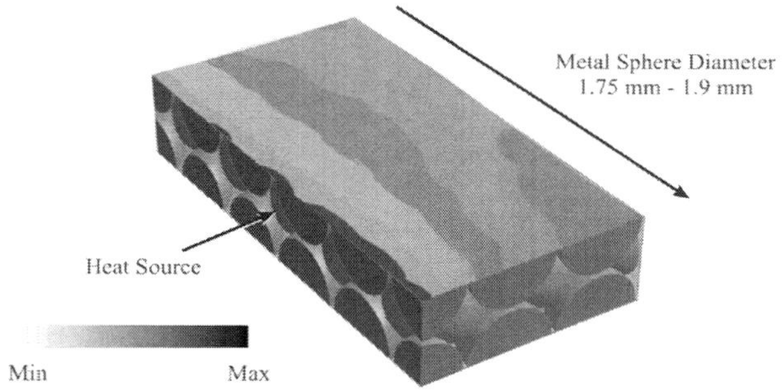

Figure 3 The simulated heat flow in the thermodynamic crystal of the graded structure with the body centered cubic pattern. The metal spheres with varied diameters were arranged in the ceramics matrix.

CONCLUSIONS

We have successfully fabricated three dimensional thermodynamic crystals with body centered cubic structures composed of pure copper spheres arrangements in alumina matrices by using stereolithography and careful optimized dewaxing and sintering processes. Periodic arrangements with high part accuracies of dense alumina and copper phases were observed and verified by using a digital optical microscope and a scanning electron microscope. Heat and stress distributions in the fabricated thermodynamic crystals were simulated theoretically and discussed though finite difference time domain methods. The heat flow transmitted through the connected copper spheres with high thermal conductivities, and the tensile stress were distributed periodically into the alumina lattices with high Young's modulus. Subsequently, graded body centered cubic structure composed of the copper spheres with varied diameters in the alumina matrix. The heat flow distributions could be curved along the graded direction for the copper spheres arrangements with the larger diameters. The thermodynamic crystals with the artificial crystal structures are considered to be applied to the effective heating and cooling devices and the useful shock and stress absorbers in the near future industrial fields.

ACKNOWLEDGMENTS

This study was supported by Priority Assistance for the Formation of Worldwide Renowned Centers of Research - The Global COE Program (Project: Center of Excellence for Advanced Structural and Functional Materials Design) from the Ministry of Education, Culture, Sports, Science and Technology (MEXT), Japan.

REFERENCES

[1] C. C. Kai and L. K. Fai, Rapid Prototyping, *John Wiley and Sons,* Singapore (1997).

[2] A. Pique, A. S. Holmes, and D. B. Dimos, Rapid Prototyping Technologies, *Materials Research Society Symposium Proceedings*, 758-764 (2003).

[3] S. Kirihara, M. Takeda, K. Sakoda, and Y. Miyamoto, Control of Microwave Emission from Electromagnetic Crystals by Lattice Modifications, *Solid State Communications*, **124**, 135-139 (2002).

[4] S. Kanehira, S. Kirihara, and Y. Miyamoto, Fabrication of TiO_2-SiO_2 Photonic Crystals with Diamond Structure, *Journal of the American Ceramic Society*, **88**, 461-1464 (2005).

[5] S. Kirihara, and Y. Miyamoto, Terahertz Wave Control Using Ceramic Photonic Crystals with Diamond Structure Including Plane Defects Fabricated by Micro-stereolithography, *The International Journal of Applied Ceramic Technology*, **6**, 41-44 (2009).

[6] D-Mec Ltd., Acculas SI-C1000, http://www.d-mec.co.jp/eng/

EFFECT OF VARIOUS FACTORS ON INTERFACE FORMATION IN MAGNETIC PRESSURE SEAM WELDING

Hisashi Serizawa, Isao Shibahara, Sherif Rashed and Hidekazu Murakawa
Joining and Welding Research Institute, Osaka University
11-1 Mihogaoka, Ibaraki, Osaka 567-0047, Japan

Shinji Kumai
Department of Materials Science and Engineering, Tokyo Institute of Technology
4259 Nagatsuta-cho, Midori-ku, Yokohama, 226-8502, Japan

ABSTRACT

The magnetic pressure seam welding is one of the candidate methods to join thin sheet multifunctional materials. In this research, to demonstrate the wavy morphology at the joint interface produced by the magnetic pressure seam welding, numerical simulations of the impact were carried out by using a commercial Euler-Lagrange coupling software MSC.Dytran (MSC.Software), where the friction between the dissimilar sheets (Fe and Al) and the oxide layer adhered to the original sheet (Al) were selected as the considerable important factors in the formation of joint interface. From the serial numerical results, it was found that the effect of friction on the success of magnetic pressure seam welding would be little although the movement of Al along the interface would be prevented and the deformation of Fe slightly increased with increasing the coefficient of friction. By considering the deformation of oxide layer, the wavy morphology at interface could be demonstrated. Also, from the comparison between the numerical analyses and microstructural observations, it can be considered that the deformation of oxide layer obtained might be related to the creation of intermediate phase layer tracing the wavy morphology.

INTRODUCTION

As a result of R&D efforts in the field of advanced ceramics and ceramics composite materials, various types of new multifunctional materials have been developed. As for the practical use of these materials, it is necessary to join them to other materials without any functional defects. The ordinary physical joining method using heat sources such as arc welding, laser welding and electron beam welding have to cause the microstructural changes at the joint interface [1,2] and might largely affect the original feature of the multifunctional materials. Also, the mechanical joining method would induce some flaws in these advanced materials. In order to overcome these problems, several joining methods have been proposed, which are, for examples, diffusion bonding [3], explosive bonding [4], friction bonding [5], magnetic pressure seam welding [6] and so on. In these joining methods, a magnetic pressure seam welding, which has been developed by Aizawa *et al.*, seems to be one of the most candidate methods for joining dissimilar thin sheet materials between the multifunctional materials and other materials, because the temperature of the joint is close to room temperature and the superior properties in the multifunctional materials seem to be preserved although the large impact is applied at the interface between the dissimilar materials.

Although there have been many experimental investigations, in which the conditions necessary for successful joining, microstructural observations of the joint interface and mechanical evaluation of the joints were discussed [7-10], the mechanism of this joining process is still not well understand. So the appropriate joining conditions have been decided from huge experimental struggles even now. The final target of this research is to examine the mechanism of the magnetic pressure seam welding from a dynamic viewpoint and to reveal the appropriate joining conditions theoretically. As a result of our previous numerical studies about the dissimilar joint between Fe and Al made by the magnetic pressure seam welding using a commercial Euler-Lagrange coupling software MSC.Dytran (MSC.Software) [11], it was found that there were two patterns of plastic strain distributions near the joint interface depending on the collision velocity and collision angle and the plastic strain pattern might be related to the success of magnetic pressure seam welding [12]. However, in these computations, the wavy morphology [7-10], which is observed at the joint interface made by the magnetic pressure seam welding as same as the explosive welding [13], could not be demonstrated although the minimum size of finite element was 1.5 μm and the cycle of the wave was in the range from 10 to 50 μm. Also, there have not been any reports to demonstrate the wavy morphology obtained by the magnetic pressure seam welding, whereas such wavy morphology observed in the explosive welding was demonstrated by using the smoothed particle hydrodynamics and its cycle was about 200 μm [14]. So, in order to simulate the wavy morphology in Fe-Al joint interface obtained by the magnetic pressure seam welding, the friction between the dissimilar sheets and the oxide layer adhered to Al were selected as the considerable important factors in the formation of joint interface and the effect of these factors on the interface formation was examined by using the commercial software MSC.Dytran.

EFFECT OF FRICTION

Model for Analysis

According to our previous numerical researches [12], only a part of two sheets was modeled for demonstrating the wavy morphology due to the cycle of wave and the limitation of computational resources. So, the model shown in Fig. 1 was employed for the precise numerical analyses where a minimum element size was 1.5 x 1.5 μm^2, and the total number of elements and nodes was 281445 and 566956, respectively. Although the Euler-Lagrange coupling software was used, these two sheets were modeled by only Lagrange finite element in order to introduce the friction between two sheets. As one typical example of the success cases in magnetic pressure seam welding, the collision velocity and the collision angle was assumed to 300 m/s and 5 degree, respectively, according to our previous

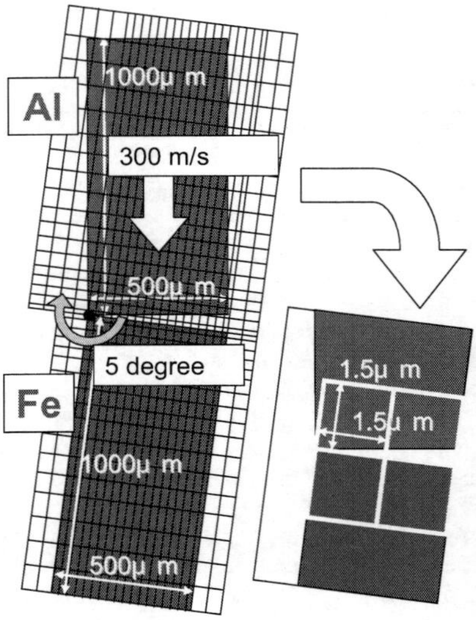

Fig. 1 Schematic illustration of numerical model for magnetic pressure seam welding.

Table 1 Material properties used for numerical analyses.

	Fe	Al
Young's Modulus (GPa)	206	70.3
Yield Stress (MPa)	500	200
Density (kg/m^3)	7.87 x 10^3	2.70 x 10^3
Poisson's Ratio	0.3	0.345
Linear Expansion Coefficient (1/K)	1.18 x 10^{-5}	2.39 x 10^{-5}
Shear Modulus (GPa)	79.2	26.0
Specific Heat (J/kg ·K)	440	900
Melting Point (K)	1808	933

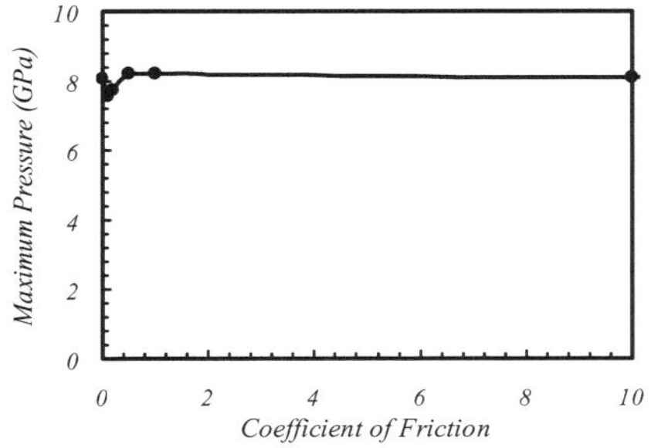

Fig. 2 Effect of friction coefficient on maximum pressure.

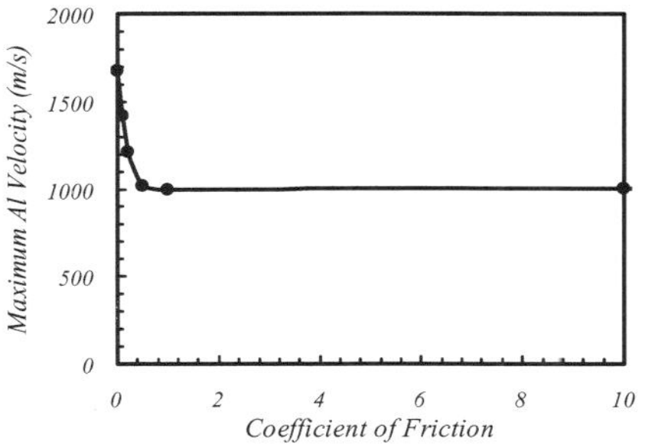

Fig. 3 Effect of friction coefficient on maximum Al velocity.

analyses [12]. The coefficient of friction was varied between 0 and 10. Table 1 shows mechanical and thermal properties used in this research.

Results and Discussions

From our previous studies [12], it was found that the collision velocity and collision angle largely affected the mean stress occurred at collision point, which could be considered as a pressure at the joint surface. Also, it was revealed that the maximum Al velocity parallel to the joint interface was influenced by the collision velocity and collision angle. Since the plastic strain pattern near the joint interface depending on the collision velocity and collision angle might be related to the success of magnetic pressure seam welding, the effect of friction on the maximum pressure and maximum Al velocity was examined and was summarized into Figs. 2 and 3. From these figures, it was found that the change of coefficient of friction would slightly affect the maximum pressure and the maximum Al velocity drastically decreased with increasing the coefficient of friction from 0 to 0.5. In the cases that the coefficient of friction was larger than 0.5, both the maximum pressure and maximum Al velocity were almost independent to the friction. So, in order to examine the effect of friction more precisely, the influence of collision angle on the maximum pressure and maximum Al velocity was studied by changing the collision angle from 0 to 10 degree in the cases that the coefficient of friction is 0 or 0.2. Figures 4 and 5 show the computed results and it was found that the influence of friction on the maximum pressure could be little regardless of the collision angle. On the other hand, although the maximum Al velocity was decreased by the friction, the collision angle, in which the maximum Al velocity had a maximum value, would not be affected. According to our previous researches [12], not the absolute value of maximum Al velocity but the inflection point seems to be related to the success of magnetic pressure seam welding since the maximum value was still keeping over 1000 m/s and was much faster than the collision velocity (300 m/s). Figure 6 shows the deformations near the contact point of two sheets. From this figure, it was found that, with increasing the coefficient of friction, the movement of Al would be prevented while the deformation of Fe slightly increased. Also, in these deformations, any wavy morphology could not be obtained. Then, as a result of these computations, it can be concluded that the effect of friction on the success of magnetic pressure seam welding would be little.

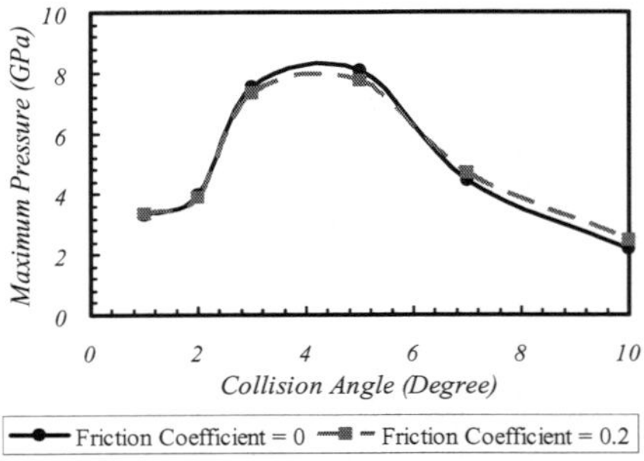

Fig. 4 Effect of collision angle on maximum pressure.

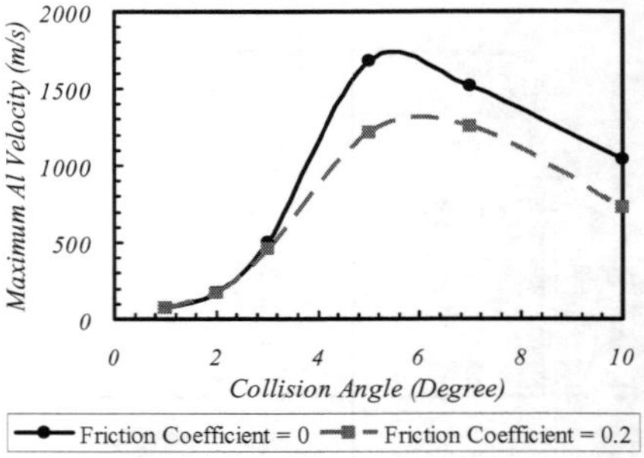

Fig. 5 Effect of collision angle on maximum Al velocity.

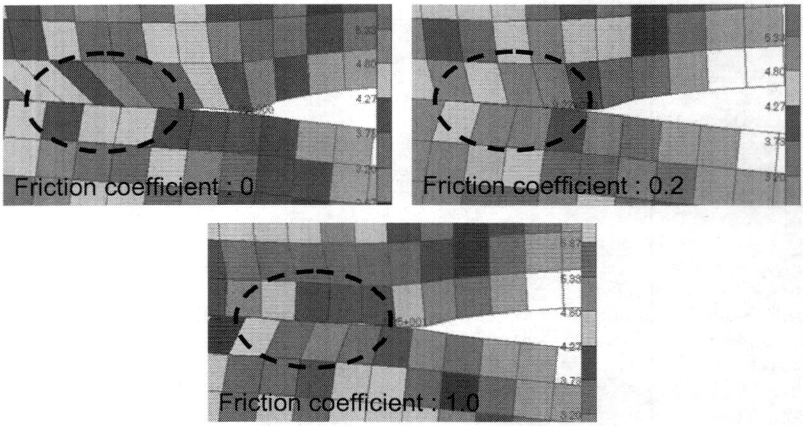

Fig. 6 Effect of friction coefficient of deformations near the contact point.

EFFECT OF OXIDE LAYER

Model for Analysis

In the previous experimental studies for the dissimilar joint between Fe and Al sheets made by the magnetic pressure seam welding, the metal jet, whose composition was mainly Al, was observed [15]. Also, in the microstructural observations of the joint interface, any oxide components could not be identified [7-9], although it is well known that Al is easily oxidized and the oxide layer would adhere to Al. From these previous researches, it could be considerable that the oxide layer might be dispersed as the metal jet by the collision between Fe and Al sheets. So, as another possible factor for wavy morphology at joint interface, the effect of oxide layer on the joining process was examined. The model for analyses was almost as same as that for the previous analyses, and a part of Fe and Al sheets was computed. These sheets, whose size was 1 x 0.5 mm^2, were modeled by Lagrange finite element. Since the specific properties of oxide layer were uncertain, the thickness of oxide layer was assumed to 1 μm in this analysis and its mechanical properties were supposed to be weaker than the original properties. In order to examine the deformation of oxide layer precisely, the oxide layer was modeled by Euler finite element, whose size was 1 x 0.5 μm^2 while the size of Lagrange element was 1.5 x 1.5 μm^2. Figure 7 shows the overview and enlarged view of model employed in this analysis. The collision velocity and collision angle was assumed to 200 or 500 m/s and 5 degree, respectively as the typical examples.

Results and Discussions

The deformed shapes in the collision velocity of 500 m/s were summarized into Fig. 8. From this figure, it was found that, at the initial contact point, the oxide layer moved along the Al surface and the Fe and Al sheets were attached directly. With increasing the contact area, a part of oxide layer was trapped between Fe and Al sheets and the oxide layer was deformed to the wavy morphology. Also, from the enlarged view of the joint interface with both Lagrange and Euler meshes in Fig. 8, it was found that the cycle of wave obtained was about 15 μm. Such a behavior was also obtained the

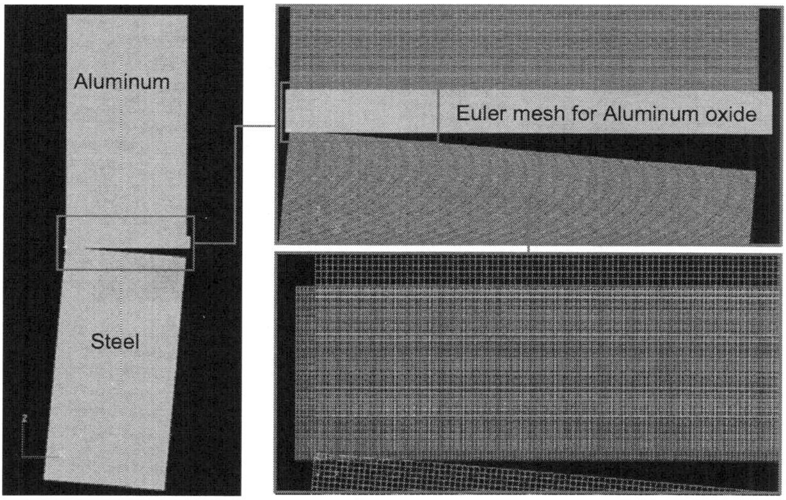

Fig. 7 Model and mesh division for numerical analysis with oxide layer.

case with the collision velocity of 200 m/s. On the other, the practical cycle of wave was observed as the range from 10 to 50 μm as shown in Fig. 9, where the breadth of intermediate phase layer tracing the wavy interface shown in Fig. 9(b) was in the range from 10 to 20 μm. From the microstructural analysis, it was reported that the intermediate phase layer consisted of an amorphous including extremely fine Al grains [9]. Since the cycle of wave computed has a very good agreement with the breadth of intermediate phase layer and the very large pressure, which was 5 and 100 times larger than the yield stress of Al, occurred at the interface, it can be considered that the deformation of oxide layer obtained might be related to the creation of intermediate phase layer and the morphology of joint interface. However, more numerical studies by changing the uncertain properties such as mechanical properties and thickness of the oxide layer have to be conducted for understanding the relationship between the wavy morphology computed and the intermediate phase layer.

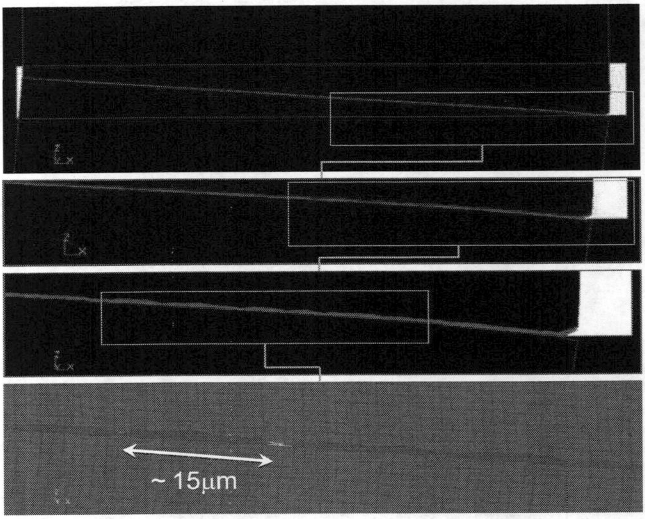

Fig. 8 Deformed shapes and mesh devision in analysis with oxide layer.

(a) Optical micrograph

(b) SEM-BEI micrograph

Fig. 9 Optical and SEM-BEI micrographs of joint interface.

CONCLUSIONS

In order to demonstrate the wavy morphology at the joint interface produced by the magnetic pressure seam welding, numerical simulations of the impact were carried out by using a commercial Euler-Lagrange coupling software MSC.Dytran (MSC.Software), where the friction between the dissimilar sheets (Fe and Al) and the oxide layer adhered to the original sheet (Al) were selected as the considerable important factors in the formation of joint interface. The conclusions can be summarized as follows.

(1) The effect of friction on the maximum pressure at the interface would be little regardless of the collision angle, while the maximum Al velocity along the interface was decreased by the friction although the inflection point of maximum Al velocity would not be affected.

(2) Although, with increasing the coefficient of friction, the movement of Al would be prevented and the deformation of Fe slightly increased, the effect of friction on the success of magnetic pressure seam welding would be little.

(3) By considering the deformation of oxide layer adhered to Al surface, the wavy morphology at interface could be demonstrated.

(4) From the comparison between the numerical analyses and microstructural observations, it can be considered that the deformation of oxide layer obtained might be related to the creation of intermediate phase layer tracing the wavy morphology.

REFERENCES

[1] H. Serizawa, Y. Kawahito, H. Ogiwara, H. Tanigawa and S. Katayama, "Weldability of Reduced Activation Ferritic/Martensitic Steel under Ultra Power Density Fiber Laser Welding", *Proceedings of the 13th International Conference on Fusion Reactor Materials*, CD-ROM (2007).

[2] H. Tanigawa, T. Hirose, K. Shiba, R. Kasada, E. Wakai, H. Serizawa, Y. Kawahito, S. Jitsukawa, A. Kimura, Y. Kohno, A. Kohyama, S. Katayama, H. Mori, K. Nishimoto, R.L. Klueh, M.A. Sokolov, R.E. Stoller and S.J. Zinkle, "Technical Issues of Reduced Activation Ferritic/Martensitic Steels for Fabrication of ITER Test Blanket Modules", *Fusion Engineering and Design*, (2008), to be published.

[3] N. Iwamoto, M. Yoshida, S. Tanabe, T. Takeuchi and M. Makino, "Diffusion Welding of Mild Steel to Aluminum", *Transactions of JWRI*, **4**, 67-70 (1975).

[4] M. Kikuchi, H. Takeda and S. Morizumi, "Bonding Interfaces in Friction-and Explosive-Welded Aluminum and Steel Joints", *Journal of Japan Institute of Light Metals*, **34**, 165-172 (1984).

[5] T. Shinoda, M. Ogawa, S. Endo and K. Miyahara, "Friction Welding of Aluminum and Plain Low Carbon Steel", *Quarterly Journal of the Japan Welding Society*, **18**, 365-372 (2000).

[6] T. Aizawa, K. Okagawa and M. Kashani, "Seam Welding Method Using Magnetic Pressure from One Side", *Proceedings of International Symposium on Joining Technologies in Advanced Automobile Assembly 2005*, 97-105 (2005).

[7] M. Watanabe, S. Kumai and T. Aizawa, "Interfacial Microstructure of Magnetic Pressure Seam Welded Al-Fe, Al-Ni and Al-Cu Lap Joints", *Materials Science Forum*, **519-521**, 1145-1150 (2006).

[8] K.J. Lee, S. Kumai, T. Arai and T. Aizawa, "Interfacial Microstructure and Strength of Steel/ Aluminum Alloy Lap Joint Fabricated By Magnetic Pressure Seam Welding", *Materials Science and Engineering A*, **471**, 95-101 (2007).

[9] S. Kumai, K.J. Lee and M. Watanabe, "Characteristic Interfacial Microstructure of Aluminum Alloy/ Steel Lap Joints Fabricated by Several Advanced Welding Methods", *Proceedings of 11th International Conference on Aluminum Alloys*, **2**, 1945-1951 (2008).

[10] M. Watanabe, S. Kumai, K. Okagawa and T. Aizawa, "In-situ Observation of Magnetic Pulse Welding Process for Similar and Dissimilar Lap Joints Using a High-Speed Video Camera", *Proceedings of 11th International Conference on Aluminum Alloys*, **2**, 1992-1997 (2008).

[11] MSC.Software, *MSC.Dytran Manuals*, MSC.Software, (2008).

[12] H. Serizawa, I. Shibahara, S. Rashed, H. Murakawa, M. Watanabe and S. Kumai, "Basic Study of Joint Interface Formation in Magnetic Pressure Seam Welding", *Ceramic Engineering and Science Proceedings*, **30**, 8, 121-130 (2009).
[13] B. Crossland, *Explosive Welding of Metals and its Application*, Clarendon, Oxford (1982).
[14] K. Tanaka, "Numerical Studies of Explosive Welding by SPH", *Materials Science Forum*, **566**, 61-64 (2008).
[15] M. Watanabe, S. Kumai, K. Okagawa and T. Aizawa, "In-situ Observation of the Magnetic Pulse Welding Process Using a High-Speed Video Camera", *Preprints of the National Meeting of Japan Welding Society*, **82**, 122-123 (2008).

PRODUCTION ENVIRONMENT LASER ASSISTED MACHINING OF SILICON NITRIDE

Federico Sciammarella [1]; Joe Santner [1]; Jeff Staes [2]; Richard Roberts [2]; Frank Pfefferkorn [3]; Stephen T. Gonczy [4]; Stefan Kyselica [1]; Ricardo Deleon [2]

1. Northern Illinois University, De Kalb, IL, USA
2. Reliance Tool and Manufacturing, Elgin, IL, USA
3. University of Wisconsin, Madison, WI, USA
4. Gateway Materials Technology, Mt. Prospect, IL, USA

ABSTRACT

This academic-industry project is aimed at transferring laser-assisted machining (LAM) of structural ceramics from a laboratory science to commercial practice on the shop floor. During LAM, the ceramic workpiece is spot heated by precisely controlled lasers and then cut with a conventional single point cutting tool. With locally lowered ceramic strength at elevated temperatures, LAM can achieve higher material removal rates, reduced tool wear, and equivalent surface quality, compared to slow and costly diamond grinding.

The LAM project team developed a unique multiple laser beam system and explored its capability for machining high strength silicon nitride in a production environment. LAM trials looked at both OD turning of 1" diameter silicon nitride rods with a CBN tool, optimizing LAM processing conditions for material removal rates, surface condition, and tool life. LAM machined rods were characterized for dimensional tolerances, surface roughness, and flexure strength to determine if LAM processing can produce strengths comparable to diamond grinding.

The results presented show that the LAM machined samples show some improvement over the as received samples. This preliminary work also shows a correlation between the mean flexure strength and mean R_a, which is that strength increases as R_a decreases. Additional work will test this relationship.

INTRODUCTION

Over the last three decades, ceramics have moved from low strength applications to high temperature, high strength applications in engines and wear, based on dramatic improvements in strength, fracture toughness and impact resistance [1, 2]. These properties enable the use of advanced ceramics (silicon nitride, silicon carbide, zirconia, etc) in engine components, such as bearings, rotors, cam roller followers, and exhaust valves, where they offer higher temperature capability, lower density, and better wear resistance, compared to metals [3, 4]. These components typically require very tight dimensional tolerances and complex shapes that require cost effective finish shaping and surface machining. There are a variety of finishing methods used for ceramics, but the most common method is diamond surface grinding, because of the high hardness of these ceramics [5]. (Conventional single point machining - turning, milling, drilling – produces brittle failure, excessive surface damage, and excessive tool wear at acceptable machining rates [6, 7].) But diamond grinding still has low material removal rates and is limited to simple contours. As a result, diamond grinding costs can be as much as 70% to 90% of the total component cost for complex components [8]. The wide application of advanced ceramics has been restrained largely because of the difficulty and high cost associated with shaping these hard and brittle materials into products. The development of a low-cost, effective machining process with high material removal rates, low tool wear, and acceptable surface finish would accelerate the commercial use of advanced ceramics for structural and wear applications.

In laser assisted machining (LAM), a high energy laser is used to locally heat a small zone ahead of a single point tool (diamond or cubic boron nitride) and then shaped using turning or milling. By preheating the ceramic zone ahead of the tool raises the ceramic temperature to point where ductile, rather than brittle, deformation occurs. Ductile deformation in the ceramic enables reduced cutting forces, high material removal rates, minimal surface damage, and long tool life. [9, 10, 11].

An illustration of a laser-assisted machining set up (with the ceramic rod, cutting tool, laser spot, and material removal plane) is illustrated in Figure 1. A good review of past laser machining research on a variety of materials using Nd:YAG and CO_2 lasers and the benefits is reported in [9].

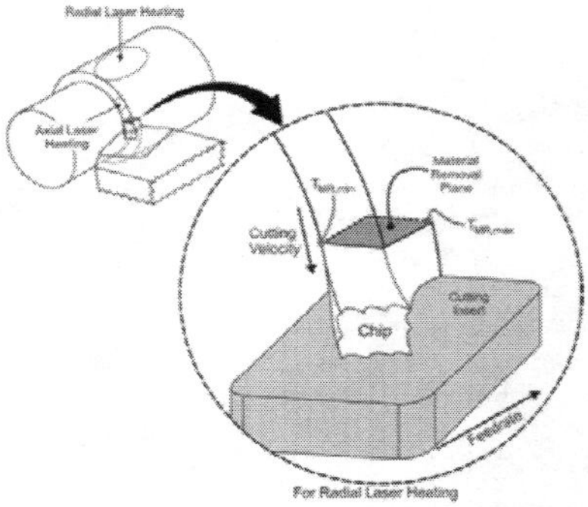

Figure 1. Laser assisted machining model illustrating the machining removal plane [11].

There has been considerable academic research in the application of LAM to high strength ceramics (silicon nitride, toughened zirconia, silicon carbide, etc) [10-18]. That research has verified the benefits of LAM -- reduced cutting forces, rapid material removal rates, and extended tool life. Numerical modeling of heating effects and temperature conditions has been an integral part of much of this work [19-23].

In the case of silicon nitride, the intense heating of the surface locally raises the temperature of the glassy ground boundary phase, which is the remainder of the oxide sintering aids used to obtain liquid phase sintering at temperatures >1300°C [4]. The composition of that sintering aid determines the temperature (~600-1200°C) at which the grain boundary phase softens. With sufficient heat, the grain boundary phase will soften and produce the desired ductile deformation. However, local overheating can introduce thermal damage (devitrification, melting, or even sublimation of the grain boundary phase, or oxidation of the silicon nitride), which is not desirable. Therefore, it is important to measure and know the surface temperature profile in the cutting zone during laser heating so that material removal can done within the critical minimum and maximum preheat temperatures .

The objective of the LAM research project is to develop a practical industrially rugged production multi beam laser assisted machining system that will handle a wide range of ceramics. The industrial set-up uses multiple high power diode fiber lasers with a much smaller equipment footprint and the flexibility of fiber-optics for laser positioning.

This project required a multi-disciplined team, consisting of: 1) the College of Engineering and Engineering Technology (CEET) at Northern Illinois University (NIU) for lab-bench studies; 2) the industrial partner, Reliance Tool & Manufacturing (RTM), located in Elgin, IL for production scale machining; 3) Multi-Scale Manufacturing Solutions, LLC (MSMS) with Frank Pfefferkorn for numerical simulation; 4) and Gateway Materials Technology (GMT) with Stephen Gonczy, for ceramic evaluation and property assessment.

The objectives of the first phase of this project were to –

1. Model and experimentally validate the laser heating parameters and machining parameters necessary to LAM a commercial high strength silicon nitride.
2. Develop and demonstrate an industrial scale LAM system.
3. Utilize a novel laser-interference method for 3-D measurement of surface roughness characterization.
4. Compare the room temperature flexural strength of the LAM machined silicon nitride with other surface finishes (as-received and ground).

EXPERIMENTAL SET UP & TESTING

Material Selection

A commercial high strength silicon nitride was chosen as the baseline material for the LAM machining study. The silicon nitride was received and LAM turned in the form of 25 mm diameter rods, 150 mm long.

This silicon nitride material was characterized (in a 2002 vintage) in a 2005 Army Research Laboratory study [24] and gave the following properties shown in Table I.

Table I. Silicon Nitride Properties Given in 2005 ARL Study [24]

PHYSICAL PROPERTIES	
Density	3.19 g/cc
Heat Capacity (RT, 1000C)	695, 1250 J/(kg-K)
Thermal Conductivity (RT, 1000C)	16, 32 W/(m-K)
Coefficient of Thermal Expansion (RT → 1000C)	3.4 ppm/K
MECHANICAL PROPERTIES	
Elastic Modulus	313 GPa
Hardness (300g, Hv) RT Hardness (300g, Hv) 400°C Hardness (300g, Hv) 800°C	13.6 GPa 11.6 GPa 9.0 GPa
Shear modulus	125 GPa
Poisson's ratio	0.253
RT Mean Tensile Strength (20 tests, SD = 51 MPa; High, Low = 643, 444 MPa) Weibull Modulus	586 MPa 16.9
RT Mean Flex Strength (4 pt, 40-80 mm) (20 tests, SD = 55 MPa; High, Low = 675, 477 MPa) Weibull Modulus	563 MPa 10.8
700°C Mean Flex Strength (4 pt, 40-80 mm) (10 tests, SD = 38 MPa; High, Low = 591, 489 MPa) Weibull Modulus	528 MPa 15.7
Fracture Toughness (RT - Chevron Notch)	8.67 MPa-m$^{1/2}$ (± 0.19)

The microstructure of this silicon nitride is illustrated in Figure 2, from Reference 25. Note the bimodal grain size with the large acicular silicon nitride grains, the light gray grain boundary phase, and the very small white inclusions.

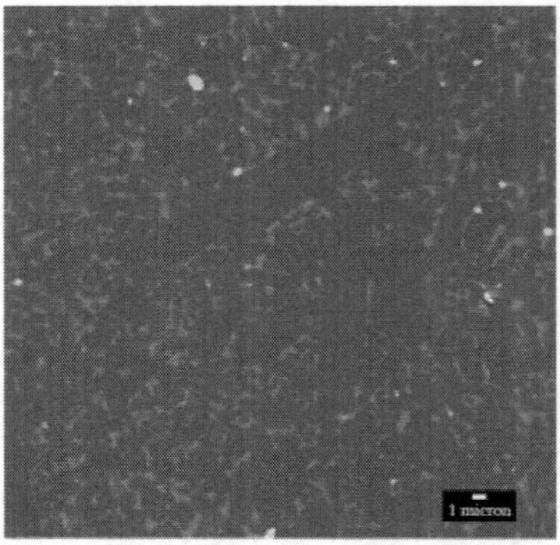

Figure 2. Microstructure of Silicon Nitride [24]

Northern Illinois University LAM System

A laboratory scale LAM system was set up at Northern Illinois University (NIU), shown in Figure 3. The system contains a Jenoptik, 250W CW, fiber-coupled diode laser integrated with a custom table top CNC lathe combined as a Class 1 laser system. The laser source delivers the beam through a 400µ fiber to a processing head which collimates and focuses the beam. The lathe, which operates up to 500 RPM, is integrated with a Delta Tau Advantage 400 CNC controller. Utilizing the CNC controller, three linear stages and servo motors are also automated, representing the X, Y, and Z axes. This allows for precise control over the tool and processing head positioning with feed rates possible up to 20 in/mm. In order to monitor the process, a FLIR A325 thermal imaging camera capable of measuring a -20-2000°C temperature range is used. The LAM turning was done with cubic boron nitride (CBN) tool inserts.

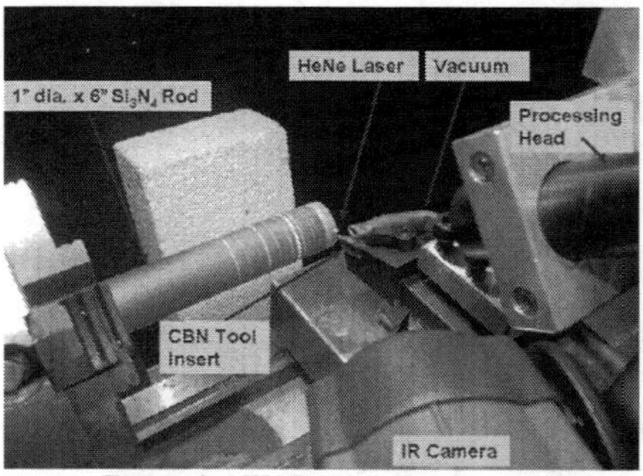

Figure 3. NIU LAM System

Reliance Tool LAM System

Based on the NIU LAM system, an industrial scale LAM system was built at Reliance Tool and Manufacturing. That system used a custom built a multi-beam diode laser system with three separate fiber diodes of 400 W each for a total of 1200 W. Each diode fiber has a six inch working distance and a 2 mm spot size. The main laser head is equipped with a built-in 2-color pyrometer system (temperature range 543-1500°C) for process monitoring and control. A FLIR A325 thermal imaging camera was added for thermal measurement. The laser system is integrated with a commercial, industrial size 5-axis 25 horsepower lathe. Dedicated hardware and software are used for data acquisition and process control. Figure 4 shows the interior working zone of the 5-axis machining station.

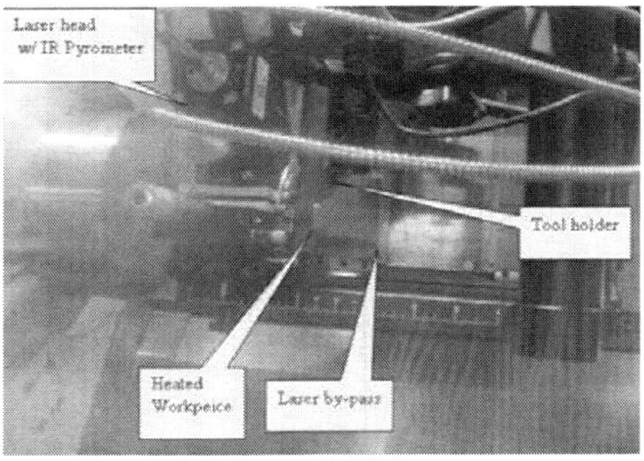

Figure 4. Reliance LAM System

Laser Surface Analysis

A laser surface inspection system was utilized to measure surface roughness of the ceramic, using an interference method similar to Shadow Moiré. The detailed theory on this optical measurement methodology is described in [25]. The laser-microscope system (Figure 5) has a capacity to resolve vertical differences to 120 nm across a 480 × 480 micron surface area. Essentially the surface of the ceramic is in contact with a grating that has 400 lines per mm. The surface is illuminated by a HeNe laser at an angle that will provide total internal reflectance. This generates a 3-D interference image that is captured by a CCD camera.

The interference patterns are analyzed with a fringe analysis software package [Holo Moiré Strain Analyzer^{TM} (HMSA) Version 2.0, developed by Sciammarella et. al. and supplied by General Stress Optics Inc. (Chicago, IL USA)]. The fringe analysis uses powerful Fast Fourier Transforms for filtering, carrier modulation, fringe extension, edge detection and masking operations, and removal of discontinuities, etc. The software produces a full statistical analysis of the interference pattern in both tabular and graphic form (Figure 6).

The laser system accuracy was cross checked against a NIST traced surface roughness calibration block. The R_a of the calibration block is 3.051 with a range of 3.018 and 3.079. The laser system gave an average R_a of 3.039 microns with multiple measurements on different regions of the calibration block.

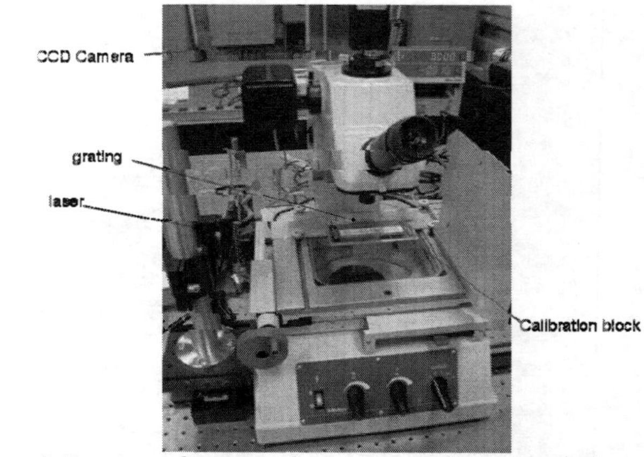

Figure 5. Experimental set up used to measure surface roughness for ceramic samples.

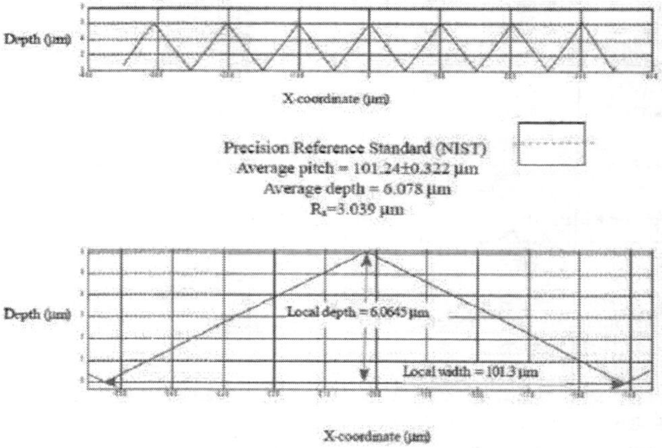

Figure 6. Experimental results from laser system given in 2D.

Flexure Strength Testing and Flaw Analysis

The 25-mm diameter 50-mm long silicon nitride rods were unsuitable (too thick and too short) for effective flexure testing. The rods were sliced into three arc segments with a smaller size (19 mm thick, 19 mm wide, and 50 mm long – Figure 7a). These arc segment specimens tested in flexure with the machined surface in tension at loads less than 4000 N (880 pounds); this geometry gave three test bars from each 25-mm diameter silicon nitride rod.

This arc segment specimen geometry has been used for other ceramics [24] and the method (specimens, fixturing, calculations) is described in detail by Quinn [26].

The test specimens were tested in 4 point- ¼ point bend test at room temperature in an Instron testing machine, using an articulated fixture (40 mm-20 mm spans) with rolling tool steel bearings (Figure 7b). The cross head rate was 0.125 mm/min. Each specimen was tested with the curved face in tension (down).

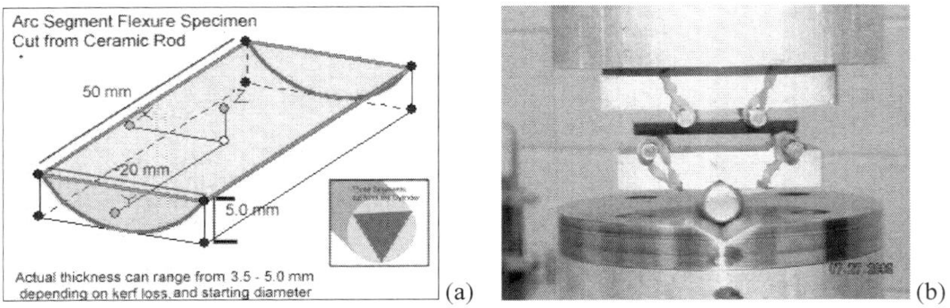

Figure 7. (a) Schematic of arc samples used for 4 point bend testing (b) view of experimental setup.

EXPERIMENTAL RESULTS AND DISCUSSION

NIU LAM Experiments

The NIU LAM parameter study used thermal modeling and bench top experiments to determine the baseline laser parameters needed for turning the silicon nitride 25-mm diameter rods, using cubic boron nitride (CBN) tool inserts. The study determined the laser power (190 W) laser spot size (2-mm), turning speed (500 RPM), and preheat time (9 minutes) needed to achieve a local temperature of ~850°C for softening the glassy grain boundary phase. This work is described in more detail in [27] and it established the baseline processing parameters for the industrial scale LAM study at Reliance Tool & Manufacturing.

Reliance LAM Experiments

The Reliance study developed proprietary baseline parameters for the successful LAM OD turning of silicon nitride rods (25-mm diameter, 150-mm long), using CBN tools. These parameters included insert geometry, spindle speed, surface velocity, laser spot-tool lead, depth of cut, and feed rate.

In a first effort to determine how LAM machining effected surface condition and mechanical strength, three different types of silicon nitride machining conditions were assessed. The first set included silicon nitride rods in the as-received conditions, to provide a baseline and to check against literature strength values. The second set of rods used a coarse (100 grit) diamond grind to baseline a high material removal rate diamond grinding that would introduce large surface flaws and reduce strength. The third -sets of rods were the laser assisted machining (LAM) samples, using the baseline Reliance LAM parameters. It should be noted at this stage of the study the baseline LAM parameters are not optimized for material removal rates, tool wear, or surface condition.

Each test set consisted of three (3) 50-mm long x 25-mm diameter silicon nitride rods. The machined (diamond ground and LAM) rods had about 0.5 mm of OD removed. The material removal rates in the phase of the study were:

Coarse Diamond Grinding – 0.15 mm^3/s

Baseline LAM machining – 1.3 mm^3/s

Each rod produced three flexure test specimens, so that there were nine flexure tests for each surface condition for a total of 27 flexure test specimens.

Surface Roughness

The surface condition of all 27 test specimens (from the nine rods) was measured with the laser surface analysis system. Each test specimen was measured in multiple locations and a statistical analysis was done on the data, producing mean, high, and low R_a values in microns for the nine rods. The mean, standard deviation (SD) and extreme values (high and low) are presented in Table II

TABLE II. Surface Roughness Measurements on Silicon Nitride A in Different Machined Conditions

CONDITION	AS-RECEIVED	DIAMOND GROUND	LAM MACHINED
Mean R_a	1.070	1.183	0.987
SD of the R_a	0.541	0.214	0.200
High, Low R_a	2.35, 0.55	2.04, 0.86	1.87, 0.63

Flexure Strength

The twenty-seven flexure test specimens broke across a range of applied loads - 1900 to 4000 Newtons (425 to 885 pounds). The specimens generally broke in 70 to 100 seconds. All the specimens broke within the inner load span. Table II shows the mean flexure strength along with the coefficient of variation (CoV) and extreme values (high and low) for the three surface conditions.

Table III. Silicon Nitride Fracture Strength Comparison with Three Different Surface Conditions

CONDITION	AS-RECEIVED	DIAMOND GROUND	LAM MACHINED
Mean - Strength (MPa)	524	469	529
CoV - Strength	11.6%	9.3%	21.6%
High, Low Strength (MPa)	604, 415	531, 408	623, 300

An analysis of this flexure data leads the following observations.

1. The mean flexure strength (524 MPa) for the as-received bars is about 7% less than the mean flexure strength (563 MPa) in the 2006 ARL study of silicon nitride A. This 7% reduction is within one standard deviation of the mean strength in the ARL study. The two sets of data had equivalent coefficients of variation (CoV) – 11.6% and 9.8%.
2. As expected the diamond ground specimens had a 10% lower mean flexure strength (469 MPa) than the as-received specimens (524 MPa), but this is within one standard deviation, so the statistical confidence of this difference is not strong. Both specimen sets had equivalent coefficients of variation – 11% and 9%.
3. The LAM specimens had an equivalent mean flexure strength (529 MPa) to the as-received specimens (524 MPa), but with a markedly larger CoV – 21% versus 11%.

The flexure data for the three specimen sets are plotted on a Weibull plot in Figure 8 [The Weibull modulus (m) was not calculated, because of the low number (9) of test specimens in each set.]

The Weibull plot clearly shows that the as-received specimens and the diamond ground specimens each have a single mode of failure, marked by all the data falling on a single line. But the data for the LAM specimens has a bimodal distribution with the three lowest values all coming from Rod #4. The bimodal behavior in LAM is a subject for future work.

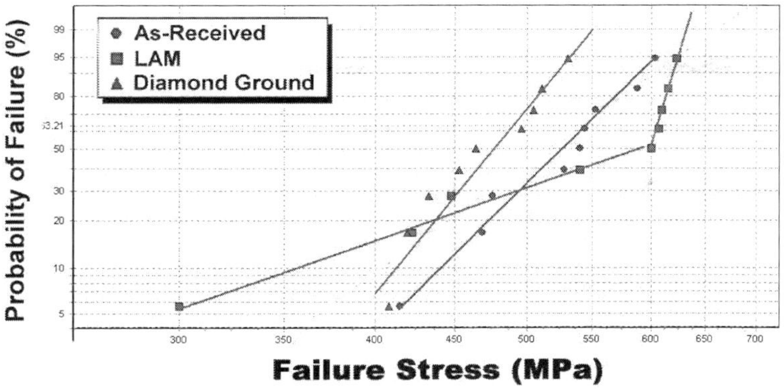

Figure 8. Weibull Plot of Flexure Strengths

By plotting the mean strength and the mean R_a (Figure 9) we can see that there is a correlation between these two quantities. The strength increases as R_a decreases. This work is preliminary and more research and investigation are necessary to see if this will hold for the other treatments and ceramics.

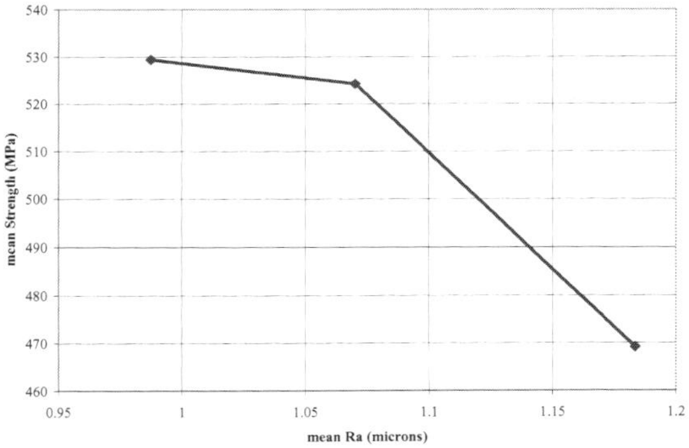

Figure 9. Mean strength (MPa) vs mean R_a (microns)

CONCLUSIONS

This work has culminated in a successful transition of LAM methods from the lab to the shop floor. Reliance Tool & Manufacturing has demonstrated its multi-beam industrial scale LAM system with the successful LAM turning of a commercial silicon nitride. The laser surface measurement system is highly sensitive, accurate, and rapid, providing detailed surface topology maps. Future work is required to confirm the correlation between strength and surface roughness taking advantage of the fact that 3-D analysis can be queried to establish other surface roughness parameters like R_t, R_q, R_z, etc. It will be also of interest to establish the mechanisms that are at the basis of this correlation and the role of the maximum flaw size.

FUTURE WORK

Future efforts will focus on six areas:

1. Optimize the LAM parameters for silicon nitride A, considering material removal rates, tool wear, surface roughness, surface flaws, and mechanical strength.
2. Assess the strength, surface condition, and material removal rates for fine diamond grinding (800 grit.) of silicon nitride A, as compared to LAM machining.
3. Determine the effect of laser surface heating on silicon nitride A, in terms of strength, surface condition, and flaw population. Determine if residual stresses play a role in strengthening.
4. Modify the LAM machining system for ID plunge turning.
5. Extend the LAM methods to silicon nitrides with different sintering aid compositions.
6. Extend the LAM methods to other high strength ceramics – zirconia, silicon carbide, mullite, and alumina.

ACKNOWLEDGEMENTS

The authors thank Tom Wagner and U.S. Army TARDEC (Contract Number W56HZV-04-C-0783) for support that made this work possible. The authors would like to express their gratitude towards Dr. Richard Johnson, Director of ROCK and Alan Swiglo, Assistant Director of ROCK, for their continued technical and management support. Special thanks to the graduate students at Northern Illinois University Mechanical Engineering Department, Sri Harsha Panuganti, Mike Matusky and Vishal Buraa for all their efforts on this project.

REFERENCES

[1]US Advanced Ceramics Market, *Fredonia Industrial Market Study*, Cincinnati, OH (December 2008).
[2]T. Abraham, U.S. Advanced Ceramics Industry—Status and Market Projections, *Ind. Ceram.*, 19 [2] 94–6 (1999).
[3]Y. Liang and S. P. Dutta, Application Trend in Advanced Ceramic Technologies, *Technovation*, 21, 61–5 (2001).
[4]F. L. Riley, Silicon Nitride and Related Materials, *J. Am. Ceram. Soc.*, Vol. 83, 245-265 (2000).
[5]I. P. Tuersley, A. Jawaid, and I. R. Pashby, Review: Various Methods of Machining Advanced Ceramic Materials,' *J. Mater. Processing Technol.*, 42, 377–90 (1994).
[6]V. Sinhoff, S. Schmidt, and S. Bausch, Machining Components Made of Advanced Ceramics: Prospects and Trends, *Ceram. Forum Int./Berichte Der Dkg*, 78 [6], E12–8 (2001).
[7]W. Konig and A. Wagemann, Machining of Ceramic Components: Process-Technological Potentials, *Machining Adv. Mater., NIST Spec. Publ.,* 847, 3–16 (1993).
[8]I. D. Marinescu, *Handbook of Advanced Ceramic Machining*, CRC Press, (2007).
[9]N.B. Dahotre and S.P. Harimkar, *Laser Fabrication and Machining of Materials*, Springer Science + Business Media (2008).
[10]W. Konig and A. K. Zaboklicki, Laser-Assisted Hot Machining of Ceramics and Composite Materials, *Int. Conf. Machining Adv. Mater. NIST Spec. Publ.*, 847, 455–63 (1993).
[11]Y. C. Shin, S. Lei, F. E. Pfefferkorn, P. Rebro, and J. C. Rozzi, Laser-Assisted Machining: Its Potential and Future, *Machining Technol.*, 11 [3] 1–6 (2000).
[12]F. Klocke and T. Bergs, Laser-Assisted Turning of Advanced Ceramics, *Rapid Prototyping Flexible Manuf., Proc. SPIE*, 3102, 120–30 (1997).
[13]J.C. Rozzi, F.E Pfefferkorn, Y.C. Shin, and F.P. Incropera, Experimental Evaluation of the Laser Assisted Machining of Silicon Nitride Ceramics, *J. of Manufacturing Science and Engineering,* Vol. 122, 666-670 (Nov. 2000).
[14]S. Lei, Y.C. Shin, and F.P. Incropera, Experimental Investigation of Thermo-Mechanical Characteristics in Laser-Assisted Machining of Silicon Nitride Ceramics, *J. of Manufacturing Science and Engineering,* Vol. 123,639-646 (Nov. 2001).

[15]S. Lei, Y. Shin, and F. Incropera, Experimental Investigation of Thermo-Mechanical Characteristics in Laser-Assisted Machining of Silicon Nitride Ceramics, *ASME J. Manuf. Sci. Eng.*, 123, 639–46 (2001).
[16]P. Rebro, Y. C. Shin, and F. P. Incropera, Laser Assisted Machining of Reaction Sintered Mullite Ceramics, *Trans. ASME, J. Manuf. Sci. Eng.*, 124, 875–85 (2002).
[17]F. E. Pfefferkorn, Y. C. Shin, Y. Tian, and F. P. Incropera, Laser-Assisted Machining of Magnesia-Partially Stabilized Zirconia, *ASME J. Manuf. Sci. Eng.*, 126, 42–51 (2004).
[18]Y. Tian and Y.C. Shin, Laser-Assisted Machining of Damage-Free Silicon Nitride Parts with Complex Geometric Features via In-Process Control of Laser Power, *J. Am. Ceram. Soc.*, 89 [11], 3397–3405 (2006).
[19]J. C. Rozzi, M. J. M. Krane, F. P. Incropera, and Y. C. Shin, Numerical Prediction of Three-Dimensional Unsteady Temperatures in a Rotating Cylindrical Workpiece Subjected to Localized Heating by a Translating Laser Source, 1995 *ASME Int. Mech. Eng. Conf. Exposition*, San Francisco, California, HTD,317 [2] 399–411 (1995).
[20]J. C. Rozzi, F. E. Pfefferkorn, F. P. Incropera, and Y. C. Shin, Transient, Three-Dimensional Heat Transfer Model for the Laser Assisted Machining of Silicon Nitride: I-Comparison of Predictions with Measured Surface Temperature Histories, *Int J. Heat Mass Transfer*, 43, 1409–24 (2000).
[21]J. C. Rozzi, F.P Incropera, and Y.C. Shin, Transient, Three-Dimensional Heat Transfer Model for the Laser Assisted Machining of Silicon Nitride: II. Assessment of Parametric Effects, *Int. J. of Heat and Mass Transfer* 43, 1425-1437 (2000).
[22]F. E. Pfefferkorn, F. P. Incropera, and Y. C. Shin, Heat Transfer Model of Semi-Transparent Ceramics Undergoing Laser-Assisted Machining, *Int. J. Heat Mass Transfer*, 48 [10] 1999–2012 (2005).
[23]Y. Tian and Y.C. Shin, Thermal Modeling for Laser Assisted Machining of Silicon Nitride Ceramics with Complex Features, *Journal of Manufacturing Science and Engineering*, Vol. 128 425-434, (May 2006).
[24]J. J. Swab, A.A. Wereszczak, J. Tice, R. Caspe, R. H. Kraft, and J. W. Adams, Mechanical and Thermal Properties of Advanced Ceramics for Gun Barrel Applications, *Army Research Laboratory Report ARL-TR-3417*, February (2005).
[25]C.A. Sciammarella, L. Lamberti, F.M. Sciammarella, G. Demelio, A. Dicuonzo, and A. Boccaccio, Application of Plasmons to the Determination of Surface Profile and Contact Stress Distribution, *Strain*, (2009).
[26]G. D. Quinn, The Segmented Cylinder Flexure Strength Test, *Ceramic Eng. and Sci. Proc.*, 27[3], 295 – 305 (2006).
[27]F.M. Sciammarella and M. Matusky, Fiber Laser Assisted Machining of Silicon Nitride, *Conference Proceedings ICALEO*, 2009 (to be published).

Net Shape Forming

GELCASTING OF HIGH PERFORMANCE CARBIDE CERAMICS WITH LARGER SIZE/COMPLEX SHAPE

Dongliang Jiang

The State Key Laboratory of High Performance Ceramics and Superfine Microstructures, Shanghai Institute of Ceramics, Chinese Academy of Sciences, 1295 Dingxi Road, Shanghai 200050, China

ABSTRACT

In this paper, the gelcasting process of silicon carbide and boron carbide ceramics with big size and complex shape are presented. The main concern is to develop a reliable process to produce homogeneous, crack free gelcasting samples. This is based on the understanding of interactions among components, the optimization of the slurry rheological properties and preventing the obvious settling of ceramic powder before gelation. The green samples with high bending strength and green density as well as homogeneous microstructure were realized. Subsequently, the optimized sintering process was also proposed for obtaining high performance ceramics with effective control of the deformation or warping during sintering. The microstructure development and resulting properties of green and final sintered sample were studied.

1. INTRODUCTION

Carbide ceramics possessing a variety of desirable properties, such as high mechanical strength, high chemical stability, good thermal conductivity, low coefficient of expansion and outstanding corrosion resistance, have found a wide range of applications in aerospace, industry, nuclear energy and high-temperature thermoelectric conversion, defense [1,2,3,4] etc. However, the widespread applications of carbide ceramics are limited by the low reliability, difficulty in machining and high manufacturing cost, especially in the fabrication of parts with big size and/or complex shape.
Gelcasting is a colloidal-forming process to yield ceramics, achieving in situ solidification of the ceramic slurry through polymerization of monomers [5,6,7,8,9,10,11,12]. The gelcasting process involves preparation of aqueous slurries with the addition of monomers, cross-linkers, initiator and sintering aids. [5].

In gelcasting process, high solid loading slurry with low viscosity should first be prepared by dispersing the ceramic powders in a premixed solution of the monomer and the crosslinker. After the addition of an initiator and a catalyst, the slurry is poured into a mold. The monomer and crosslinker then copolymerize in situ to create a macromolecular network, holding ceramic particles very close to each other. Homogeneous ceramic green bodies with uniform microstructure and excellent mechanical property are therefore obtained. Gelcasting can be used to prepare ceramic parts with complex shape, which reduce the huge machining cost. In recent years, gelcasting has been widely studied to produce ceramic materials, including Al_2O_3[6,7,8, 13], Si_3N_4[6,7,13,14], SiC[15,16,17,18,19,20,21], PZT[22], Al_2O_3-ZrO_2[23], SiC -Si_3N_4[24], $BaTiO_3$[25] etc.
In glecasting, the solids loading is a key issue because higher solid content could reduce the shrinkage during drying and firing and increase the wet and dry green strength. This is usually realized through the selection of dispersant and its content, the optimization of slurry properties including pH, rheological behavior etc. [26,27,28,29,30]. Drying of 'green' gelcast part is a complex, multiphase transport process need to be well studied [31,32,33].

The objective of this paper was to introduce the recent advance in our lab about the development of reliable gelcasting process for forming SiC and B_4C ceramics with larger size and complex shape and to examine the properties of the green and sintered samples. The rheological properties and the gelation process were also characterized.

2. EXPERIMENTAL PROCEDURES

2.1. Starting materials

The raw materials used in gelcasting process are listed in Table 1. Based on the study in our lab, an initiator system composed of 2,2'-azobis[2-(imidazolin-2-yl) propane]dihydrochloride(AZIP) and other components were proposed.

Commercially available SiC powder (FCP-15, Saint-Gobain's Silicon Carbide department, Norway) produced by the Acheson method was used in this study. The average particle size and the specific surface area are 0.58μm and 15.24m^2/g respectively. Boron carbide (Mudanjiang Jingangzuan Boron Carbide Co., Ltd.) with the average particle size as 0.93 μm and a specific surface area as 10.78 m^2/g were used as the sintering additives. Dextrin was used as the C source (sintering additives).

2.2. Gel-casting and sintering

The schematic of gelcasting process is described in Fig. 1. Initially, SiC powder was suspended in a premix solution, which had been prepared by dissolving 12 wt% N,N-dimethyl acrylamide (DMAA) and N,N'-methylenebisacrylamide (MBAM) in a 13:1 ratio in de-ionized water. The slurries were prepared by adding ceramic powder to the premix solutions under mechanical agitation. To improve the dispersion of SiC powder in the premix solution and the fluidity of the suspension, PEI (at the rate of 0.456 mg/m^2 of SiC) was used as the dispersant [26,34]. All the particulate slurries were degassed for 10-30 min after the addition of initiator in order to eliminate the air bubbles trapped inside before casting into moulds. Afterward, the slurries were cast into non-porous molds, which were then allowed to set in water bath at 45-60 °C for 30-90 min in order to gel the system. The gelled green bodies were de-molded and dried under controlled humidity conditions to avoid cracking and non-uniform shrinkage due to rapid drying. The green blocks were heated to 600°C at a heating rate of 1.0 °C/min and held several times in the range of 300-400°C to burn out the monomers and other volatiles, followed by pressureless sintering at 2200 °C for 1 h in an inert atmosphere for densification purpose.

The gelcasting process of B_4C is similar to that of SiC except that only TiC was used as the sintering additives. The pressureless sintering of B_4C is performed at 2250°C for 30min in Ar atmosphere.

Rheological measurements were performed on a stress controlled rheometry (SR-5 Rheomeric scientific instrument company, U.S.A.). Measurements were carried out at 25°C.

2.3. Characterization of gel-casting green bodies and as-sintered samples

The density of green pieces was determined by Hg intrusion porosimetry in a Micromeritics ASAP2010 porosimeter. Bulk densities of as-sintered samples were determined by Archimedes' principle. Flexural strength was measured by three-point bending with a span of 30mm and a cross-head speed of 0.5 mm/min, using Instron 5566 universal testing machine. The size of the samples was normally 3mm×4mm×36 mm. Hardness and toughness were measured by indentation test on Wilson-wolpert Tukon 2100B (INSTRON), the load and loading time were 50N and 10 s respectively. Microstructures of SiC and B_4C samples were investigated by field emission scanning electron microscopy (JSM-6700F, Jeol, Japan).

3. RESULTS AND DISCUSSIONS

3.1 Aqueous gelcasing of SiC

The effect of dispersant, solid content, composition on the rheological properties of SiC slurries as well as the consolidation process has been well studied in our lab [18, 26,27]. Based on the study before, the monomer content was kept as 12wt% and the DMAA/MBAM mass ratio as 13:1.

A suspension for gelcasting should always meet two requirements: good fluidity and high solid content. On the one hand, a suspension with good fluidity and homogeneity is required for the casting process; on the other hand, higher solid content can result in higher green density as well as less deformation and defects of the green and sintered parts. The effects of monomer and crosslinker on the rheological properties of SiC slurries were characterized. Fig.2 showed the rheological properties of the slurries in the absence and presence of monomers. It can be found that the monomer and cross-linker have limited influence on the viscosity of 50 vol% SiC slurries. The slurries have enough fluidity and low viscosity for the gelcasting process.

Fig.3 showed the polymerization of acrylamide monomers using AZIP mixture as the initiator. It was found that with the increase in temperature from 40 to 60°C, the idle time increased from 15min to about 23min. In our experimental range, the temperature is usually controlled in the range of 40-50°C.

SiC green bodies with smooth surface and no visible defects can be produced by gelcasting. The wet part contains moisture about 15-17wt% of its mass. The drying process was conducted in oven through the temperature and humidity control. After drying in ovens, the linear shrinkage is in the range of 2-2.5%, the samples are nearly uniform in all dimensions and no region is dried faster than the other. The relative density and flexural strength of green bodies are around 57% and 40MPa, respectively. The microstructure of the green body as observed by SEM is shown in Fig.4. It shows that powders in green body compact closely and homogeneously, and that the pores decrease greatly with the size less than 1μm. Green bodies with the size up to 180mm were successfully prepared and showed in Fig.5.

The linear shrinkage and the relative density of the as sintered SiC samples are 16% and 98% respectively. Microstructure of the as sintered sample is shown in Fig.6. The fracture surface of SiC samples exhibited a homogeneous microstructure, few pores were observed. Most of the SiC crystal grains grow up to about 1-5 μm. Based on the SEM observation, the fracture mode is a mixture of intergranular and intragranular type.

3.2 Aqueous gelcasting of B_4C

The gelcasting of B_4C has been reported previously [35,36]. Based on the results, TiC is selected as sintering additives for the pressureless sintering of B_4C. The content of TiC was fixed at 10wt% after screening. The morphology and particle size distribution of B_4C powder was shown in Fig.7. The formulation of B_4C slurries was shown in Table 2. It is more feasible for preparing B_4C slurries with the solid content as high as 57vol% with enough fluidity. The stability of B_4C slurries in the presence of initiator was characterized by rheological measurement. Results are shown in Fig.8. Similar to that for SiC slurries, well stabilized B_4C slurry can be obtained, which was quite stable and no obvious gelation was observed after holding at room temperature for 24h. After degassing, the concentrated B_4C gelcasting slurries have good fluidity and can be solidified with AZIP mixture as the initiator at 45-50°C. The gelation process at 50°C was shown in Fig.9. A gradual increase in the storage modulus was observed, which are different from that for SiC samples. This might be due to the presence of carbon on B_4C powder surface, which could inhibit the gelation reaction. The drying of B_4C samples was similar to that for SiC. After drying, well-shaped green bodies were obtained. The relative density of gelcasted B_4C green body is around 64vol%. The microstructure of B_4C green samples are shown in Fig.10.

The green samples exhibited high mechanical properties comparable to that of SiC samples. The bending strength of green samples is as high as 56MPa. This high strength comes from the cross-linked gel network, which adhered to B_4C particles and bound them together. Pores with different sizes in the green bodies were also observed. The pore diameter distribution, obtained by Hg intrusion porosimetry, showed a monomodal distribution type with the peak around 156 nm.
The relative density of sintered samples at 2250°C was 96.2%. The XRD patterns showed the pure phase of B_4C. The fracture strength and hardness are 253.3±20.0MPa and 24.01±3.27GPa respectively. Micrographs of the fracture surface of B_4C sintered samples are shown in Fig.11. B_4C samples without TiC as the sintering additives were also prepared and sintered under the same condition for comparison. The relative density of as-sintered samples at 2250°C was 91.6%. The fracture strength is 261±39MPa, which is comparable to that of B_4C samples with TiC. While the hardness was 19.04±2.27 GPa. Microstructure of the samples showed a lot of trapped pores in the fracture surface, Fig.11. As pure B_4C ceramics tend to have more satisfied mechanical properties, further work on the sintering of pure B_4C is underway.

4. CONCLUSIONS

SiC and B_4C samples with complex shape and/or larger size were successfully prepared using aqueous gelcasting process with a newly developed initiator system. The slurries showed shear thinning rheological behavior and good stability for gelcasting. The green strength of SiC bodies was over 40MPa, suitable for green machining. After gel-casting, drying and sintering at 2200°C, SiC samples can be densified to a relative density of 98%. The mechanical properties of the obtained pieces are satisfying. The relative density of as-sintered B_4C samples was 96.2%. The gelcasting process appears to be suitable for manufacturing complex shaped components requiring minor/no post sintering machining.

Table 1 Raw materials used in gelcasting process

Raw materials	Abbreviation	Function
Silicon carbide	SiC	Powder
Boron carbide	B_4C	Sintering additives
Dextrin	Dextrin	C source
Polyethyleneimine	PEI	Dispersant
Deionized water	H_2O	Solvent
N,*N*′-Dimethylacrylamide	DMAA	monomer
N,*N*′-Methylenebisacrylamide	MBAM	Cross linker
2,2'-azobis[2-(2-imidazolin-2-yl)propane] mixture	AZIP	Initiator

Table 2 Formulation for the gelcasting of B_4C

Materials and reagent	Function	Content (vol%)
B_4C	Powder	55-53
TiC	Sintering additives	2.9-3.1
Deionized water	Solvent	34-36
TMAH	Dispersant	1.8-1.9
DMAA	monomer	4.5-4.8
MBAM	Cross linker	0.9-1.0
AZIP mixture	Initiator	0.02

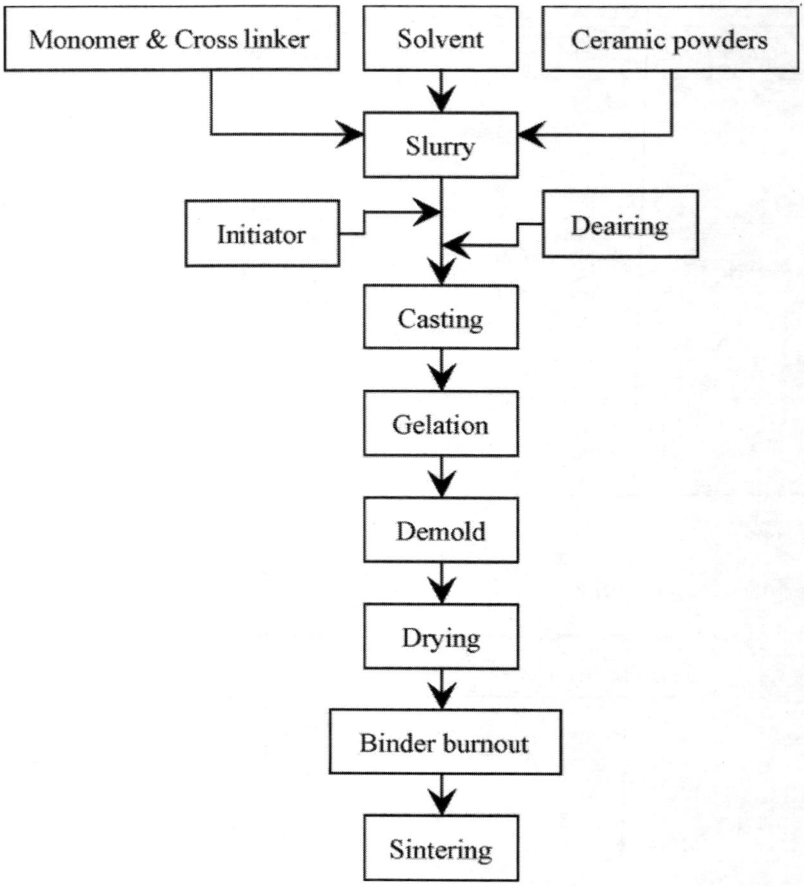

Fig. 1 Schematic of gelcasting process.

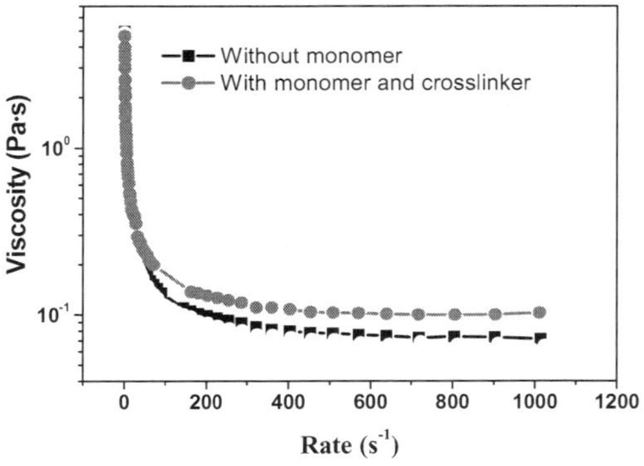

Fig.2 Rheological properties of 50vol% SiC gelcasting slurries

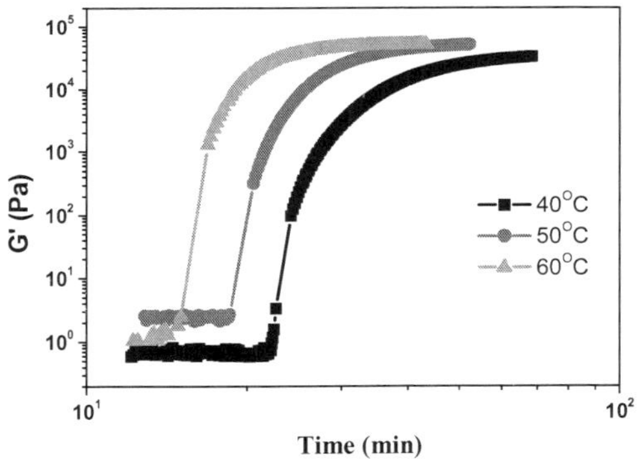

Fig.3 Variation of storage modulus of SiC slurries with time and temperature

Fig.4 Microstructure of green SiC samples

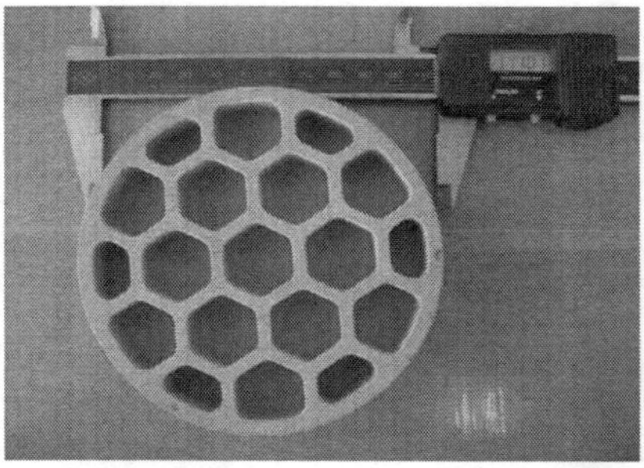

(a) ϕ110 green samples before machining

(b) ϕ180 green samples after machining

Fig. 5 Optical pictures of SiC samples after gelcasting and drying

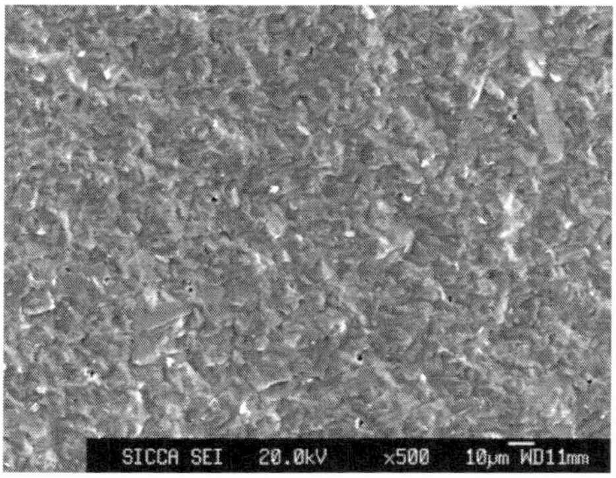

(a) Fracture surface at low magnification

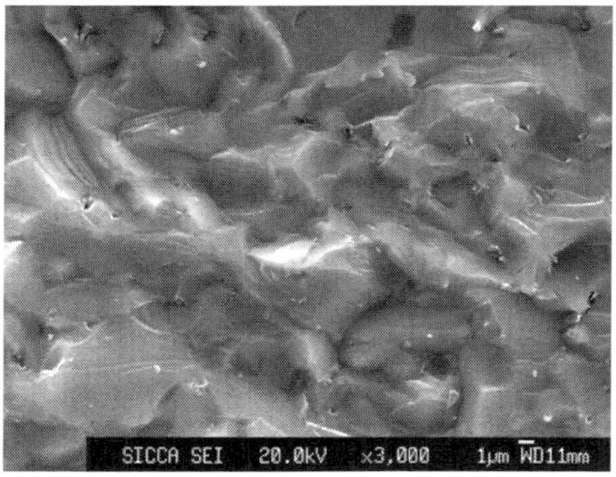

(b) Fracture surface at high magnification

Fig. 6 Micrographs of SiC samples after sintering

(a)

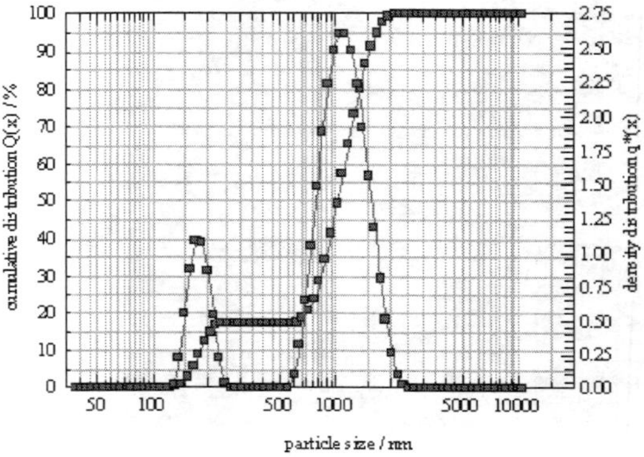

(b)

Fig.7 Morphology and particle size distribution of B_4C

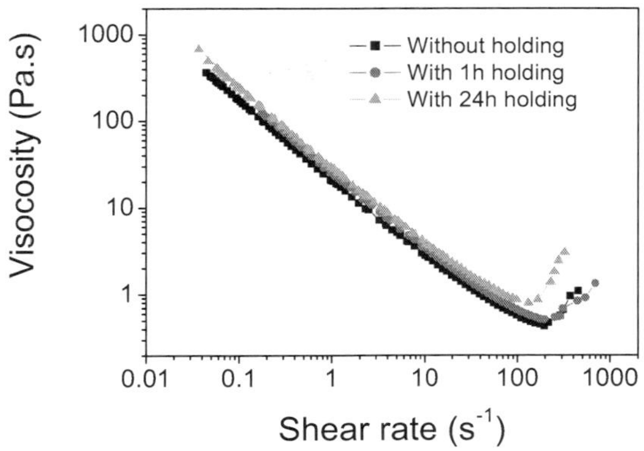

Fig. 8 Rheological of B_4C slurries before and after holding

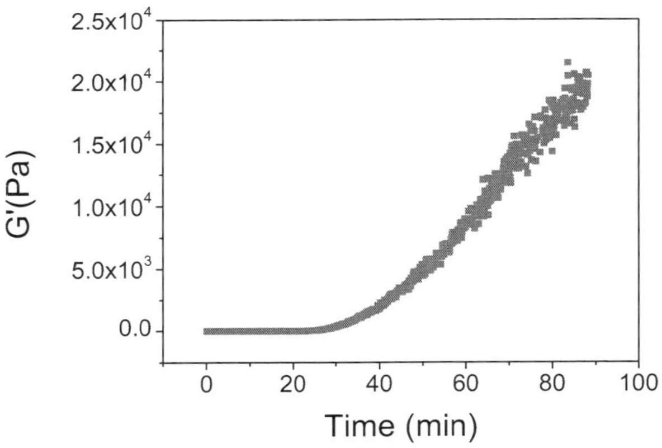

Fig. 9 The gelation of B_4C slurries at 50°C

(a)

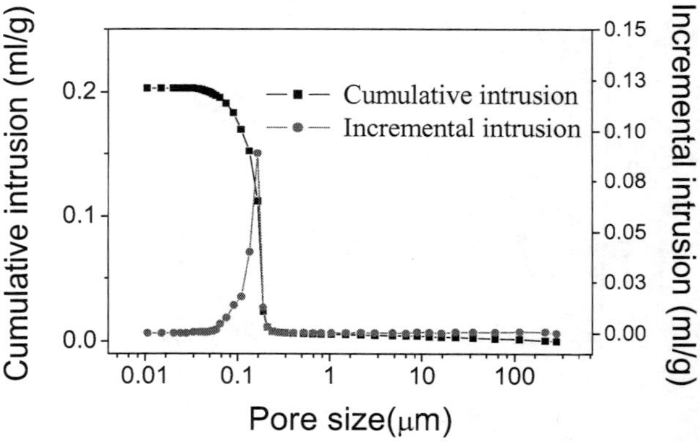

(b)

Fig. 10 Microstructure and pore size distribution of green B_4C samples

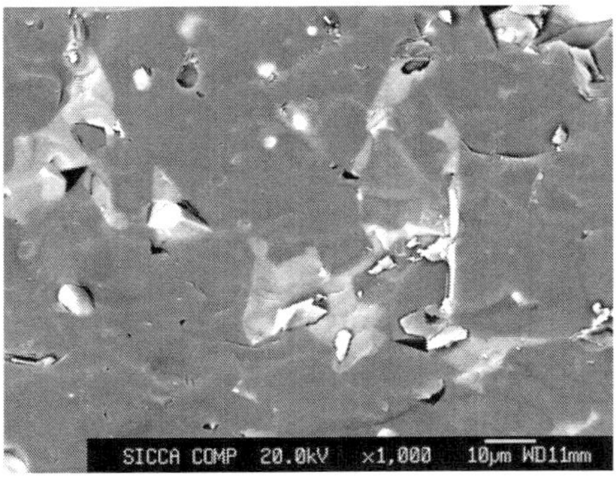

(a) with TiC

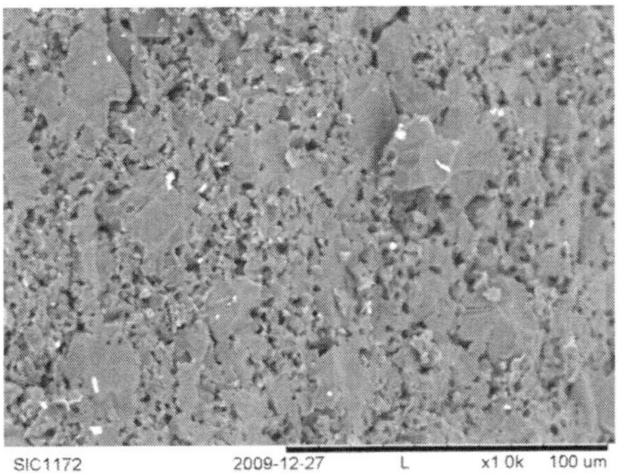

(b) without TiC

Fig.11 microstructure of B_4C samples after sintering

REFERENCES

[1] W. C. Johnson, Advances in Polishing Plate Bases, Cup Rims, *Am. Ceram.Soc. Bull.*, **80**, 64-65(2001).

[2] K. Y. Chia, W. D. G. Boecker, and R. S. Storm, Silicon Carbide Bodies Having High Toughness and Fracture Resistance and Method of Making Same, *US Patent*, 5,298,470 (1994).

[3] J. Robichaud, Silicon Carbide Replaces Berylium for Airborne and Space-Based Optics, *MRS Bulletin*, 914(2004).

[4] K. Komeya, M. Matsui, in: R.W. Cahn, P. Haasen, and E.J. Kramer (Eds.), Materials Science and Technology: A Comprehensive Treatment; in: M.V. Swain (Ed.), Structure and Properties of Ceramics, vol. 11, VCH,

Weinheim, 1994, p. 517.

[5] A. C. Young, O. O. Omatete, M. A. Janney, and P. A. Menchhofer, Gelcasting of Alumina, *J. Am. Ceram. Soc.*, **74**, 612-8 (1991).

[6] O.O. Omatete, M.A. Janney, and R.A. Strehlow, Gelcasting-A New Ceramic Forming Process, *Am. Ceram. Soc. Bull.*, **70**, 1641-9 (1991).

[7] O.O. Omatete, M.A. Janney, and S.D. Nunn, Gelcasting: From Laboratory Development Toward Industrial Production, *J. Eur. Ceram. Soc.*, **17**, 407-13 (1997).

[8] K. Prabhakaran, and C. Pavithran, Gelcasting of Alumina Using Urea–Formaldehyde II. Gelation and Ceramic Forming, *Ceram. Int.*, **26**, 67-71 (2000).

[9] M.A.Janney, and O.O.Omatete, Method for molding ceramics powders using a water-based gel-casting process, U. S. Patent, 5145908.

[10] R. Gilissen, J.P. Erauw, and A. Smolders et al., Gelcasting, a near net shape technique, *Mater. Design*, **21**, 251-257(2000).

[11] M.A. Janney, Method for molding ceramic powders, US Patent, 4894194(1990).

[12] M.A. Janney, and O.O. Omatete, Method for molding ceramic powders using a water-based gelcasting, U.S.Pat., 5,028,362(1991).

[13] K. Prabhakaran, and C. Pavithran, Gelcasting of alumina from acidic aqueous medium using acrylic acid, *J. Eur. Ceram. Soc.*, **20**,1115-1119(2000).

[14] L.J. Zhou, and Y. Huang, et al., Preparation of Si_3N_4 ceramics with high reliability via a processing strategy, *J. Eur. Ceram. Soc.*, **22**, 1347-1353(2002)

[15] F. Li, and H.-Y. Chen, et al., Effect of polyethylene glycol on the surface exfoliation of SiC green bodies prepared by gelcasting, *Mater. Sci. Eng.*, **A368**, 255-259(2004).

[16] Z.Z. Yi, Z.P. Xie, and Y. Huang, et al., Study on Gelcasting and Property of Recrystallized Silicon Carbide, *Ceram. Inter.*, **28**, 369-376(2002).

[17] Ibram Ganesh, D. C. Jana, Shamshad Shaik, and N. Thiyagarajan, An Aqueous Gelcasting Process for Sintered Silicon Carbide Ceramics, *J. Am. Ceram. Soc.*, **89**, 3056-3064 (2006)

[18] T. Zhang, Z.Q. Zhang, J.X. Zhang, D.L. Jiang, and Q.L. Lin, Preparation of SiC ceramics by aqueous gelcasting and pressureless sintering, *Mater. Sci. Eng. A*, **443**, 257-261 (2007).

[19] M. D. Vlajic, and V. D. Krstic, Strength and machining of gelcast SiC ceramics, *J. Mater. Sci.*, **37**, 2943-2947(2002).

[20] Z.-Z. Yi, Z.-P. Xie, and J.-T. Ma, et al., Study on gelcasting of silicon nitride-bonded silicon carbide refractories, *Mater. Lett.*, **56**, 895- 900 (2002).

[21] L. Zhou, Y. Huang, and Z. Xie, Gelcasting of concentrated aqueous silicon carbide suspension, *J. Eur. Ceram.*

Soc., **20**, 85-90(2000).

[22] D. Guo, K. Cai, and L. Li, et al., Application of gelcasting to the fabrication of piezoelectric ceramic parts, *J. Eur. Ceram. Soc.*, **23**, 1131-1137(2003).

[23] O.O. Omatete, A. Bleier, and C.G. Westmoreland et al., Gelcast zirconia - alumina composites. *Ceram. Eng. Sci. Proc.*, **12**,2084-2094(1991).

[24] Z.Z. Yi, Z.P. Xie, and Y. Huang et al., Preparation of High Concentrated Suspension and Gelcasting Process for Silicon Nitride Bonded Silicon Carbide Refractories, *Key Engineering Materials*, **224-226**, 685-690(2002).

[25] Y.X. Hu, D.X. Zhou, and D.L. Zhang et al. , PTCR characteristics of gelcast $BaTiO_3$ ceramic thermistor, *Sensors Actuators A*, **88**,67-70(2001).

[26] J. X. Zhang, D. L. Jiang, S. H. Tan, L. H. Gui, and M. L. Ruan, Aqueous Processing of SiC Green Sheets I: Dispersant, *J. Mater. Res.*, **17**, 2012-8 (2002).

[27] J.X. Zhang, Q. Xu, F. Ye, Q.L. Lin, D.L. Jiang, and M. Iwasa, Effect of citric acid on the adsorption behavior of polyethylene imine (PEI) and the relevant stability of SiC slurries, *Colloids and Surfaces A: Physicochemical and Engineering Aspects*, **276**,168-175(2006).

[28] B. J. Briscoe, A. U. Khan, and P. F. Luckham, Optimising the Dispersion on an Alumina Suspension Using Commercial Polyvalent Electrolyte Dispersants, *J. Eur. Ceram. Soc.*, **18**, 2141-7 (1998).

[29] V. A. Hackley, Colloidal Processing of Silicon Nitride with Poly (Acrylic Acid): I, Adsorption and Electrostatic Interactions, *J. Am. Ceram. Soc.*, **80**, 2315-25 (1997).

[30] R.R. Rao, H.N. Roopa, and T.S. Kannan, Effect of pH on The Dispersability of Silicon Carbide Powders in Aqueous Media, *Ceram. Int.*, **25**, 223-30 (1999).

[31] A. Barati, M. Kokabi, and N. Famili, Modeling of liquid desiccant drying method for gelcast ceramic parts, *Ceram. Int.*, **29**, 199-207(2003).

[32] A. Barati, M. Kokabi, M. Hossein, and N. Famili, Drying of gelcast ceramic parts via the liquid desiccant method, *J. Eur. Ceram. Soc.*, **23**, 2265-2272 (2003).

[33] S. Ghosal, A. Emami-Naeini, Y.P. Ham, B.S. Draskovich, and J.P. Pollinger, A physical model for the drying of gelcast ceramics, *J. Am. Ceram. Soc.*, **82**, 513-520(1999).

[34] W. Li, P. Chen, M. Gu, and Y. Jin, Effect of TMAH on Rheological Behavior of SiC Aqueous Suspensions, *J. Eur. Ceram. Soc.*, **24**, 3679-84 (2004).

[35] J.X. Zhang, D.L. Jiang, and Q.L. Lin, Aqueous processing of boron carbide, *Journal of the Ceramic Society of Japan*, **116**, 681-684(2008).

[36] D.L. Jiang, Gelcasting of carbide ceramics, *Journal of the Ceramic Society of Japan*, **116**, 694-699(2008).

PROCESSING OF COMPLEX-SHAPED MICRO PARTS BY REACTION-BONDING AND SINTERING OF SILICON NITRIDE

M. Müller, W. Bauer, R. Knitter
Karlsruhe Institute of Technology, (KIT), Institute for Materials Research III,
Hermann-von-Helmholtz-Platz 1, 76344 Eggenstein-Leopoldshafen, Germany.

J. Rögner
Karlsruhe Institute of Technology, (KIT), Institute for Materials Science and Engineering I,
Kaiserstr. 12, 76131 Karlsruhe, Germany.

ABSTRACT

Paraffin-based feedstocks for low-pressure injection molding containing silicon and varying sintering additive percentage were prepared to mold micro-bending test specimens as well as prototypic components of a micro turbine. After thermal debinding samples were nitrided in forming gas atmosphere (N_2 + 5% H_2) at 1390°C and subsequently sintered pressureless at temperatures between 1700 °C and 1800 °C for periods of 1 or 2 hrs. It was stated that the additives Y_2O_3, Al_2O_3, and MgO not only control the sintering behavior of Si_3N_4, but also influence the nitridation reaction of silicon. After sintering densities up to 98% were achieved. The investigation of the mechanical properties – based upon micro-bending bars – revealed values for σ_0 between 500 and 1200 MPa. It was found that the micro bending strength is rather limited by the presence of crystalline secondary phases (mainly melilite, $Y_2Si_3O_3N_4$) at the surface of the micro specimens, than by the residual porosity.

As micro samples exhibit a large surface-to-volume ratio they are in particular affected by decomposition of Si_3N_4 and volatilization of sintering additives. To obtain micro components with high strength and reliability the occurrence of secondary phases especially at the sample surface needs to be avoided within the processing framework.

INTRODUCTION

Among the numerous modifications of Si_3N_4 sintered reaction-bonded silicon nitride (SRBSN) can be considered as a favorable compromise between the high-performance sintered Si_3N_4 (SSN) and the low-cost reaction-bonded Si_3N_4 (RBSN). Analogous to the RBSN process, SRBSN silicon nitride is generated in-situ by direct nitridation of Si. A following post-sintering step at higher temperatures with the presence of sintering aids diminishes the considerable porosity which still exists after reaction-bonding. By this densification 93-99% of theoretical density can be realized and thus mechanical properties which are comparable to hot-pressed (HPSN) or conventionally sintered (SSN) silicon nitride are achieved[1-3]. But in contrast to SSN, cost-effective (metallurgical) silicon qualities can be used as raw material instead of expensive Si_3N_4 powder.

As a consequence of a volume increase of about 22 % during RBSN formation the "green density" is significantly higher than for green bodies prepared from Si_3N_4 powder. The resulting sintering shrinkage for full densification is correspondingly lower in the case of SRBSN. This aspect renders SRBSN interesting for near net-shape manufacturing where a reduced sintering shrinkage should enable narrower dimensional tolerances[4]. Especially micro components or micro-patterned parts demand high precision from the manufacturing process as final machining is excluded for economical and practical reasons.

It was the aim of the present work to evaluate if (low-pressure) injection-molding of sintered reaction-bonded silicon nitride is a suitable processing technology for the manufacturing of high strength *micro* parts with narrow tolerances. To adopt a processing technology to smaller dimensions some micro specific features have to be considered. An important factor when changing from the macro to the micro world is the increase of surface-to-volume ratio with decreasing size. For macroscopic or thick-walled components the diffusion of nitrogen into the inner regions might become limiting especially when the pore cross section becomes smaller in the course of the reaction. Also the exothermic character of silicon nitride formation, which may cause melting of the powder compact, has to be tackled[5]. Both pore diffusion and dissipation of heat become minor problems when small components with high surface area have to be formed, and consequently faster processing is possible. On the other hand the higher surface-to-volume ratio is responsible for the fact that the mechanical properties of micro parts are increasingly dominated by the defect density of the surface region[6,7]. It is therefore essential that the mechanical properties are determined on the basis of samples with correspondingly reduced dimensions.

EXPERIMENTAL

Powder and feedstock preparation

As starting material Si powder (SiMP grade AX05, HC Starck, Germany) with a mean diameter d_{50} of 3.9 μm was used. To reduce the particle size and for homogeneous distribution of the sintering aids, Si powder was planetary ball milled together with varying amounts of Y_2O_3 (Aldrich, 99.99%, d_{50} = 3.4 μm), Al_2O_3 (Ceralox SPA05, d_{50} = 0.35 μm) and MgO (Merck, p.a. d_{50} = 0.25 μm) for 20-48 h in 2-propanol using zirconia beaker and balls. After ball milling the powder blends were dried, sieved, and characterized according to their average particle size (Dispersion Technology, DT1200) and BET specific surface area (Flow Sorb II, Micromeritics), as shown in Table 1.

Low-pressure injection moulding requires low-viscosity feedstocks and therefore a paraffin wax (Terhel 6403, Schüman Sasol) was used as main binder component. The following dispersants were added to the molten paraffin to ease wetting of the powder blend by the nonpolar binder, to stabilize the feedstock against segregation, and to control the rheological behaviour of the feedstock: ODTMS (Octadecyltetramethoxysilane, 90% in ethanol, Aldrich), Hypermer LP1, (Uniquema) and N-alkyl-succinimide (OLOA 1200, Chevron).

Moulding and thermal treatment

One of the main advantages of LPIM is the possibility to use soft moulds due to the moderate injection pressure of only 0-3 bar. Even silicone rubber moulds will work and their preparation is simple, fast, and inexpensive as they are obtained within hours just by copying a master model made of

Table 1: Properties of ball-milled powder mixtures and feedstock composition

Composition	Y_2O_3 wt%	Al_2O_3 wt%	MgO wt%	BET m^2/g	solids loading vol%
YA-1	8.0 (5.0)	3.2 (2.0)	-	6.8	56.5
YA-2	8.0 (5.0)	3.2 (2.0)	-	8.8	61.0
YA-3	12.3 (7.7)	3.0 (1.9)	-	6.3	56.4
YA-4	12.3 (7.7)	3.0 (1.9)	-	9.1	56.2
YA-5	12.3 (7.7)	1.7 (1.1)	-	7.8	56.6
YM-1	12.4 (7.8)	-	0.7 (0.4)	10.5	57.1
YM-2	12.3 (7.7)	-	3.0 (1.9)	7.1	55.6
YM-3	12.3 (7.7)	-	3.0 (1.9)	7.1	60.0
YAM-1	12.3 (7.7)	1.5 (0.9)	1.5 (0.9)	6.3	56.8
YAM-2	12.3 (7.7)	1.5 (0.9)	1.5 (0.9)	10.1	56.9

* Numbers in brackets refer to the composition after Si has been converted into Si_3N_4

any material. Due to the low feedstock viscosities in the order of 10 Pa•s complete mould filling was achieved just by casting the feedstock into preheated silicone moulds at 90-100 °C. Entrapped air was evacuated afterwards from the filled mould while the feedstock was still liquid. High mould elasticity and low adhesion forces facilitate the demolding even of complex-shaped parts after cooling down to room temperature. To evaluate the dimensional changes during reaction-bonding and sintering simple geometries like rectangular and disc samples were chosen as well as more complex components were prepared to assess the micro moldingability of the feedstocks (see also Fig.4). Fig.1 shows the "column array" termed test structure which primarily serves as a supply for a large number of identically prepared specimens for micro-bending tests. For the master model a brass block with dimensions of 20x15x10 mm^3 has been micromachined on one side to leave an array of 15x15 columns, approx. 260 μm in length and width, and 1500μm in height.

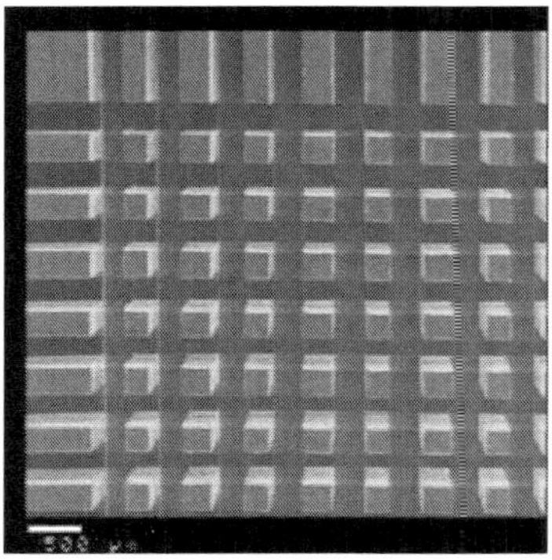

Fig.1: Master model of column array test structure (brass).

The paraffin binder was burned-out in air atmosphere at a maximum temperature of 500 °C where about 95% of the binder is withdrawn and significant oxidation of the Si powder is still avoided; small amounts of residual carbon preserve a sufficient mechanical stability for handling the samples. After debinding the samples were placed in BN crucibles and reaction-bonded in a furnace equipped with tungsten heating elements. The standard program comprised a heating rate of 10 K/min up to 1390 °C, dwell time of 4 h, and cooling rate of 20 K/min. A static atmosphere with 95% nitrogen and 5% hydrogen (forming gas) at an overall pressure of 0.16 MPa was applied. Due to the Si_3N_4 formation the pressure decreased continuously during dwell time; when it dropped below 0.15 MPa new forming gas was added until 0.16 MPa was reached again. The amount of Si conversion in the reaction-bonded samples was calculated by determining the weight increase the debinded samples gained during nitridation and by correlating the weight increase with the theoretical value of 66.5% (1-(mol. weight Si_3N_4/(3x mol. weight Si)) and the fraction of Si in the starting powder blend. In contrast to the nitridation treatment the sintering step was conducted in pure nitrogen under static atmosphere at a nearly pressureless condition (0.16 MPa). To reduce Si_3N_4 decomposition the samples were placed in a powder bed (Si_3N_4/BN mixture, BN crucible) and sintered at temperatures between 1700 and 1790°C with varying dwell times. Heating and cooling rates were 10 K/min and 20 K/min, respectively.

The mechanical properties of the micro specimens were determined by means of a micro universal testing machine equipped with an electro-dynamical actuator. Specimens used for the micro mechanical testing were obtained by detaching columns from the array; their average dimensions were approx. 240 x 240 x 1400 μm^3 in size. The micro-three-point bending device was provided with lower bending supports with a distance of 0.8 mm and a radius of 50μm, and an upper support with a radius of 100μm. From the fracture forces the ultimate bending strengths were calculated. The statistic scattering was analysed using the Weibull theory, which gives σ_0 as an estimate of strength for a 63.2 % failure probability and m as an exponent growing with decreasing scattering.

RESULTS AND DISCUSSION

Effect of sintering additives on Si_3N_4 formation

Varying additions of the following sintering aids to the Si starting material were investigated according to their influence on nitridation: Y_2O_3 + Al_2O_3 (YA), Y_2O_3 + MgO (YM), and Y_2O_3 + Al_2O_3 + MgO (YAM). To complete these variations the amount of the inherent sintering additive SiO_2 was altered as the specific surface area and particle size of the Si powder was modified by varying the time for ball milling. Fig 2 summarizes the Si conversion of the different feedstock compositions in dependence of the dwell time. From Table 1 the corresponding amounts of the different additions can be taken. As expected, with increasing dwell time the Si_3N_4 formation also increases; however, substantial differences between the different powder compositions are observed. Obviously the MgO containing compositions reveal higher reaction rates. Already after 0.5 h dwell time more than 90% conversion was obtained for YM-3 samples. In contrast YA-samples exhibit merely 80 % conversion although treated at 1390 °C up to 8h.

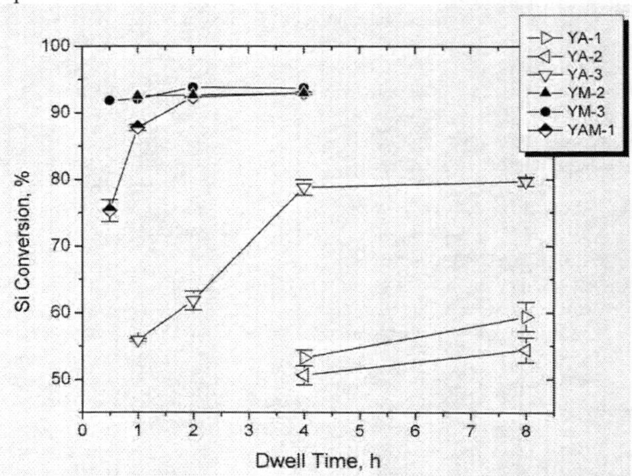

Fig.2: Influence of sintering additives on the nitridation behaviour; Si conversion of "column array" samples in dependence of dwell time at 1390°C.

To distinguish the influence of the single sintering additive those compositions have to be compared where only one factor was changed while the others preferably not. Fig. 3(a) illustrates the positive effect of higher Y_2O_3 contents on Si nitridation. By increasing the Y_2O_3 content from 8 wt.% to about 12 wt.% Si conversion after 4h is increased by the half.

The opposite effect is observed for the SiO_2 content, represented by the specific surface area of the ball-milled powder mixture in m^2/g. As Fig. 3(b) shows, that powders of the same nominal composition but with different surface area behave different during nitridation. Although the influence is not as pronounced as for the Y_2O_3 content, the clear trend is found that higher SiO_2 amounts inhibit Si_3N_4 formation slightly.

Directly opposed influence on the Si conversion arises from the two sintering additions Al_2O_3 and MgO (Fig. 3(c)); increasing the Al_2O_3 content from 0 to 3 wt.% results in a decrease of Si_3N_4 formation rate while Si conversion is promoted by the addition MgO. To understand the effect sintering additives impose already on the nitridation process one must not forget that nitridation temperature (1390 °C) lies above the lowest melting temperatures in the Y-Si-Al-O-N (1280 °C) and the Y-Si-Mg-O-N (~1200 °C) system, respectively. The occurrence of an – at least intermediate – liquid phase is very likely. A higher volume of liquid phase may cause blocking of the pore channels before the interior of the Si body has been completely converted[8]. With increasing specific surface area more SiO_2 will contribute to the formation of a liquid phase and thereby retarding nitridation (Fig 3(b)).

On the other hand by increasing the Y_2O_3 content Si conversion is supported twofold by the formation of crystalline secondary phases: (i) the native silica layer covering the Si particles is removed[9] which is a prerequisite for direct reaction between silicon and nitrogen[10]; (ii) the total amount of liquid phase is reduced (densification is retarded to higher temperatures).

The different behavior of Al_2O_3 and MgO during nitridation is not fully understood, yet. However, it is assumed that the accessible pore structure plays a decisive role as samples with high Al_2O_3 (and SiO_2) content exhibit substantial linear shrinkage of up to 3% during nitridation (~9 % in volume) which is more than twice as much as MgO containing compositions reveal[11]. Thus the loss of open porosity is considered to be the main factor for the lower reaction rate of the Al_2O_3 containing samples.

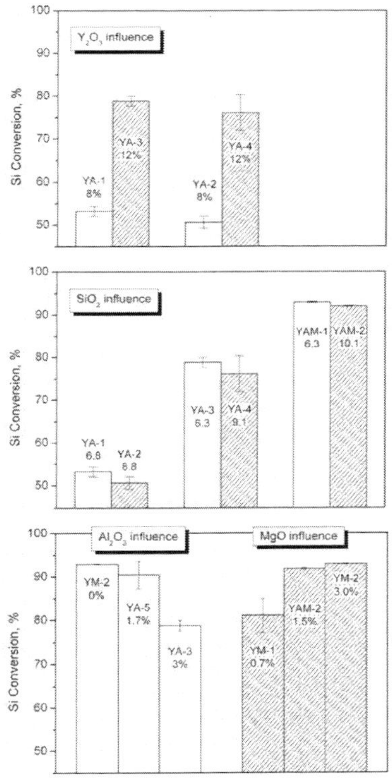

(a)

(b)

(c)

Fig.3: Influence of the different sintering additions; (a) Y_2O_3, (b) Al_2O_3, MgO, and (c) SiO_2 on the Si conversion after 4h at 1390°C (N_2/H_2 atmosphere).

Effect of samples size on Si_3N_4 formation

For the nitridation of the silicon powder the nitrogen transport into the compact might become a limiting factor for parts with larger wall thickness and/or smaller pore channels. Therefore it was expected that a *decrease* in sample size would be beneficial for the reaction rate of Si_3N_4 formation[12]. To reveal the influence of the specimen size on the Si conversion different sample geometries were evaluated. In Fig. 4 exemplarily the Si conversion of YAM-1 samples of different size and shape is plotted in dependence of the sample mass (after debinding). Contrary to the expectations it was observed that a decrease of sample size was accompanied by a decrease of Si conversion. The same correlation was also detected for the compositions YM-2, YM-3, YAM-2 and to a lower extent also for YA-3 (sample size has not been varied for the other compositions).

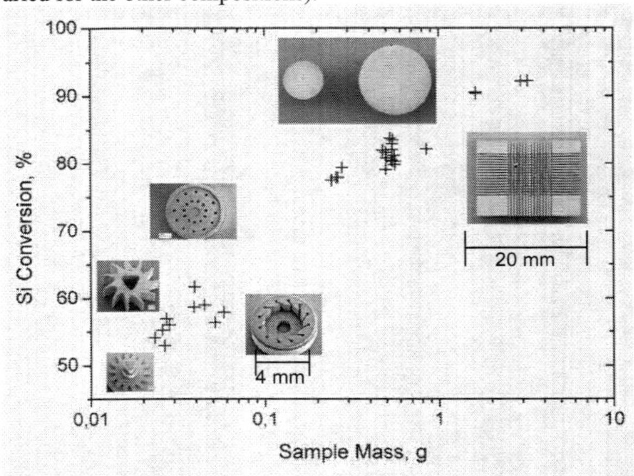

Fig.4: Si conversion in dependence of sample mass after nitridation at 1390°C, 2h dwell time, feedstock composition YAM-1.

To understand the positive correlation between sample size and nitridation rate it has to be envisioned that the reaction between silicon and nitrogen (eqn. 1) is exothermal[13] and that the reaction rate always increases with temperature.

$$3Si(s)+2N_2(g)\rightarrow Si_3N_4(s) \qquad \Delta G = -723+0.315T/K\cdot(kJ/mol) \qquad (1)$$

Owing to a decreasing surface-to-volume ratio with increasing size the heat of reaction will not be dissipated as fast as for smaller samples witch are rather in thermal equilibrium with the environment. Thus higher local temperatures are responsible for the improved Si conversion of bigger samples. It can be expected that this effect has its limitations when the heat of reaction causes Si melt formation. In contrast to the MgO containing compositions, only a weak increase of Si conversion with sample mass was found for the alumina blended feedstock YA-3. This observation can be explained by the generally lower reaction rate of the feedstocks YA-1 to YA-4 where less heat is released within a certain period of time, and heat transfer from the sample surface into the environment is effective enough so that the self-accelerating effect vanishes.

Effect of sintering additives on sintering behavior

After nitridation and reaction-bonding the samples were sintered in a separate processing step in N_2 atmosphere at a constant pressure of 0.16 MPa. Sintering was performed at 1700 °C, 1750 and about 1800 °C with 1 and 2h dwell time. The degree of densification was found to depend considerably on the type and amount of sintering additions. What had begun to show after nitridation, i.e. the higher shrinkage of the Al_2O_3 containing and finer particle size compositions, is even more pronounced after sintering. In Fig. 5 the open porosity of column array samples sintered at 1750°C for 2h is taken as an indicator to illustrate the positive effect of higher Al_2O_3 and – indirectly – higher SiO_2 contents on densification. Although the closed porosity cannot be detected by mercury intrusion porosimetry and hence some uncertainty regarding the total porosity remains, these results are in good agreement with the measured cumulated shrinkage.

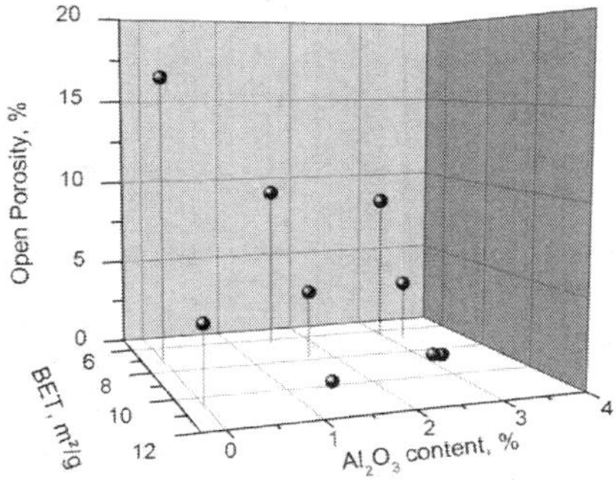

Fig. 5 Open porosity after sintering at 1750°C / 2h in dependence of Al_2O_3 content and specific surface area of the Si starting powder mixture.

Process – microstructure – property correlations

In Tab.2 the characteristic bending strength σ_0 and the modulus m obtained of the micro-three-point bending tests are presented, together with sintering conditions and the open porosity determined at the corresponding array socket plate. The values of σ_0 cover the range from about 500 MPa up to 1200 MPa, while m – which is a measure for the scattering of bending strength – varies between 10 and 20 for the majority of the samples. Although it might stand to reason to ascribe the broad distribution of the mechanical properties to the differences in porosity, an obvious correlation however could not be found. As shown in Fig. 6 the lowest porosities represented by samples YA-1, YA-4 and YAM-2, do not correspond to the highest strength values. Rather samples from composition YM-1 and YA-5 exhibit promising σ_0 values of 1200 MPa and 1000 MPa, respectively, although significant open porosities from 4 to 7% were detected for these samples. In contrast to the experiences gained from ZrO_2 micro-specimens[6,7], in the present study porosity appears to have only a secondary influence on the micro-bending strength of the SRBSN micro specimens.

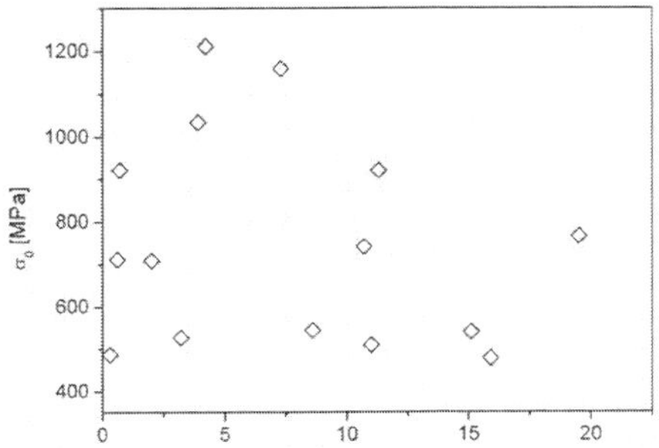

Fig. 6: characteristic strength σ_0 of micro specimens in dependence of open porosity of the corresponding bulk material.

Table 2 Sintering conditions and mechanical properties of micro bending specimens

	T_{sint} °C	Dwell time h	σ_0 MPa	m -	Open Porosity %	peak ratio $Y_2Si_3O_3N_4/Si_3N_4$
YA-1	1750	1	712	11	0.6	0.13
YA-3	1700	1	540	14	15.1	0.61
	1750	1	527	6	3.2	0.70
	1750	2	544	13	8.6	0.51
YA-4	1750	1	921	12	0.7	0.08
	1750	2	953	7	-	0
	1790	1	873	12	-	0.31
YA-5	1750	1	920	13	11.3	0.10
	1790	1	1035	12	3.9	0
YM-1	1750	2	1161	19	7.3	0.02
	1790	1	1212	10	4.2	0.03
YM-2	1750	2	478	24	15.9	0.41
YM-3	1750	2	767	13	19.5	0.04
YAM-1	1750	1	742	11	10.7	0.46
	1750	4	509	5	11	0.40
YAM-2	1750	2	487	5	0.3	0.76
	1790	1	709	7	2.0	0.08

As porosity could not explain the observed strength very well, research was conducted regarding the primary factors determining the (micro) mechanical properties. One approach was to take a deeper look at the phases which are formed during sintering by means of XRD-analysis (step scan mode using Cu Kα radiation). First of all it was confirmed that Si_3N_4 has completely changed from α- into β-modification. Furthermore it was found that substantial differences in the composition occurred between surface and sample interior. As Fig. 7 illustrates only minor amounts of crystalline secondary phases are detected from the cross sections (mainly YN-apatite $Y_5Si_3O_{12}N$). From the surface scan however, quite strong peaks of the melilite phase ($Y_2Si_3O_3N_4$) can be identified which are not present in the sample interior. The origin of melilite formation is considered to be the high temperature reaction of SiO_2 and Si_3N_4 (Eqn. 2).

$$Si_3N_4(s) + 3SiO_2(l) \longleftrightarrow 6SiO(g) + 2N_2(g) \tag{2}$$

It is generally accepted that this reaction between SiO_2 and Si_3N_4 forming SiO contributes substantially to the weight loss of Si_3N_4 during sintering[14-16]. The reduction of SiO_2 will take place especially at the surface which is the site where the gaseous products SiO and N_2 escape and no equilibrium is established. As the composition of the oxynitride liquid phase becomes poor of SiO_2, instead of the SiO_2-rich apatite phase the melilite phase is formed.

Tab. 2 lists also the ratio of the melilite and silicon nitride main peaks (height ratio of (211)-reflex of $Y_2Si_3O_3N_4$ to average of (200)/(101)/(210)-reflexes of β-Si_3N_4). The peak ratio assigned to a set of micro-bending specimens is not considered as a real quantitative measure for the actual concentration of melilite. It is remarkable that the degree of melilite enrichment at the surface ranges from negligibility to nearly 0.8 without showing an obvious relationship to the composition or the sintering temperature and time.

The question which therefore needs to be addressed is the origin of the large differences of the melilite content observed in the different samples. As discussed above the enrichment of secondary phase at or near the surface is initiated by the loss of SiO_2. It is noted that although all the samples were treated under uniform nitrogen conditions, the same powder bed was typically used several times for subsequent sintering runs. Apparently the efficiency of the BN/Si_3N_4 powder bed at supplying a high nitrogen partial pressure in the environment of the samples decreases with every run. This is probably due to the fact that the Si_3N_4 fraction is diminished as a result of decomposition, and the original α-phase is transformed into the more stable β-phase. As a consequence the amount of melilite at the surface is continuously increasing. In fact, for those cases in which the powder bed was omitted during sintering, the melilite phase at the surface was found to be the dominant phase (relative peak height up to 1.6).

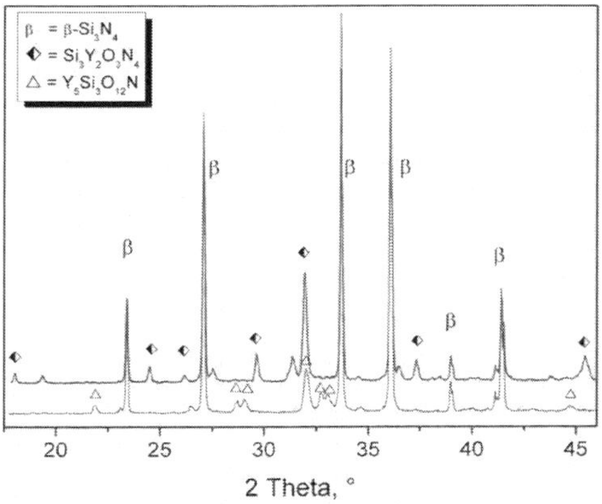

Fig.7 XRD-patterns of sintered YAM-1 sample; top line: sample surface, bottom line: cross section

Actually it is the amount of melilite phase formed during sintering which allows a correlation with the mechanical properties, namely that with increasing amount of melilite the micro bending strength is decreasing (see Tab. 2). Two reasons might be responsible for the observed effect: (i) due to the large surface-to-volume ratio of the micro specimens even thin layers of secondary phases at the surface potentially weaken the bending strength significantly; (ii) as melilite formation is coupled with the decomposition of Si_3N_4, it can be expected that additional defects are introduced into the microstructure at very sensitive sites at the specimens' surface. Following the second explanation melilite rather would act as an indicator for the occurrence of degradation processes than to be the primary cause for the strength decrease.

CONCLUSION

Micro components cannot be surface-machined to remove any disturbed sintered skin of undefined constitution. This fact is in particular critical as the surface region occupies an increasing fraction of the volume with decreasing component dimensions, and thereby becomes the dominating factor for the strength. It is therefore of great importance to suppress the decomposition reactions of Si_3N_4 by applying appropriate preparative measures. In case of pressureless sintering special attention has to be paid to the state of the powder bed. In doing so characteristic strength values of at least 1200 MPa are achievable with SRBSN. In future work also the evidence has to be provided that a decreased sintering shrinkage of about 10 % actually allows for the realization of higher production accuracy and better property reproducibility.

ACKNOWLEDGEMENT

The financial support provided by the German Research Foundation (DFG) in the framework of the SFB 499 is greatly appreciated. We want to thank M. Deuchert, wbk Institute of Production Science, KIT, for the micro machining of the master models.

REFERENCES

[1]G. Petzow, M. Herrmann, Silicon Nitride Ceramics, Structure and Bonding, Vol 102, Springer Verlag, Berlin-Heidelberg,47-167 (2002).
[2]S. Hampshire, Nitride ceramics, in: R. W. Cahn, P. Haasen, E. J. Kramer (Eds.), Material Science and Technology Vol. 11, Structure and Properties of Ceramics, VCH Verlagsgesellschaft, Weinheim, 119-171 (1994).
[3]A. Giachello, P. Popper, Post-sintering of reaction-bonded silicon nitride, *Ceram. Int.* **5** (3) 110-114 (1979).
[4]P. Greil, Near net shape manufacturing of ceramics, *Mater. Chem. Phys.*, **61,** 64-68 (1999).
[5] N. Kondo, H. Hyuga, H. Kita, T. Kaba, Fabrication of thick silicon nitride by reaction bonding and post-sintering, *J. Ceram. Soc. Jap.*,**115**, 285-89 (2007).
[6]B. Kasanická, M. Müller, M. Auhorn, V. Schulze, W. Bauer, T. Beck, H.J. Ritzhaupt-Kleissl, D. Löhe, Correlations between production process, states and mechanical properties of microspecimens made of zirconia, *Microsyst. Technol.*, **12**, 1133-41 (2006).
[7]M. Müller, J. Rögner, B. Okolo, W. Bauer, H.J. Ritzhaupt-Kleissl, Factors Influencing the Mechanical Properties of Moulded Zirconia Micro Parts, in: J.G. Heinrich and C. Aneziris, Proc. 10th ECerS Conf., Göller Verlag, Baden-Baden, 1291-96 (2007).
[8]J.S. Lee, J.H. Mun, B.D. Han, H.D. Kim, B.C. Shin, I.S. Kim, Effect of raw-Si particle size on the properties of sintered reaction-bonded silicon nitride, *Ceram. Int.*, **30**, 965-76 (2004).
[9]X. Zhu, Y. Zhou, K. Hirao, Processing and thermal conductivity of sintered reaction-bonded silicon nitride I: Effect of Si powder characteristics, *J. Am. Ceram. Soc.*, **89**, 3331-39 (2006).
[10]M.N. Rahaman, A.J. Moulson, the removal of surface silica and its effects on the nitridation of high-purity silicon, *J. Mat. Sci.*, **19**,189-94 (1984).
[11]M. Müller, J. Rögner, B. Okolo, W. Bauer, R. Knitter, Processing of micro-components made of sintered reaction-bonded silicon nitride (SRBSN). Part 2: Sintering behaviour and micro-mechanical properties, *Ceram. Int.*, **36**, 707-17 (2010).
[12]M. Müller, W. Bauer, R. Knitter, Processing of micro-components made of sintered reaction-bonded silicon nitride (SRBSN). Part 1: Factors influencing the reaction-bonding process, *Ceram. Int.*, **35**, 2577-85 (2009).
[13]A.J. Moulson, Review: Reaction-bonded silicon nitride: its formation and properties, *J. Mater. Sci.*, **14**, 1017-51 (1979).
[14]F.F. Lang, Volatilization associated with the sintering of polyphase Si_3N_4 materials, *J. Am. Ceram. Soc.*, **65,** C120-21 (1982).
[15]K. Yokohama, S. Wada, Solid-gas reaction during sintering of Si_3N_4 ceramics (part 1), *J. Ceram. Soc. Jap.*, **108**, 6-9 (2000).
[16]K. Yokohama, S. Wada, Solid-gas reaction during sintering of Si_3N_4 ceramics (part 4), *J. Ceram. Soc. Jap.*, **108**, 357-364 (2000).

THERMOPLASTIC CERAMIC INJECTION MOLDING OF ZIRCONIA TOUGHENED ALUMINA COMPONENTS

F.Kern*, M. Abou El-Ezz, R. Gadow
University of Stuttgart, IFKB
Allmandring 7b, D-70569 Stuttgart, Germany

ABSTRACT

Zirconia toughened alumina ceramics are interesting materials for engineering and biomedical applications due to their high strength, hardness, fracture toughness and abrasion resistance. For the same reason final machining is rather costly, making near net shape manufacturing technologies such as ceramic injection molding even more attractive to produce ZTA components economically. ZTA containing 10 vol-% of zirconia reinforcement of different yttria content was produced from submicron size powders. Injection molding feedstocks were prepared using a commercial binder system, homogenized at high torque. The rheological properties of the feedstocks were investigated in order to optimize the recipes and predict the molding behavior. Molded test samples – bars and plates – were produced. The debinding process was characterized by thermal analysis, the sintering process was optimized by determination of the sintering kinetics. The sintered samples in this study were characterized by SEM, bending test, hardness test and fracture toughness investigation. To clarify the transformability of zirconia, polished and grinded ZTA samples were analyzed and evaluated by x-ray diffraction. Samples had a homogeneous and defect free microstructure. The bending strength ranged between 350-700 MPa and was noticed to decline gradually as the yttria content increased. The fracture toughness varied from 5.6 to 6.2 MPa·√m at optimum sintering temperatures.

INTRODUCTION

Zirconia toughened alumina (ZTA) shows considerable advantages compared to alumina ceramics with respect to toughness, bending strength and abrasion resistance. The fraction of zirconia reinforcement usually ranges from 3-20 vol-% for different applications. Cutting tools are usually reinforced with ~ 5% of monoclinic zirconia, components for biomedical applications have higher reinforcement contents up to the percolation point of 16-17 vol-%. 10 vol-% zirconia ZTA is a standard recipe suitable for a wide range of applications in engineering and biomedical applications.

The toughening effect in ZTA is associated with two mechanisms, microcrack toughening induced by large monoclinic zirconia grains[1] which transform from tetragonal to monoclinic during cooling down from sintering temperature introducing microcracks which act as energy dissipation centers. Such ZTAs show excellent toughness but usually only moderate strength. Transformation toughening requires tetragonal zirconia grains in the sintered ceramic which must be considerably smaller in the range of < 0.5 μm [2]. Therefore special care has to be taken to limit grain growth. Starting powders should be very fine (<0.5 μm) and sinterable. Densification should be performed at temperatures of < 1500°C. This is quite challenging in case of pressureless atmospheric sintering very often pressure assisted sintering is used either as hot-pressing or sinter-HIPing in order to eliminate residual porosity and obtain fully dense materials. It has been shown that by careful choice of the sintering parameters this can also be achieved in pressureless sintering cycles[3].

At extremely low amounts < 3 vol-% zirconia acts mainly as a grain growth inhibitor with little additional contribution to strength, however recently Chevalier et. al[4] recently showed that a ZTA nanocomposite with only 1.7 vol-% zirconia exhibits extremely high resistance to slow crack growth.

* e-mail: frank.kern@ifkb.uni-stuttgart.de, Tel. +49-711-685-68233, Fax: +49-711-685-68233

New material developments in ZTA based ceramics include the use of in situ platelet reinforcements[5]. Burger could show that the formation of strontium aluminate platelet coupled with the addition of chromia leads to ultra high strength > 1300 MPa and enhanced hardness ~ 19 GPa enabling the use of such materials in hip and knee implants.
In case of such biomedical applications the ZTA has a big advantage in the field stability in contact with body fluids compared to zirconia[6]. Below the percolation threshold a hydrothermal degradation can only take place at the surface and, as the zirconia grains are not linked can not penetrate into the volume.
For parts of complex shapes in large lot sizes press to shape methods should be used if possible to reduce production costs and obtain components in near net shape quality. In such cases where this is impossible, due to design features such as undercuts or threads, ceramic injection molding (CIM) is the method to be preferred as it enables producing high surface qualities and very flexible designs. Compared to pressing processes the process cycle of CIM technology is more complex. Feedstock preparation requires intensive and homogeneous mixing of powder, binder and additives. Rheological properties have to be adjusted in such a way that the mold filling process becomes possible without shear cracks induced by too high viscosity or binder segregation caused by too high binder contents. The binder added to plastify the feedstock has to be removed before sintering which can be very time consuming and difficult in parts with high wall thickness. A process immanent problem is the low green density compared to pressed or slip cast green bodies resulting in even higher requirements to sinterability of the powders. Form filling is an in instationary process concerning temperature and viscosity, powder particles orient in the flowing binder and are frozen in this position when the injection is finished leading to flow textures which can even be amplified during sintering[7].
In earlier works it was shown that the use of extremely ultrafine powders causes processing problems while too coarse recipes are problematic due to poor sinterabilityof the compounds[8]. Therefore in this study a standard recipe of ZTA with 10 vol-% zirconia was produced from submicronsize alumina and zirconia and a commercial thermoplastic binder system in order to investigate and determine the basic processing parameters in feedstock preparation, molding, debinding and sintering. Alumina powders and yttria content of zirconia powders were varied.

EXPERIMENTAL

The powders used for compounding of the ZTA feedstocks were α-alumina (Sasol North America, Ceralox Division Tucson, Arizona) and unstabilized and partially stabilized zirconia (Tosoh, Japan), details and recipes are given in table 1 and 2. The binder used was Licomont EK 583 G (EmBe products & service GmbH, Thierhaupten, Germany). Alumina pastes were produced as a reference. Both alumina pastes were doped with 500 ppm MgO added as Spinel (S30 CR Baikowski, France). The binder system is polyethylene wax based and partially water soluble allowing combined debinding processes.

Table 1 : Powders used for preparing ZTA feedstocks

Powder	Material	manufacturer	Mean particle size * [nm]	Specific surface area * [m²/g]
SPA 0.5	α-alumina	Sasol North America	400	8
APA 0.5	α-alumina	Sasol North America	300	8
TZ-0	m-zirconia	Tosoh	350 (agglomerate)	15
TZ-3YSE	t-zirconia	Tosoh	350 (agglomerate)	7

* manufacturer's data

5 kg batches of powders and binder were mixed at 140°C in a sigma blade mixer (Hermann Linden Maschinenfabrik, Marienheide Germany) at 25 rpm for 1 h to obtain a well premixed thermoplastic feedstock. The premix was extruded and subsequently homogenized by passing it over a twin screw extruder (Thermo Electron, Karlsruhe-Durlach, Germany) with a screw diameter d= 16mm and a length to diameter ratio of L/d = 25 at 300 rpm. The homogenizing temperature was 140°C. The ready to process feedstock was obtained in form of extruded rods of 3mm diameter which were crushed into pieces of 5-15 mm length. Rheological properties of the feedstocks were measured on the same twin screw extruder with a mounted capillary rheometry die. In the relevant processing temperature range the apparent viscosity of the feedstocks was determined at apparent shear rates ranging from 1-1000 [1/s]. Injection molding was carried out on a standard injection molding machine (BOY50M, Dr. Boy, Neustadt-Fernthal, Germany). The mold mounted on the machine (Graveurbetrieb Leonhardt, Hochdorf, Germany) has two cavities for bending bars (30mm x 3mm x 4mm) and plates (45 mm x 32 mm x 3 mm). In the plate shape cavity a 2 mm diameter pin can be inserted to introduce a weld line. Debinding of samples was carried out as a combined cycle with debinding in distilled water at 60°C overnight and subsequent thermal debinding. Presintering was carried out at 800°C in air for 12 h. Debinding an presintering were carried out subsequently in the same furnace.

Table 2. Recipes of the ZTA and alumina feedstocks

Feedstock	SPA 0.5 [m-%]	APA 0.5 [m-%]	TZ-0 [m-%]	TZ-3YSE [m-%]	Binder [m-%]
ZTA-1	72	0	13	0	15
ZTA-2	72	0	6.5	6.5	15
ZTA-3	0	72	6.5	6.5	15
ZTA-4	72	0	3.2	9.8	15
ZTA-5	72	0	0	13	15
AL-1	85	0	0	0	15
AL-2	0	85	0	0	15

The debinding behaviour was investigated by DTA-TG measurements (STA 429 C, Netzsch Gerätebau, Selb, Germany), the sintering behaviour was determined by dilatometry (DIL 402 C, Netzsch Gerätebau, Selb, Germany). Debinding and sintering kinetics were calculated from dynamic TG/DTA and dilatometry experiments carried out with heating rates of 2, 5 and 10 K/min in air, the modelling was carried out using Netzsch thermokinetics software (Netzsch Gerätebau, Selb, Germany). The procedure has been repoted previously[9,10]. Final temperatures were 800°C in case of the thermogravimetry studies and 1590°C for the dilatometry. Microstructures of the materials were investigated by SEM (LEO VP 438, Leo, Cambridge UK).
Samples were ground and polished for mechanical characterization. Bending strength was determined by 3-point bending according DIN EN 6872 with a 16 mm span on polished and in some cases also as-fired bending bars (Hegewald&Peschke, Nossen, Germany). Hardness HV0.1 and Elastic modulus of the ceramics were measured by universal hardness method with a microindenter (Fischerscope, Fischer, Sindelfingen, Germany). HV10 values were measured with a standard hardness tester (KB Prüftechnik, Hochdorf-Assenheim, Germany). Fracture toughness was measured by indentation, toughness values were calculated according to the formula of Niihara[11]. Pore size distributions of presintered samples were measured by mercury porosimetry (Pascal 140/440, Porotec, Hofheim, Germany). Densities were determined by Archimedes method. Phase composition of zirconia was determined by analysing XRD-data according the calibration curve determined by Toraya[12].

RESULTS AND DISCUSSION

Rheology

Figure 1 shows the viscosity curve of ZTA-3 determined with a 2 mm diameter capillary with a length of 20 mm. At low shear rates the feedstock shows shear thinning behaviour for all tested plastification temperatures. At intermediate shear rates of ~ 100 s^{-1} the behaviour turns to newtonian. The onset of the newtonian behaviour is shifted to higher shear rates with rising plastification temperatures. Due to the high solid loading of the feedstocks the behaviour at very high shear rates could not be measured as the maximum pressure in the measuring setup is limited to 10 MPa. For the same reason measuring with longer capillaries was impossible thus only apparent viscosities can be shown. A Bagley correction was not possible. The behaviour of ZTA-3 is typical for all ZTA and alumina feedstocks tested.

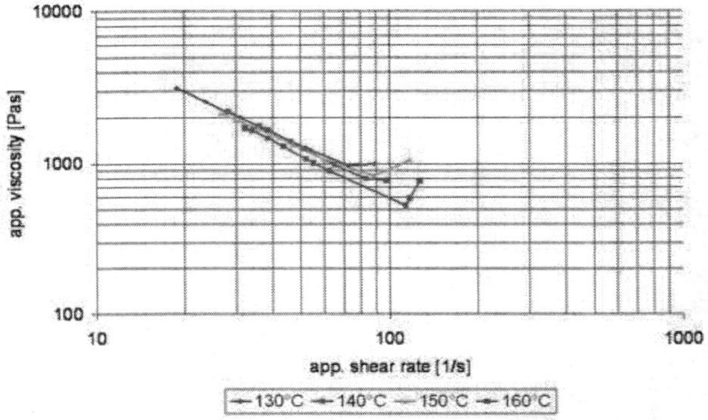

Fig. 1: Viscosity curves of ZTA-3 in the temperature range between 130-160°C

Resulting from the rheological characterization we expected good processing properties in the injection process, with one limitation. Injection speed has to be either low in order not to enter the Newtonian range or very high to surpass the newtonian field. At very high injection speed there is however a high risk of binder segregation leading to structural defects.

Molding was carried out under very moderate pressure and temperature conditions. The details of molding conditions are shown in table 3. Plastification temperature did not exceed the maximum processing temperature of 160°C recommended by the binder manufacturer. Despite the fact that we used a standard thermoplast screw setup, the injection pressures did not exceed 1200 bar. Packing pressure was 800 bar in all cases. In most of the cases the pressures were lower. Mold temperatures were slightly higher than recommended (60°C). It was quite interesting to see that the feedstocks ZTA-4 and ZTA-5 with the highest yttria contents required higher injection pressures than the alumina and ZTA-feedstocks with less than 1.5 % yttria in TZP. This fact may hint at an influence of the stabilizer on rheology. One may conclude that the yttria which is a basic oxide, even if mainly located in the bulk of the TZP, can interact with additives such as carboxylic acids (e.g. oleic or stearic acid) which are frequently used in commercial binder systems. This would lead to protonation of yttria and formation

of yttria hydroxide which can negatively affect rheological behaviour. High amounts of the monoclinic zirconia TZ-0 with a high surface area (ZTA-1) did not cause any rheological difficulties.
Filling studies with the plate-shape mold and a flow obstacle (figure 2) showed that the mold filling through a film gate is associated with a rippled flow front. The two branch currents unite quite fast and lead to a well embedded insert. At a distance of 2/3 of the maximum mold length the feedstock loses wall contact at the sides. The ripples originally located at the flow fronts are pushed aside by fountain flow. While the corners and the central region of the components are well filled this causes microstructural defects at the sides. The inserted pin leads to the formation of a weld line ranging from the circular hole to the bottom of the component. Apparently this weld line is well annealed.

Table 3: Injection molding parameters for different feedstocks

Feedstock	Plastification temperature [°C]	Injection pressure [bar]	Injection speed [cm³/s]	Mold Temperature [°C]
ZTA-1	150-160	800-1000	30-50	65-75
ZTA-2	150-160	800-1000	30-50	65-75
ZTA-3	150-160	800-1000	30-50	65-75
ZTA-4	150-160	1000 -1200	30-50	65-75
ZTA-5	150-160	1000 -1200	30-50	65-75
AL-1	150-160	800-1000	30-50	65-75
AL-2	150-160	800-1000	30-50	65-75

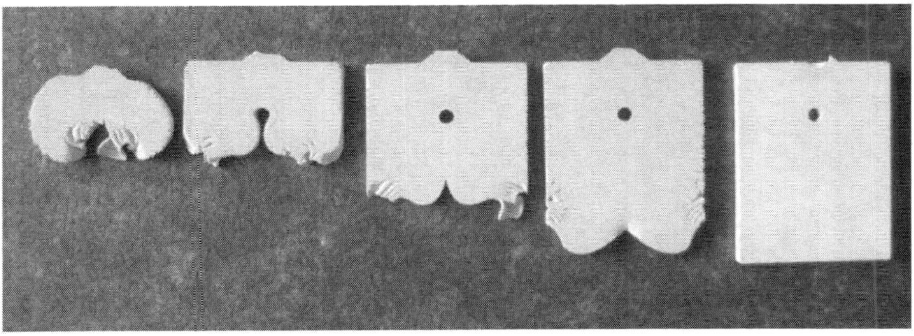

Fig. 2: Mold filling study , injection molding around a round mold insert

Debinding

Water debinded bars molded from Feedstock AL-1 were pre-dried at 105°C till all residual moisture was eliminated. The bars were cut into slabs of 9 mm length for dilatometry. For TG/DTA samples of ~ 0.8 g each were taken from crushed predebinded green bodies. Figure 3 shows the thermogravimetry curve measured. At low heating rates the paste can take up some oxygen and gain weight before decomposing. At higher heating rate this was not observed. The binder burnout starts at 200°C and is completed at 460-600°C depending on heating rate. The mass loss during debinding takes place in three stages. A DTG plot (fig. 5) of the first derivative makes this more clear. In order to chose a meaningful model first a Ozawa-Flynn-Wall[13,14] model-free estimation of the activation energy was carried out. The activation energy declines from 100 kJ/mol at the beginning to 40 KJ/mol in the end indicating a shift from reaction controlled behaviour of the initial stage to diffusion controlled behaviour in the final stage as can be expected.

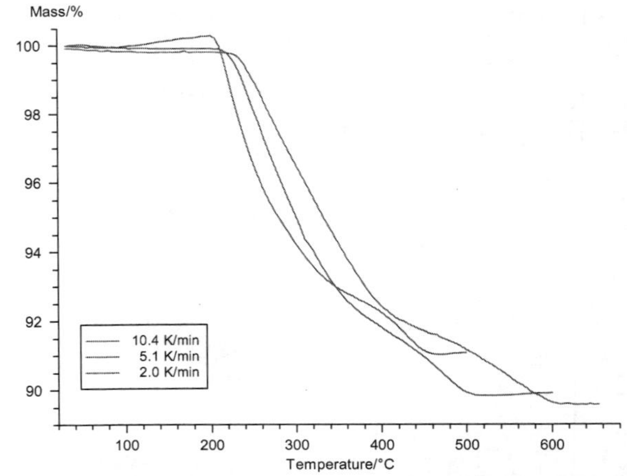

Fig. 3: thermogravimetric analysis of water-debinded Al-1 at three heating rates

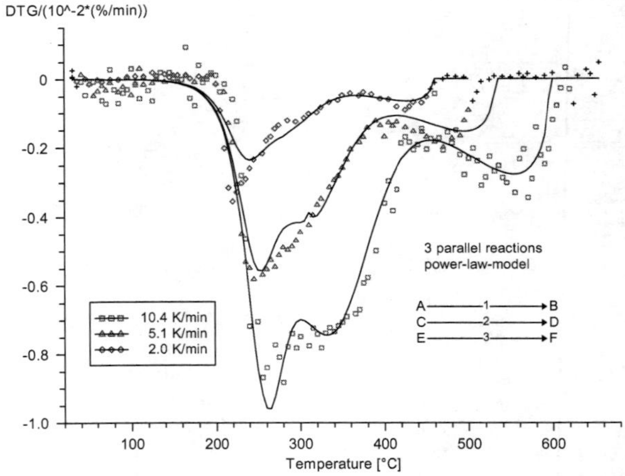

Fig. 4 DTG-curve with fitted modelling data

Describing the debinding behaviour in a macrokinetic model by a set of three parallel reactions led to the highest prediction quality. The kinetic parameters determined are given in table 4. The values for activation energies are in the typical range of oxidation reactions, the reaction orders are in good accord with the above made statements, high orders are typical for reaction control (reaction 1 and 2) whereas low orders are typical for diffusion control (reaction 3). With respect to the measured data the debinding curve was defined as follows: 10K/h to 280°C, 12h dwell, 18K/h to 800°C, 12h dwell. Free cooling to room temperature.

Table 4: Kinetic parameters of the debinding reaction model

Parameter	Unit	Calculated value	Standard Deviation
Preexponential factor 1	[1/s]	11.7	3.1
Activation energy 1	[kJ/mol]	139.7	29.6
Reaction order 1	[-]	1.56	0.52
Preexponential factor 2	[1/s]	3.0	0.77
Activation energy 2	[kJ/mol]	63.6	8.15
Reaction order 2	[-]	1.26	0.27
Preexponential factor 3	[1/s]	0.366	0.147
Activation energy 3	[kJ/mol]	49.7	1.86
Reaction order 3	[-]	0.31	0.24
Fraction reaction 1	[-]	0.21	0.057
Fraction reaction 2	[-]	0.54	0.073

Sintering

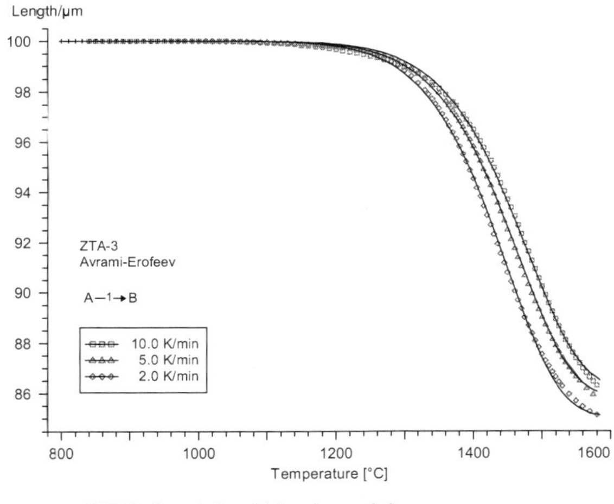

Fig. 5 Dilatometry curve of ZTA-3 and fitted kinetic model

Figure 5 shows the dilatometry curve of ZTA-3 and the fitted kinetic model. Unfortunately the densification was not complete within the accessible temperature range of < 1590°C. The sintering behaviour of ZTA-3 can be satisfactorily explained with a simple one-stage n-dimensional Avrami-Erofeev nucleation and growth model. The prediction quality is 99.97 %. Adding a further stage of power-law kinetics does slightly improve the prediction quality. A critical revision showed however that this stage explains less than 3 % of the length change and leads to incorrect predictions concerning the sintering behaviour.
An isothermal prediction of the sintering behaviour using the proposed model is shown in figure 6. This graph assumes an infinitely fast heating to final temperature and a dwell of up to 2 h. It is obvious that in order to fully densify the sample a temperature of 1500°C and a dwell of 2h are necessary. This is well within the boundary conditions defined to obtain a fine grained zirconia dispersion and preserve the tetragonal phase after cooling. All other ZTA-types showed a similar sintering behaviour.

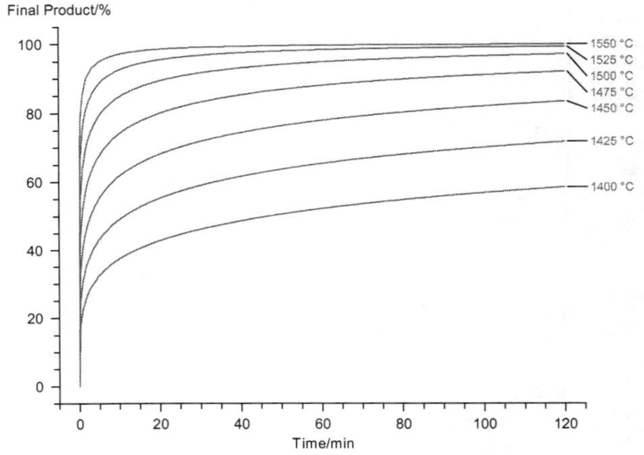

Fig. 6 Prediction of sintering behaviour of ZTA-3

Table 5: Kinetic parameters of the sintering model for ZTA 3

Parameter	Unit	Calculated value	Standard Deviation
Preexponential factor 1	[1/s]	37.7	0.91
Activation energy 1	[kJ/mol]	1335.9	30.7
Dimension 1	[-]	0.263	0.005

Microstructure

ZTA specimen were sintered according to the defined sintering schedule, final temperatures were chosen in the range between 1475°C and 1525°C, dwell was 1-2 h. The heating rate to final temperature was 5 K/min. Fracture surfaces were studied by SEM after bending tests. The observations made in SEM confirm the predictions made by sinter modelling. To obtain full density a sintering temperature of 1475°C and 1h dwell are insufficient. Figure 7 shows the evolution of density for the different materials with respect to sintering temperature. ZTAs require sintering temperatures of 1500°C to become completely dense. In some cases 1475°C is sufficient to obtain >99% of theoretical density. AL-1 which is based on SPA powder requires higher sintering temperature than its ZTA composites, AL-2 based on APA powder is dense at 1450°C (not shown in Fig. 7) and shows a better sinterability than its composite.

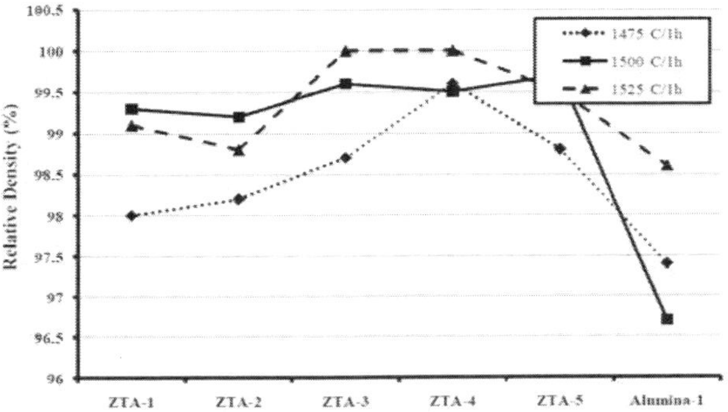

Fig. 7 Density of the ZTA and Alumina ceramics depending on sintering conditions

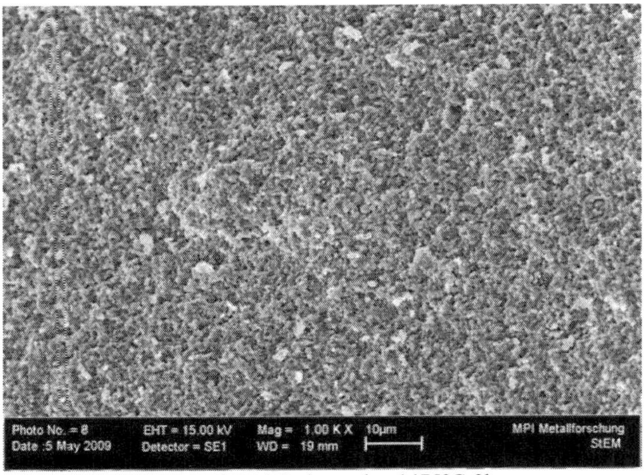

Fig. 8 Overview of the microstructure of ZTA-2 sintered at 1475°C-2h

Figure 8 shows that the microstructure of the components produced was very homogeneous. Flaws or large pores cannot be detected. A more detailed look at higher magnification shows the distribution of residual porosity after sintering and the size of the alumina and zirconia grains. Moreover the homogeneity of zirconia distribution into the alumina matrix can be evaluated. Figures 9 and 10 show the microstructure of ZTA-2 sintered at 1475°C/2h and 1500°C/2h respectively. At 1475°C there are a still lot of pores and the facets of the alumina grains are not well developed. The sample sintered at 1500°C shows a well developed and dense microstructure. Fracture is intergranular for both alumina and zirconia grains at 1475°C while at 1500°C the fracture of zirconia grains becomes transgranular indicating an efficient toughening mechanism. Grain sizes of alumina are approximately 1 μm, zirconia grains are < 0.5 μm at both sintering temperatures. Zirconia is well dispersed.

Fig. 9 Detail of the microstructure of ZTA-2 sintered at 1475°C-2h

Fig. 10 Detail of the microstructure of ZTA-2 sintered at 1500°C-2h

XRD analysis

As fired and ground samples were analyzed by XRD. Figure 11 shows the XRD analysis for ZTA-1 with only monoclinic zirconia and ZTA-2 with 1.5Y-TZP reinforcement in the relevant 2-theta range between 20-40°. The monoclinic content estimated according the formula of Toraya indicates that zirconia in ZTA-1 as a transformability of 16 % while in ZTA-2 a higher value of 18% is obtained, thus the contribution of transformation to toughening is higher in ZTA-2 than in ZTA-1.

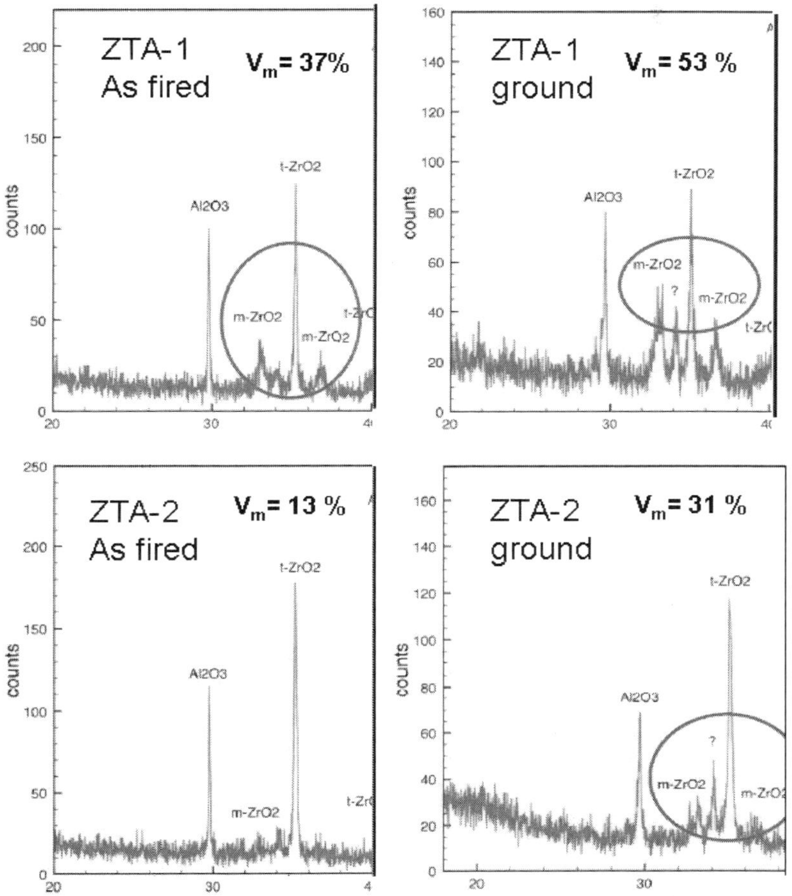

Fig. 11 XRD analysis of ZTA-1 and ZTA-2 in as-fired and ground condition, in the 2theta-range 20-40°, determination of the volumetric content of monoclinic phase V_m acc. Toraya.

Mechanical properties

The results of bending tests (fig. 12) showed that bending strength of the samples varied strongly with yttria content of zirconia, alumina powder and sintering temperature. Maximum strength for the SPA alumina based ZTAs was reached at low sintering temperatures and yttria contents. Overstabilization of ZTA results in poor mechanical properties which fall from 720 MPa for the 0Y to 420 MPa for the 3Y-TZP reinforced ZTA. ZTAs with high yttria content are almost invariant in strength at different sintering temperatures. APA based ZTA-3 (1.5Y-TZP) showed a higher strength of 720 MPa than SPA-based ZTA-2 with 570 MPa. Even in as-fired condition ZTA-3 still had a strength of 620 MPa. Obviously the surface chemistry of the alumina derived from different alumina production processes strongly influences the strength of ZTA composites. Aluminium chloride based SPA seems incompatible with yttria stabilized powders. Both reference aluminas showed excellent strength of 600-620 MPa. ZTAs with inappropriate processing conditions and recipes can thus not surpass the alumina in strength.

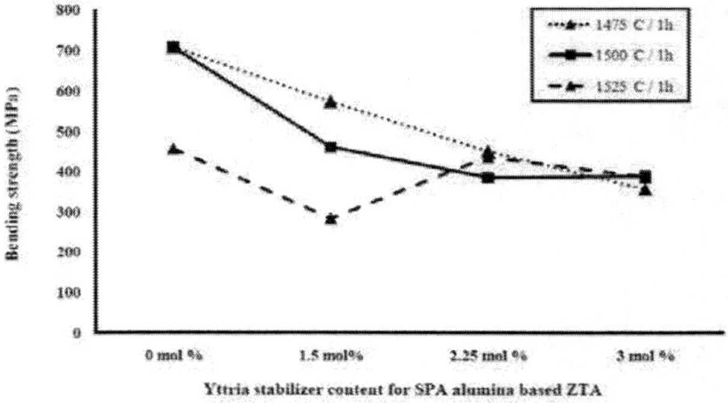

Fig. 12: Bending strength of SPA0.5-based ZTAs

The dependency of fracture toughness on recipe and processing conditions is shown in in Fig. 13. As expected the 0Y and 1.5Y-TZP reinforced materials sintered at low temperature had the best toughness values of 5.7-5.9 MPa·√m. With rising yttria content an inversion of this behaviour can be detected 2.25Y-TZP reinforced ZTA-4 shows invariant toughness of 5.7 MPa·√m, overstabilized ZTA-5 (3Y) showed an inverse correlation. Toughness of ZTA-5 rises with increasing sintering temperature (from 5.3 to 5.6 MPa·√m). Again APA 0.5 based ZTA-3 showed the best toughness value of 6.2 MPa·√m at a sintering temperature of 1475°C.

The correlations found can be easily explained. At low yttria contents grain growth of zirconia grains proceeds faster than at high stabilizer contents. Once a critical grain size is reached, in understabilized ZTA-1 tetragonal zirconia transforms to monoclinic during cooling leaving only a small amount of zirconia for transformation toughening, the main contribution to toughness should then come from microcracks. These microcracks are however too small in zirconia grains of ~ 0.6 μm to function efficiently for energy dissipation.

At higher stabilizer content the grains ought not to be to small to be transformable at all. Therefore rising sintering temperatures lead to tougher materials. The optimum mechanical properties can be achieved with an APA based alumina matrix, stabilizer contents of ~ 1,5 mol-%Y-TZP, low sintering temperatures of 1475-1500°C and a dwell of 1-2 h. Hardness values HV_{10} showed a good correlation with density. For all tested ZTAs hardness rose from 17 GPA at 1475°C/2h to 18 GPA at 1500°C/2h to finally 19 GPa at 1525°C/2h. The youngs modulus rose from 350 to 385 GPA with increasing sintering temperatures.

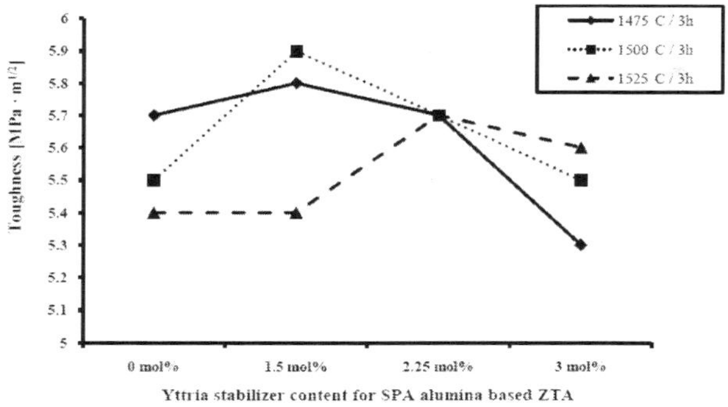

Fig. 13: Fracture toughness of SPA0.5-based ZTAs

Properties of ideal samples vs. real components

Testing of defect-free bending bars which were molded against a withdrawing plunger shows that the mechanical properties of the sintered ZTAs are very good provided that the recipes and processing parameters are well adjusted. In real components with more complex geometry, design features such as weld lines, changes in width or dead zones in the mold may lead lead to significant loss in mechanical strength. The plates molded with a round flow obstacle (see. Fig. 2) apparently had well cured weld lines at first sight.

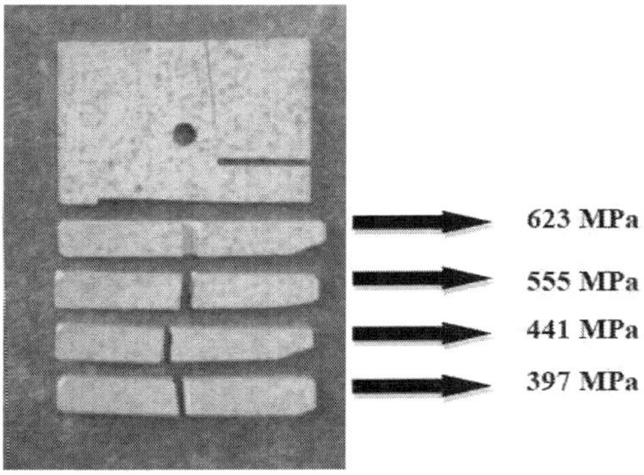

Fig. 14 Test plate with weldline manufactured from ZTA-4

Plates of such geometries were cut into strips and tested by 3-point bending. In injection direction with rising distance from the hole a steep decline in strength can be observed (Fig. 14). This behaviour is related to the instationary mold filling process. During the mold filling the feedstock simultaneously cools down. The viscosity increases and the ability to cure weak zones by diffusion of binder molecules decreases. With falling temperatures induced by heat transfer from the paste to the cooled mold, the weld line becomes weaker and weaker. In a similar reference experiment we molded test rings with two gates and two resulting weld lines located in 90° position to the gates. Mechanical testing of these rings revealed that their strength was only ~ 50 % compared to test bars. Weld lines located in positions with high mechanical load have to be avoided by any means. It is therefore mandatory to optimize the mold geometries, gate design and gate position with respect to critical design features and check the results on sintered parts. Mechanical properties determined on defect-free test geometries might lead to mistakes in the design and dimensioning of components for critical applications.

SUMMARY AND CONCLUSION

The lay out of a complete process chain for ceramic injection molding is a very complex task for a ceramics engineer. In addition to the basics in materials science many processing related boundary conditions have to be respected in order to obtain a product with tailored properties and a production cycle which is reliable and economically performing.

Compounding of feedstocks includes the choice of the powders and binders in the correct ratio, and their perfect homogenization in order to obtain a basis for successful molding processes. For feedstocks produced from sub-µm powders, high intensity mixing devices such as twin-screw extruders are required to achieve high solid contents and good flowability and avoid binder segregation. Rheology measurements are very helpful to determine the optimum binder content and to validate the efficiency of additives such as surfactants.

Molding processes can be performed on modified standard machines with abrasion resistant plastification units (screws with adjusted slope and cermet-clad barrels). The lay out of the mold requires a lot of experience which can today be supported by form filling simulations. Identification and elimination of critical design features by a proper mold and gate design are crucial for the technical and economic success of a CIM production line. As molds are precision tools with a high cost, design adaptations or complete redesigning are very expensive and should be avoided.

Debinding is a key technology in processing of CIM parts. The preservation of a defect free microstructure after debinding is the basis for high quality sintered parts. Binders with a soluble polymer fraction can be debinded in two stages. The water debinding step creates an open porosity through which volatile binder components can easily escape in the subsequent thermal debinding cycle. Thus parts with a larger wall thickness can be safely processed. Modelling the decomposition behaviour of the binder in the green bodies by analysing DTA/TG data is very helpful to lay out an optimized thermal processing and save furnace time while retaining a reliable debinding process.

The optimization of the sintering process is usually done based on sintering experiments. Including sinter modelling into the engineering process helps saving development time and experimental effort. Engineering tools like Netzsch thermokinetics software may not fulfil the academic standards of materials science as they work with black box macrokinetic models and do not respect the microkinetic aspects of sintering processes. They are however very powerful in determining the time-temperature curves of sintering processes. The model created should however be experimentally checked. While the sintering model can safely predict the densification behaviour other microstructural features to be developed during sintering such as reaching a defined size target for e.g. zirconia in ZTA cannot be predicted. This still has has to done conventionally e.g. using the modelling results as the starting point for a design-of experiments based optimization process.

Concerning our ZTA composites derived from submicronsize powders it was shown that application of the above mentioned engineering tools is very efficient. Materials with high strength, toughness and microstructural integrity were produced which can compete with pressed materials of the same composition. Including a subsequent post-HIP cycle may further improve the good properties.

ACKNOWLEDGEMENTS

We would like to thank Andreas Vogel and Willi Schwan for their support in testing and characterization, Thomas Theye of IMI and Johannes Rauch for XRD measurements and analysis and our students Matthias Häusler and Michael Wilhelm who did a part of the investigations in their seminar papers.

REFERENCES

[1] N. Claussen, Fracture toughness of Al_2O_3 with an unstabilized ZrO_2 dispersed phase, J. Am. Ceram. Soc. 59(1-2), 59-51, 1976

[2] P.F. Becher, Slow crack growth behaviour in transformation-toughened Al_2O_3-ZrO_2(Y_2O_3) ceramics, J.Am. Ceram. Soc. 66(7) 485-488, 1983

[3] R. Gadow, F. Kern, Pressureless sintering of injection molded zirconia toughened alumina nanocomposites, J. Cer. Soc. Jap. 114 (1335), 958-962, 2006

[4] J. Chevalier et. al., Nanostructured ceramic oxides with a slow crack growth behaviour close to covalent materials, Nanoletters, 5(7) 1297-1301, 2005

[5] W. Burger and H.G. Richter, High strength and toughness alumina matrix composites by transformation toughening and in situ platelet reinforcement (ZPTA) – the generation of bioceramics, Key Eng. Mat., 192-195, 545-548, 2001

[6] J. Chevalier, S. Grandjean, M. Kuntz, G. Pezzotti, On the kinetics and impact of tetragonal to monoclinic transformation in an alumina/zirconia composite for arthroplasty applications, Biomaterials, 30 5279–5282 (2009)

[7] R. Gadow, R. Fischer, M. Lischka, Computer assisted colorimetric optical texture analysis for CIM components, Int. J. Appl. Ceram. Tech., 2 (4) 271–277 ,2005

[8] F. Kern, M. Häusler, R. Gadow, Thermoplastische Formgebung und rheologisches Verhalten von ZTA Micro- und Nanocomposites: in: H.G. Fritz, C.D. Eisenbach (eds.) : Proceedings of 21st Stuttgarter Kunstoffkolloquium 2009, Paper Nr. 3.3

[9] F. Kern, R. Gadow, A thermokinetic analysis of two step sintering of submicron alumina. In: A. Bellosi, G.N. Babini (eds.) Proc. 2nd Conf. on Ceramics, 2008, Verona, Italy, Document Nr. 6-P-067

[10] M. Frischbier, F. Kern and R. Gadow, Determination of sintering kinetics of zirconia toughened alumina ceramics by analysis of dilatometry data, In: J.G. heinrich, C. Aneziris (eds.) Proc. 10th Int. Conf. And Exhib. of the Eur. Cer. Soc., ECERS 2007, Berlin. Göller-Verlag, 1099-1104

[11] K. Niihara, A fracture mechanics analysis of indentation-induced Palmqvist crack in ceramics, J. Mat. Sci. Let., 2 (1983) 221-223

[12] H. Toraya, M. Yoshimura, S. Somiya: Calibration Curve for Quantitative Analysis of the Monoclinic-Tetragonal ZrO_2 System by X-Ray Diffraction, J. Am. Ceram. Soc., 67, 6, C119-121 (1984)

[13] T. Ozawa: A new method of analyzing thermogravimetric data. Bull. Chem. Soc.Japan, 38 1881-1886, 1965,

[14] J.H. Flynn, L.A. Wall: A quick direct method for the determination of activation energy from thermogravimetric data, Polym. Lett., 4, 232-328, 1966

FABRICATION OF ALUMINA DENTAL CROWN MODEL WITH BIOMIMETIC STRUCTURE BY USING STEREOLITHOGRAPHY

Mitsuyori Suwa and Soshu Kirihara
Smart Processing Research Center
Joining and Welding Research Institute
Osaka University
11-1 Mihogaoka, Ibaraki, Osaka Japan 567-0047

Taiji Sohmura
Dental Technology Institute
School of Dentistry
Osaka University
1-8 Yamadaoka Suita, Osaka 565-0871, Japan

ABSTRACT

Our research group has been succeeded in manufacturing all alumina dental crown models by using stereolithography of a computer aided designing and manufacturing (CAD/CAM) process. In order to imitate a natural tooth structure with a surface enamel layer, La_2O_3-B_2O_3-Al_2O_3-SiO_2 glass for the dental use was infiltrated into open pores into the fabricated ceramic objects. The alumina ceramic and acrylic resin particles of 1.8 and 8.0 μm in diameter were mixed each other at 7:3, 6:4 and 5:5 in volume ratios, respectively. These mixed particles were dispersed into the photosensitive acrylic resin at 70 volume percent. The resin paste was spread on a metal substrate with 60 μm in layer thickness. The ultra violet laser beam of 355 nm in wave length was scanned on the surface to create cross sectional thin layers through the photo polymerization. After the layer stacking process, the solid objects were processed. These precursors were dewaxed and sintered at 600 and 1700 °C in an air atmosphere, and the alumina rectangle pieces of 1.2×4.0×20 mm in dimensions composed of the porous structure were obtained successfully. The slurry paste including the glass powder was plastered on the sintered objects. Through the heat treatment at 1100 °C in the air, the molten glass was infiltrated into the opened porous. In the microstructures of the test specimens, the residual pores were not observed by scanning electron microscope (SEM). Maximum relative density and flexural strength measured as approximately 95 % and 328± 27 MPa thorough Archimedes method and a bending test apparatus, respectively.

INTRODUCTION

All ceramic dental crowns have been focused in the region of dental restorations and dentures during recent years. Their aesthetics and biocompatibility are superior to traditional metallic ones. A number of methods have been contrived to form a ceramic dental prosthesis. [1-3] In particular, cutting work is currently a predominant method of forming the ceramic crowns. However, it is difficult to cut high hardness ceramics. In addition, only a single crown can be fabricated during one operation by this method. By contrast, stereolithography of layer lamination process makes it possible to form a complicated structure. Our research group has succeeded in fabricating complex three dimensional ceramic configurations by using stereolithography.[4-5] In previous study, three dimensional alumina dental crown models have been fabricated by using the stereolithography. Top graphics in Fig. 1 show the CAD data of the dental crown model of upper molar tooth obtained from CT (Computerized Tomography) scanning and lower pictures are fabricated alumina dental crown models before and after thermal treatment. Complex groove structures at the top surface are recognized to be precisely shaped. Flexural strengths of the formed alumina crowns have been measured by using

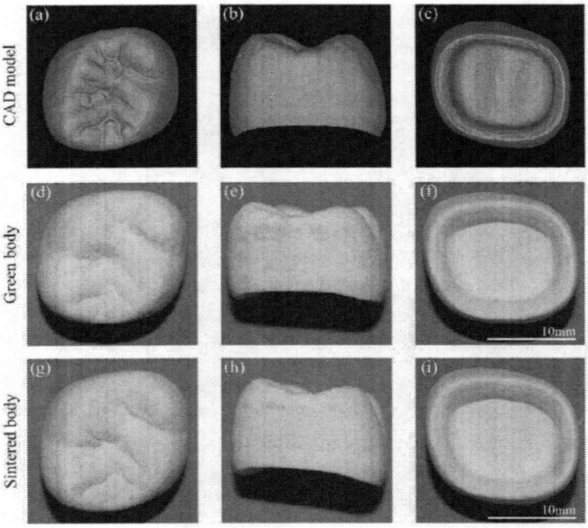

Fig. 1 Different directional views of dental crown models. A computer graphic (a) and photograph (d) · (g) are top view, (b) · (e) · (f) side and (c) · (f) · (i) bottom, respectively.

bending tests specimens of 1.2×4.0×20 mm in the previous study. Through infiltrating glass ceramic powders into the specimens, flexural strengths have been improved about 462 MPa. This obtained value is exceeds the required level of 400 MPa in clinical use.

In this investigation, we aimed to control porous ratios in sintered structures and infiltrate glass ceramic powders into the porous structures in order to imitate a surface enamel layer of a natural tooth. Moreover, we have estimated the mechanical properties of flexural strengths of the ceramic dental crown models with biomimetic structures.

EXPERIMENTAL PROCEDURE

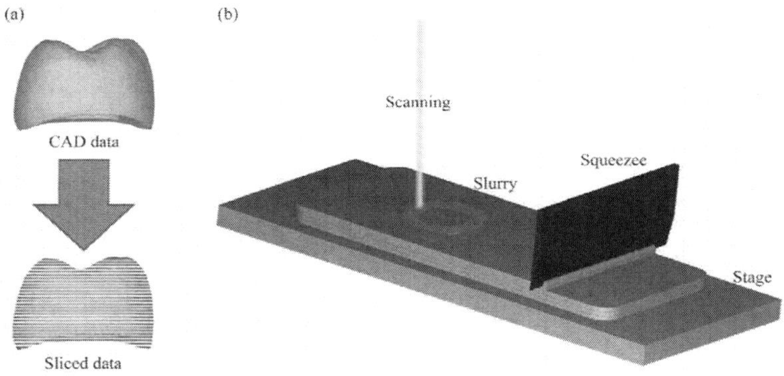

Fig. 2 Schematic illustrations of stereolithographic processes. The graphics (a) and (b) show model data slicing and layer stacking processes, respectively

Flexural test specimens of 1.2×4.0×20 mm in dimensions were designed and fabricated by the stereolithography of the CAD/CAM system as shown in Fig. 2. Three dimensional graphic models designed through the CAD software were converted into the stereolithography format (STL-Format) and sliced into a series of plate data with uniform thickness by the CAM applications. These numerical data were transformed directly into the stereolithography equipment (SCS-300P, D-MEC Co. Ltd., Japan). Alumina particles and spherical acrylic resin beads of 1.8 and 8.0 μm in diameter were mixed each other at 7:3, 6:4 and 5:5 in volume ratios, respectively. Photo sensitive acrylic resins dispersed these mixed particles at 70 % in volume content were spread on a metal platform and smoothed with 60 μm in layer thickness though a mechanical knife edge moving. An ultraviolet laser beam of 355 nm in wavelength was scanned on the slurry surfaces according to the cross sectional images of sliced plates. The laser beam was 100 μm in diameter size and 100 mW in spot power. The photo sensitive slurries were polymerized with the mixed particles. By the layer stacking process, solid objects were successfully fabricated. These samples were dewaxed at 600 °C for 2hs with the heating late of 0.5 °C/min. The dispersed acrylic beads were evaporated and the micro porous structures were formed. These precursors were sintered 1700 °C for 2hs with the heating rate of 8.0 °C/min in the air. Twelve flexural test specimens were manufactured at the each mixing ratio of the alumina and acrylic resin particles, and the six of them were coated with La_2O_3-B_2O_3-Al_2O_3-SiO_2 glass ceramic powders (In-Ceram-Alumina, Zhanfabrik, Co. Ltd., Germany) and annealed at 1100 °C for 2hs in the air. This glass powder is widely used for the ceramic coating on metal teeth in dentistry. The excess glass powders on the surface were eliminated by sand blasting. Flexural strength and relative density were measured for these six kinds of test bars. The all fabricated specimens were named as shown in Table 1. The specimen names of S73, S64 and S55 indicate sintered alumina specimens made from the acrylic resin precursors including the alumina particles and the acrylic resin beads with 7:3, 6:4 and 5:5 in volume ratios, respectively. Moreover, the code names G73, G64 and G55 indicate the glass infiltrated samples into the S73, S64 and S55 of sintered alumina bulks, respectively.

Table 1 Names of sintered and glass infiltrated specimens

Sample names	Alumina particles:Acryl beads 7:3	6:4	5:5
Sintered samples	S73	S64	S55
Glass infiltrated samples	G73	G64	G55

RESULTS AND DISCUSSION

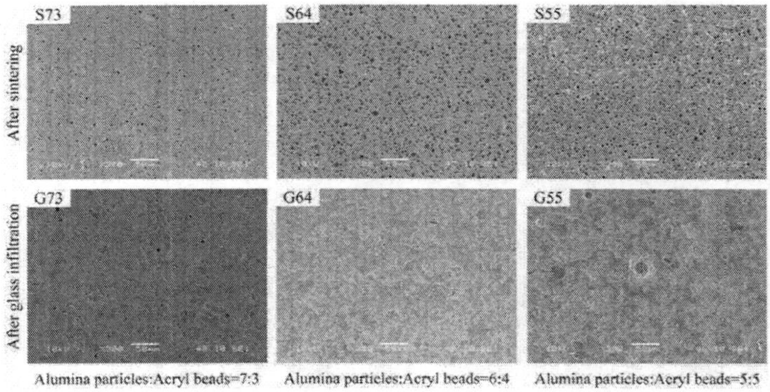

Fig. 3 Surface microstructures of sintered alumina specimens with or without glass infiltrations. Pores arrangements were observed by using scanning electron microscope.

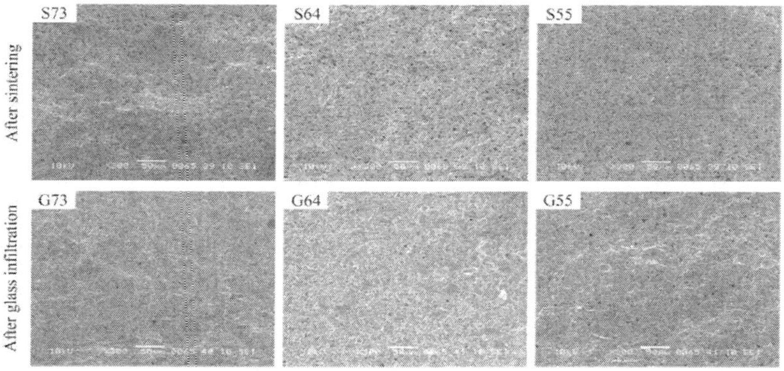

Fig. 4 Cross sectional microstructures sintered alumina specimens before and after glass infiltrations. Broken surfaces were observed by the scanning electron microscope.

Figure 3 shows the surface microstructures of the S73, S64, S55, G73, G64 and G55 specimens. Arrangements densities of the open air paths were increased according to the acrylic beads contents in the formed precursors. Figure 4 shows the cross sectional microstructures of the S73, S64, S55, G73, G64 and G55 specimens. The opened air holes were filled successfully with the molten glass. The remained holes of the closed pores are considered to reduce the relative densities of the bulk samples.

The relative densities of sintered and glass infiltrated specimens were indicated in Fig. 6. Relative densities of the sintered samples S73, S64 and S55 were 74, 66 and 64 %, respectively. And, the densities of the glass infiltrated samples G73, G64 and G55 reacted at 93, 95 and 94 %, respectively. The flexural strengths variations through the glass infiltrations into the alumina models were shown in Fig. 5. Average flexural strengths of the sintered specimens were increased from 102 MPa to 156 MPa according to the volume ratios of alumina particles. In the case of glass infiltrated ones, G73 and G55 showed approximately 294 MPa, and G64 showed the maximum value of 328 MPa. These results are comparable with the tendencies of relative densities variations as shown in Fig. 4. In the coating process, the molten glass could infiltrate into the open air paths in theS64 sample comparing with the S73 samples including closed pores. The S55 sample with larger amount of the pores inevitably showed lowest flexural strength.

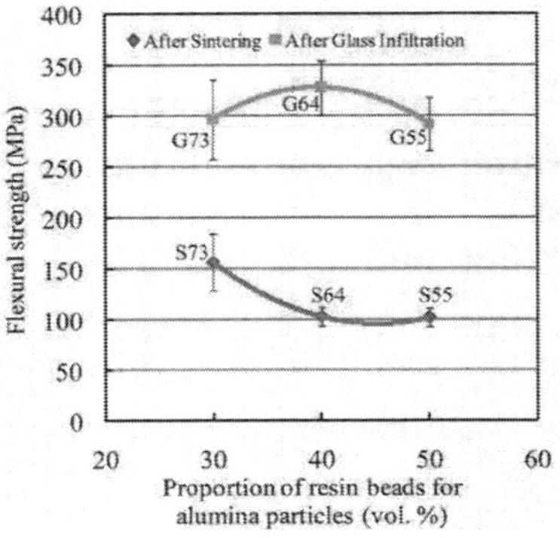

Fig. 5 Flexural strengths before and after the glass infiltrations into alumina model.

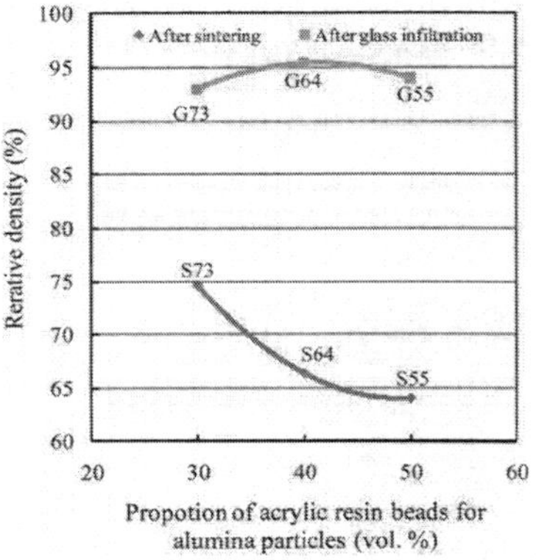

Fig. 6 Relative densities variations before and after the glass infiltration techniqu.

CONCLUSION

Composite precursors composed of photo polymerized acrylic resin with alumina particles and micro resin beads at 70 % in total volume contents were fabricated by using stereolithography of a CAD/CAM process. After dewaxing 600 °C for 2hs and sintered 1700 °C for 2hs, test specimens of

1.2×4.0×20 mm in dimensions with porous alumina structures were obtained. In order to imitate a natural tooth structure having a surface enamel layer, La_2O_3-B_2O_3-Al_2O_3-SiO_2 glass for dental use was infiltrated into open pores of the sintered alumina specimens. Maximum relative density and flexural strength reached approximately 95 % and 328± 27 MPa, respectively.

ACKNOWLEDGEMENTS

This study was supported by Priority Assistance for the Formation of Worldwide Renowned Centers of Research - The Global COE Program (Project: Center of Excellence for Advanced Structural and Functional Materials Design) from the Ministry of Education, Culture, Sports, Science and Technology (MEXT), Japan.

REFERENCE

[1] J. Stampfl, H. C. Liu, S. W. Nam, K. Sakamoto, H. Tsuru, S. Kang, A. G. Cooper, A. Nikel, F. B. Prinz, Rapid prototyping and manufacturing by gelcasting of metallic and ceramic slurries, *Materials Science and Engineering*, **A334**, 187-192 (2002).

[2] H. H. Tnag, F. H. Liu, Ceramic laser gelling, *Journal of the European Ceramic Society*, **25**, 627-632 (2005).

[3] R. Zauner, Micro powder injection moulding, *Microelectronic Engineering*, **83**, 1442-1444 (2006).

[4] S. Kirihara, Y. Miyamoto, K. Takenaga, M. Takeda, K. Kajiyama, Fabrication of electromagnetic crystals with a complete diamond structure by stereolithography, *Solid State Communication,* **121**, 435-39 (2002).

[5] M. Ishikawa, S. Kirihara, Y. Miyamoto, T. Sohmura, Freeform fabrication of alumina dental-crown model by using stereolithography, *Advanced Processing and manu- facturing technologies for Structural and Multifunctional Materials II, Ceramic Engineering and Science Proceeding* **Volume 29**, Issue 9 (2008).

Author Index